Water Resources and Development

'The book provides a critical and thoroughly referenced introduction to water resource availability and management. The authors highlight the importance of the stewardship of the water environment and give a balanced treatment of the role of social and political priorities in the allocation of water. This is an ideal text for first and second degree courses.'

J A [Tony] Allan *Professor at King's College London and the School of Oriental and African Studies, London.*

'The authors provide valuable perspectives on the increasingly complex business of managing water. This book will serve both as a valuable reference to practitioners and students in water-related disciplines as well as a guide to those from other fields who need to understand how water and its management impact on their worlds.'

Mike Muller *Visiting Professor at the Wits University Graduate School of Public and Development Management, Johannesburg.*

Since the start of the twenty-first century there has been an unprecedented focus upon water as a key factor in the future of both society and environment. Water management lies at the heart of strategies of development to which has been added the hazard of climate change.

Water Resources and Development, provides a stimulating interdisciplinary introduction to the role of water resources in shaping opportunities and constraints for development. The book begins by charting the evolution of approaches to water management. It identifies an emerging polarisation in the late twentieth century between 'technical' and 'social' strategies. In the past decade these two axes of policy debate have been further intersected by discussion of the scale at which management decisions should be made: the relative effectiveness of 'global' and 'local' gover-nance of water. A variety of case studies elaborate this analytical framework, exemplifying four key development challenges: economic growth, poverty reduction, competition and conflict over water, and adaptation to climate change. Current 'best practice' for water management is exam-ined, addressing strategies of water-supply augmentation, the ecological implications of intensified use and strategies of demand management guided by economic or political principles. It is argued that defining 'successful' water management and best practice requires first the establishment of development goals and the implicit trade-offs between water consumption and conservation.

This engaging and insightful text offers a unique interdisciplinary analysis by integrating scientific, engineering, social and political perspectives. This is an essential text for courses on development studies, geography, earth sciences and the environment.

Clive Agnew is a Professor of Physical Geography at the School of Environment and Development, University of Manchester, UK.

Philip Woodhouse is Senior Lecturer in Environment and Rural Development, Institute for Development Policy and Management, University of Manchester, UK.

Routledge Perspectives on Development

Series Editor: Professor Tony Binns, *University of Otago*

The *Perspectives on Development* series will provide an invaluable, up to date and refreshing approach to key development issues for academics and students working in the field of development, in disciplines such as anthropology, economics, geography, international relations, politics and sociology. The series will also be of particular interest to those working in interdisciplinary fields, such as area studies (African, Asian and Latin American Studies), development studies, rural and urban studies, travel and tourism.

If you would like to submit a book proposal for the series, please contact Tony Binns on j.a.binns@geography.otago.ac.nz

Published:

Third World Cities, 2nd edition
David W. Drakakis-Smith

Rural–Urban Interactions in the Developing World
Kenneth Lynch

Children, Youth and Development
Nicola Ansell

Theories and Practices of Development
Katie Willis

An Introduction to Sustainable Development, 3rd edition
Jennifer A. Elliott

Environmental Management and Development
Chris Barrow

Gender and Development
Janet Henshall Momsen

Tourism and Development
Richard Sharpley and David J. Telfer

Southeast Asian Development
Andrew McGregor

Population and Development
W.T.S. Gould

Postcolonialism and Development
Cheryl McEwan

Conflict and Development
Andrew Williams and Roger MacGinty

Disaster and Development
Andrew Collins

Non-Governmental Organisations and Development
David Lewis and Nazneen Kanji

Cities and Development
Jo Beall

Gender and Development Second Edition
Janet Henshall Momsen

Economics and Development Studies
Michael Tribe, Frederick Nixson and Andrew Sumner

Water Resources and Development
Clive Agnew and Philip Woodhouse

Forthcoming:

Global Finance and Development
David Hudson

Africa: Diversity and Development
Tony Binns and Alan Dixon

Politics and Development
Heather Marquette

Local Knowledge, Environment and Development
Tony Binns, Christo Fabricius and Etienne Nel

Participation and Development
Andrea Cornwall

Food and Development
E.M. Young

Natural Resource Extraction
Roy Maconachie and Gavin M. Hilson

Health and Development
Hazel Barrett

Theories and Practices of Development, 2nd edition
Katie Willis

Water Resources and Development

Clive Agnew and Philip Woodhouse

Routledge
Taylor & Francis Group

LONDON AND NEW YORK

First published 2011
by Routledge
2 Park Square, Milton Park, Abingdon, Oxon, OX14 4RN

Simultaneously published in the USA and Canada
by Routledge

270 Madison Avenue, New York, NY 10016

Routledge is an imprint of the Taylor & Francis Group, an informa business

Typeset in Times New Roman by Glyph International
Printed and bound in Great Britain by TJ International Ltd, Padstow, Cornwall

British Library Cataloguing in Publication Data
A catalogue record for this book is available from the British Library

Library of Congress Cataloging in Publication Data
Agnew, Clive, 1955-
Water resources and development / Clive Agnew and Philip Woodhouse.
p. cm.
Includes bibliographical references and index.
1. Water resources development. 2. Water-supply. 3. Water resources
development–Environmental aspects. I. Woodhouse, Philip. II. Title.
HD1691.A365 2010
333.91–dc22   2010017017

ISBN: 978-0-415-45137-6 (hbk)
ISBN: 978-0-415-45139-0 (pbk)
ISBN: 978-0-203-84065-8 (ebk)

Contents

Plates

Figures

Tables

Boxes

Preface

Efforts to understand and manage water resources have generated a vast literature of books and journals devoted to the topic's many facets, from the underlying science of hydrology and engineering options available, to the geopolitical problems generated by water scarcity and the adaptive strategies needed to tackle floods and pollution. The literature has demonstrated many different ways of organising our thinking on water. While some take a global perspective, others focus on key regional problem areas, while yet others emphasise distinct water use sectors. To this burgeoning print literature may be added a rapidly growing number of web pages through which national and international agencies and advocacy organisations present current, or at least recent, information on water-resource issues and priorities. It is perhaps a wonder, then, that we ever considered starting this book, and we could be excused for being intimidated by both the volume of existing work and by the challenge of addressing 'water resources' coherently, never mind adding the complication of using a development perspective to organise our analysis.

We were encouraged by four factors. First, and undeniably, water is an important topic, not just because it is vital for life, but because it is important in the development of society. This can be demonstrated historically, from the early water-based cultures along the Tigris, Euphrates and the Nile, to the construction of the Dutch polders or the Hoover Dam in the USA.

Second, there is a widely held perception of 'water in crisis', an impending Armageddon that threatens an outbreak of water wars as scarcity becomes more acute. While acknowledging the many parts of the world where lack of access to water is a key constraint on human activity or wellbeing, we need to recognise the many examples where this is not the case or where people have successfully adapted and innovated. We wanted to understand better the reasons for this variability, and the nature of spatial and temporal uncertainties surrounding the impact of water constraints.

A third motivation for this book was to assess critically the extent to which paths followed in the past to tackle water-resource problems in today's 'developed' economies may not be suitable for developing countries today. There are important contrasts between the 'underdevelopment' contexts of the past and those of today. Population pressure, poverty and inequality are all higher, demand for irrigation water is greater and water increasingly crosses international frontiers, both as virtual water and through exogenous flows. Yet it is also the case that strategies of the past have been heavily modified, with greater emphasis now being placed on stakeholder engagement and economic tools rather than relying heavily or solely upon technology. Ideas of what constitutes good practice in water management have evolved, just as the contexts and challenges of development cannot be simply mapped to a simple linear path from 'underdevelopment' to 'development'. We felt it important, therefore, to examine how approaches to water resources have evolved to embrace the 'ecological', 'instrument' and 'institutional' principles widely advocated in contemporary water policy, and to consider: what is next?

Finally, we chose not to write about water-resource management, but, rather, to write a book on water-resource development. Here we see water as both a facilitator and a constraint on development, but also development as an important force shaping the nature of water resources. Harnessing and using water resources has been one of the most effective means by which societies have inhabited the planet. In doing so, development has created new meanings for water: irrigation has been fundamental to the increase in food supply during the twentieth century, the provision of hydroelectric power continues to be a major motivation for the construction of mega dams, chlorination to create safe drinking water is believed to have substantially increased life expectancy. In the future, increasing urbanisation will mean greater emphasis on the re-use of wastewater, and changing desalination technology may mean water supplied as an industrial product, rather than natural resource.

Presenting all four elements in a coherent fashion has proved a challenge and we are indebted to the advice from colleagues and reviewers of an earlier version of the book, which we substantially changed. It became clear that we should not try to write an encyclopaedia of water problems and solutions, nor could we hope to capture in a relatively short book all facets of water and development. This has meant that our treatment is unavoidably uneven. We have noted the particular importance,

for example, attached to gender roles, and have given special consideration
to urban and rural water-supply problems. However, we have not delved
deeply into water infrastructure, nor waste-water treatment techniques.
This selectivity will be apparent in the chapter structure and is explained
in Chapter 1. It is important to stress here that our main goal has been to
identify and explain the different ways in which water has combined
with human development over the last hundred years. In doing so, we
make clear that today the engagement of local populations of 'water
users' is being addressed in ways that would have been inconceivable
in the early parts of the twentieth century.

Following an introduction to the nature of the global water crisis in
Chapter 1 and an examination of water within economic development
in Chapter 2, the book is structured around two key dimensions.
Technological approaches to tackle water scarcity, and socio-economic
approaches that address water governance. Because environmental
conditions are fundamental but are beset with uncertainties over a
variety of scales (temporal and spatial) we commence with the highly
contemporary topic of climate change and water resources in Chapter 3,
before addressing agricultural exploitation of water in Chapter 4 and
then technologies for water-supply enhancement in Chapter 5. Many
of the topics covered are technical or require detailed explanation.
We have separated out some of the technical and detailed information
to aid the flow of the discussion, and presented these as 'boxes' which
the reader can skip and still cover the key arguments. The governance
of water needs to be considered across various scales, so in
Chapter 6 we focus upon the individual and the community in an
examination of demand management, then, at a coarser scale, we
give consideration to catchment-scale governance and geopolitical
conflicts in Chapter 7.

Finally, Chapter 8 reflects upon the lessons learned for management and
development of water, concluding that technological solutions, whether
guided by engineering or ecological principles, are not enough. This is
because, while they may determine the overall quantity of water available, distribution of benefits from water development is a function of the
political dynamics of society. This raises questions about the role of the
state in water development and the goals and priorities that determine
outcomes. Thus we have also sought to look at the changes in attitudes
towards water-resources development through the twentieth century. We
have examined the neo-liberal arguments emphasising water pricing and
private property rights, together with the more recent promotion of

engagement of stakeholders as a means of ensuring more democratic and accountable water management. However, the evidence is that neither approach is unproblematic. In the absence, then, of a universal strategy or solution, we argue that the key is to understand the tensions and interactions between these principles in specific development contexts.

Acknowledgements

We would like to thank Tony Binns for his encouragement to write this book in the first place, and for the invaluable advice received from the reviewers of the first draft. The diagrams have all been produced by the drawing office in the School of Environment and Development at the University of Manchester, and we thank Graham and Nick for their patience and skill. The book draws upon our experiences of working on water resources in many parts of the world over several decades and would not have been possible without collaboration from colleagues across Africa, Middle East, USA and Europe, and we would like to make special mention of the advice and motivation provided by Professor Ewan Anderson (Emeritus, Durham University), Mike Muller (former Director General of Water Affairs and Forestry, Pretoria) and Professor Tony Allan (Kings College, University of London). Finally, our thanks to our families for helping us keep a sense of perspective.

1 Water-management best practice in the twenty-first century

Introduction

There is perhaps no more graphic illustration of the centrality of water to human existence than the emphasis placed on the search for water on Mars. The confirmation by NASA in July 2008 that its Phoenix lander on Mars had identified ice in a soil sample analysed in its onboard laboratory prompted a newspaper opinion poll in which 92 per cent of respondents said they believed there was life elsewhere in the universe (*Guardian*, 2008). Beyond this fundamental association of water with 'life', however, water plays a major role in humanity's social and economic existence. Not only is the management of water needed to enable almost all productive activity, but the need to manage water has historically imposed organisational requirements on human society.

Viewed from a perspective of aggregate water use, however, this imperative to manage water resources seems unwarranted, since human activity withdraws less than 10 per cent per cent of the available water resources on the planet (Table 1.1 and Figure 1.1). 'Available water' means here the water in streams, lakes and groundwater – sometimes referred to as 'blue' water – after discounting the 99 per cent of the world's water as 'unavailable' because it is contained in the oceans or polar ice, or falls as rain but is absorbed by vegetation and returned to the atmosphere (transpiration – 'green' water) before it can drain into aquifers or streams (see Box 1.1 for a fuller account of such water classifications).

Table 1.1 *Water availability and use*

Region	Renewable (internal) freshwater resources (km³y⁻¹)	Total freshwater withdrawal (km³y⁻¹)	Freshwater withdrawals (km³y⁻¹)						Withdrawal as % of renewable freshwater
			agriculture		industry		municipalities		
			amount	%	amount	%	amount	%	
Africa	3,936	217	186	86	9	4	22	10	5.5
Asia	11,594	2,378	1,936	81	270	11	172	7	20.5
Latin America	13,477	252	178	71	26	10	47	19	1.9
Caribbean	93	13	9	68	1	9	3	23	14.4
North America	6,253	525	203	39	252	48	70	13	8.4
Oceania	1,703	26	19	72	3	10	5	18	1.5
Europe	6,603	418	132	32	223	53	63	15	6.3
World	43,659	3,830	2,664	70	785	20	381	10	8.8

Sources: based on FAO-AQUASTAT database, 2009; Molden, 2007: 70

Even within the less than 1 per cent of water that is 'available', it appears that water is so abundant relative to human needs that little effort should be needed to manage it. Across Europe total rainfall (precipitation) is ten times the amount withdrawn for human activities (EEA 2005, cited by Carter 2007). However, as Table 1.1 and Figure 1.1 suggest, the availability of water is not uniform across all parts of the globe. Moreover, the intensity of human activity and settlement is not governed

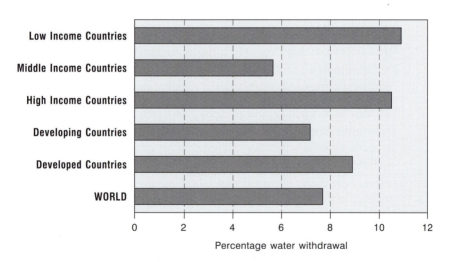

Figure 1.1 Percentage of water withdrawal compared to renewable supply using income groups

Source: after WRI, 2009, withdrawal data 2000, supply data 2007

Box 1.1

Definitions of different categories of water

A primary division is between 'saline' – predominantly seawater – and 'freshwater'. Viessman and Hammer (1998: 487) state that seawater contains 35,000 mgl⁻¹ of total dissolved solids (tds), of which 30,000 mgl⁻¹ is typically sodium chloride. By comparison, drinking water in the USA must be less than 500 mgl⁻¹ tds (Kebbekus and Mitra, 1998: 249). 'Freshwater' is not scientifically defined other than as 'non saline' as listed below (after Huggett et al., 2004: 36):

Oceans	Saline	1,350,000,000 km³	97.37%
Snow and ice	Freshwater	27,500,000	1.98
Groundwater	Freshwater	8,200,000	0.59
Atmospheric	Freshwater	460,000	0.033
Lakes and rivers	Freshwater	207,000	0.015
Soil moisture	Freshwater	70,000	0.005

A widely used terminology distinguishes between 'blue' freshwater as the naturally renewable water in streams, lakes and groundwater, and 'green' freshwater utilised by plants (see, for example, Marget, 2009: 583 on water resources in the Mediterranean).

This colour-coded terminology may be confused with a different classification of marine waters as brown, green or blue, referring to coastal, littoral or deep ocean zones, respectively.

A further colour coded nomenclature is used to indicate quality of waste-waters, to identify the differences between potable (i.e. uncontaminated freshwater) and water that carries human wastes (after Abbassi and Al Baz, 2008):

Grey: untreated household wastewater from washing and bathing which does not contain human wastes and should have low risk of contamination.

Black: sewage from household toilets containing faeces and high organic levels. There are concerns over pathogens and pharmaceutical products.

Yellow: contains urine with high levels of nitrogen and phosphorus.

Also:
Green non potable, but treated, waste water which should have low turbidity, low biochemical oxygen demand and free of pathogens. Hence can be used with low risk for irrigation and other watering uses, (WWUK, 2010).

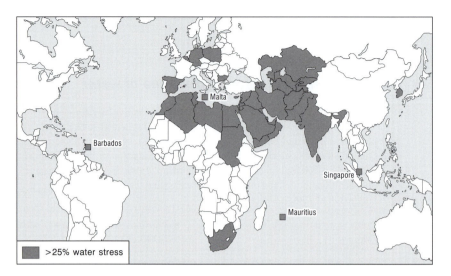

Figure 1.2 Countries that are facing high water stress as indicated by the ratio of withdrawal to renewable resources ($m^3c^{-1}y^{-1}$) where water stress is taken to be a withdrawal of more than 25% of resources

Source: based on data from WRI, 2007, renewable resources, 2000 withdrawal rate

solely by water availability, so that 'demand' is much higher relative to available supply in, say, Asia than in Latin America. Figure 1.2 displays regions of high water stress as measured by rates of water withdrawal compared to annually renewable resources. The top quartile of countries have been selected in this figure and this corresponds to at least 25 per cent of renewable resources being consumed annually, (note that data was unavailable for some parts of Eastern Europe and many island states). At a more local scale still, this unevenness is even more marked, so that in particular locations the intensity of water demand may exceed the local supply (see Box 1.2. concerning magnitudes of available water resources and levels of consumption).

For many countries, especially those in the tropics, water withdrawals for irrigated agriculture dominate consumption, with values of 70 per cent typical but rising to over 80 per cent in places. For example, 70 per cent of Australia's water is used in rural areas but almost all for irrigation, compared to 21 per cent consumed in urban areas for domestic and industrial uses (Environment Australia, 2003). Table 1.1 shows that in North America and Europe demand from industry and energy generation is significant (48 and 53 per cent respectively) with municipal supplies accounting for much less (13 per cent and 15 per cent), although much of this is available for recycling. Globally, industrial water withdrawal accounts for only 20 per cent and domestic water accounts for 10 per cent (see Figure 1.3 for examples from the Middle East).

Although agriculture dominates total water abstraction across the globe, it is the local intensity of demand for potable supplies from urban areas that has stimulated past development of water resources (see Chapter 2) and that has focused much of water-management effort. Thus, although 'domestic consumption' is much less significant in terms of the overall water budget, it has received priority due to the immediacy and relative inflexibility of demand for potable supplies. Furthermore, providing household water supplies can have much wider social and economic benefits, as exemplified by the following list of relationships charted by Moriarty and Butterworth (2003: 12) between water development and human livelihoods:

- Health (hygiene improvements, disease reduction)
- Labour availability (health and less effort required to collect water)
- Cheaper water (compared to commercial vendors)
- Improved well-being (more security, less stress)
- Education (more time, especially for children)
- Empowerment (through greater community decision making)
- Community capacity (strengthened through water-project engagement)
- Income (health and labour improvements raise other opportunities)
- Food security and nutrition (household-production opportunities)
- Investment (other improvements increase expenditure in other areas)

Nonetheless, it seems clear that, despite some two hundred years of rapid growth in economic and scientific capacity, there is widespread concern that humanity is falling short of what is needed to manage water effectively. This failure, most often characterised in terms of the 1.1 billion people who do not have access to clean water and the 2.6 billion who live without adequate sanitation, has particular resonance for many because of the centrality of water to human health and dignity: '"Not having access" to water and sanitation is a polite euphemism for a form of deprivation that threatens life, destroys opportunity and undermines human dignity' (HDR, 2006: 5). It is therefore unsurprising that aggregate statistics showing an abundance of water at regional or global scales belie an intensity and violence in local struggles whose trigger or pretext is water. Water thus plays an emblematic role in contestation over inequality in human society.

While mindful of this dimension to water-resource management, in this book our intention is to step back and take a broader view. In particular, we seek to consider the status of 'best practice' in managing water

Box 1.2

Water units and magnitudes[a] (Annual unless stated otherwise)

Column	A	B	C	D	E	F (= 100*D/B)	G	H	J	K
	Internal renewable resources	Actual renewable resources	Actual renewable resources per capita	Annual water withdrawals	Annual water withdrawals per capita	Withdrawal as % of actual renewable resources	Actual renewable	withdrawal	withdrawal L per captia per day (lcd)	Average water consumption L per captia per day (lcd)
Units	km^3	km^3	m^3c^{-1} or $km^3 \times 10^{-9}c^{-1}$	km^3	m^3c^{-1}	%	$m^3c^{-1}d^{-1}$	$m^3c^{-1}d^{-1}$	L per captia per day (lcd)	L per captia per day (lcd)
Summed total	43,219	55,273	8,549	3,803	633	6.9	23.42	1.73	1734	(300–350)[b]
Asia	11,192	14,582	4,079	2,147	631	14.7	11.18	1.73	1729	
China	2,812	2,830	2,206	630	494	22.3	6.04	1.35	1353	86
India	1,261	1,897	1,754	646	635	34.1	4.81	1.74	1739	135
USA	2,818	3,069	10,333	479	1,682	15.6	28.31	4.61	4608	575 (600–800)[b]
UK	145	147	2,474	10	163	6.8	6.78	0.45	447	149
Egypt	2	58	794	69	1,013	118.9	2.18	2.78	2,775	

(a) from UNDP 2006 Human Development Report: Beyond scarcity: Power, poverty and the global water crisis: http://hdr.undp.org/en/reports/global/hdr2006/
(b) Twort et al. (2000) for urban areas

Column A: internal renewable freshwater resources in km³, excludes saline waters but includes renewable surface waters and groundwater *recharged by precipitation occurring within the borders of that country or region*.

Column B: actual renewable freshwater resources total in km³ is a theoretical amount that might be available *taking account of upstream and down-stream water agreements*, for example the Nile flowing into Egypt. This can be higher than the internal resources, as in the case of Egypt or India which have a high dependency ratio (Actual/Internal). Where most water is derived within the country, as in the case of China, then internal and actual water are similar. Note that the figures used in Table 1.1 refer to internal renewable freshwater resources. Note also that summed, i.e. 'global', actual values exceed internal renewable resources by 28%. This is because 'global' figures are obtained by summing those for individual countries. This generates a greater amount of water in the 'actual' total than can be generated by precipitation across the globe because there is double-counting of water flowing across international frontiers.

Column C: actual renewable resources per person or per capita (based on 2000 population data), in units of m³ (numbers in terms of km³ would be too small for convenience). Note the differences in comparative magnitudes between column C and columns A and B: on a per capita basis the USA, in particular, appears as a water-rich environment. Note: *water stress* can be defined as renewable resources less than 1700 m³c⁻¹y⁻¹, and *water scarcity* defined as less than 1000 m³c⁻¹y⁻¹.

Column D: annual withdrawals (including conveyance losses) of water for direct use or storage for agricultural, domestic and industrial consumption.

Column E: withdrawals per capita. Note how Egypt, because of exogenous flows, appears to be comparatively water rich for the population directly supported.

Column F: the percentage of water withdrawal compared to actual renewable resources. Note these values are slightly different to those in Table 1.1, which is based upon internal and not actual water resources.

Column G: actual renewable water per person (cf column C) but expressed as m³ per person per day.

Column H: water withdrawals per person (cf Column E) but expressed *as* m³ per person per day.

Column J: the same data as Column H, but expressed as litres per person per day.

Column K: water consumption data (litres per capita per day – lcd). The difference between columns J and K is that the latter are average values based on the amount metered or estimated at the point of consumption, whereas the former concerns the amount of water withdrawn at its source. The difference is not just due to conveyance losses and storage, but also to recycling and re-use of water in urban and industrial supplies. WHO (Howard and Bartram, 2003) suggest water availability below 20 lcd can only assure basic consumption and hygiene requirements, and 50 to 100 lcd is required for all household water needs including laundry and bathing.

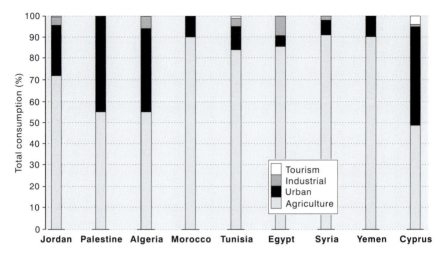

Figure 1.3 Sectoral water consumption in the Middle East

Source: after Magiera et al., 2006

resources in order to meet development goals. In doing so we will need to interrogate the specific meanings of 'scarcity' of water in particular contexts, and the extent to which these constitute constraints to the future content and direction of development.

In this chapter we will review the emergence of a water 'sector' within development thinking, and how current ideas of international best practice have come to be formulated and adopted by development agencies. We will draw upon these recent debates to set out a framework for the more detailed discussions in the remainder of the book. First, however, we will review the ways in which water supply and demand are measured, and the basic models used to characterise conditions of water scarcity.

Water scarcity: a water crisis?

As we observed earlier (Table 1.1), aggregate data suggest an abundance of water resources at a global level but mask large disparities in water distribution between geographical regions. Climate (aridity) clearly affects water availability and these disparities are accentuated when water resources are calculated per head of population (Figures 1.4 and 1.5). The relatively high populations of Asia and the Middle East mean that the renewable water per person (m^3/capita/yr) in those regions is reduced, and tends to decline as population grows.

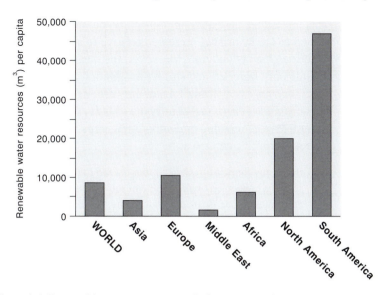

Figure 1.4 Renewable water resources relative to population

Source: WRI, 2005 data

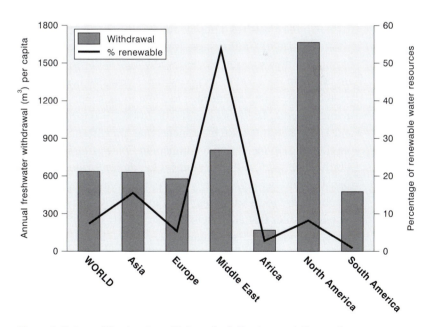

Figure 1.5 Annual freshwater withdrawal relative to population and as percentage of renewable water resources

Source: WRI, 2005 data

This underlines the rather obvious point that measurements of water resources are invariably presented relative to a defined geographical area or a given population. Moreover, the measured resource may include only water arising (from precipitation and storage) within a specified area (as in Table 1.1), or also water generated outside that area but accessible (via rivers, aquifers) within it. Yet another way of relating water use to geographic or demographic units is through the idea of a 'water footprint', defined as

> the total volume of freshwater that is used to produce the goods and services consumed by the people of the nation. Since not all goods consumed in one particular country are produced in that country, the water footprint consists of two parts: use of domestic water resources and use of water outside the borders of the country
>
> (Hoekstra and Chapagain, 2007: 36)

This is similar to the 'virtual water' concept, defined as the volume of water required to produce a commodity or service (Allan, 2001). Considerable care is therefore needed to ensure that measures of water supply and demand are appropriate to the scale and nature of the water management issue being considered (see Box 1.2).

In aggregate, global renewable water resources amount to about 8000 m^3 per person per year for a population of 6 billion people, but UNDP estimates that 700 million people in 43 (46 in Figure 1.6) countries are living in 'water-stressed' societies in which water resources per person are less than the 1700 m^3 per person per year that is conventionally treated as the minimum needed to grow food, support industry and maintain ecological functions (HDR, 2006: 135) (see Figures 1.6 and 1.7). The same source estimates that, on current trends, growing population will increase the numbers in 'water stressed' societies to 3 billion by 2025, with 14 countries becoming 'water scarce', with less than 1000 m^3 per person per year. As before, regional aggregates obscure more local variation. Thus, even today, some populations have access to water that is far below these criteria of water scarcity. UNDP notes, for example, Palestinians in Gaza who have on average 320 m^3 per person per year, compared to a Middle East average of 1200 m^3 per person per year (HDR, 2006:135). Clearly, there are factors other than natural distribution that determine water availability to human populations and their economic activity, such as the extent of investment in order to extract and use the local water resources. This is illustrated by the case of the Middle East, where relatively high rates of per capita water use are achieved by intensive exploitation

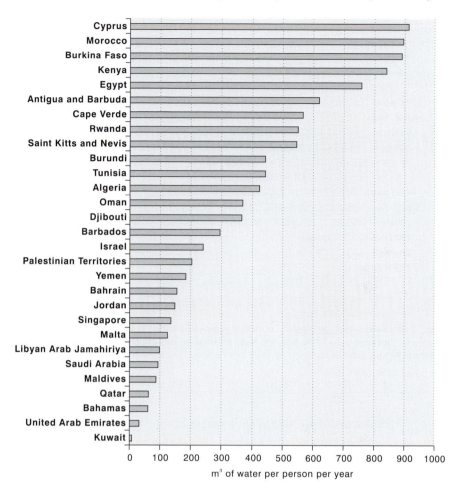

Figure 1.6 Water-scarce countries, i.e. actual renewable water resources are less than 1000 m³c⁻¹y⁻¹

Source: based on WRI 2007 data

of existing fresh-water resources. This may be compared to North America where high per capita water use can be sustained at a much lower percentage use of fresh water, and Africa, where low per capita use results from low rates of investment and consequently low rates of exploitation of fresh-water resources.

Water scarcity is determined by demand (population) and supply, and at sub-national levels demand can be expected to be highest in large and industrialised urban areas. Typical urban values of water

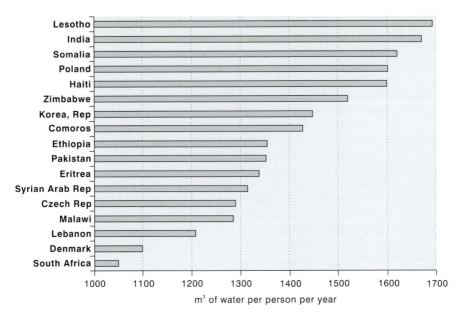

Figure 1.7 Water-stressed countries, i.e. actual renewable water resources are greater than 1000 m³c⁻¹y⁻¹ but less than 1700 m³c⁻¹y⁻¹. Note alternative designation of water-stressed countries in Figure 1.2

Source: based on WRI 2007 data

demand of 200 to 300 lcd are found in Europe but Figure 1.8 displays enormous variation of household consumption levels reflecting behaviour and environmental conditions, plus the difficulties of obtaining comparable numbers. Twort et al. (2002) suggest typical levels of consumption are:

600 to 800 lcd in industrial cities in USA
300 to 550 lcd in major cities across the world
90 to 150 lcd where water is provided by street standpipes or wells

Note that these are much lower than the figures of renewable water shown in Figure 1.6 where 1000 m³ per person per year is equivalent to 2,700 lcd because the latter refers to water available not the amount supplied to customers (see Box 1.2 for further explanation).

Swyngedouw (2006) presents urban per capita consumption values of Latin American cities of rarely less than 100 lcd and normally in excess of 300 lcd with 630 lcd for Buenos Aires, 500 lcd for Havana and similar magnitudes for Mexico City, Rio de Janeiro and Santiago. Household consumption levels are lower than this as they exclude industry and commerce. Average value for example in England and Wales is around

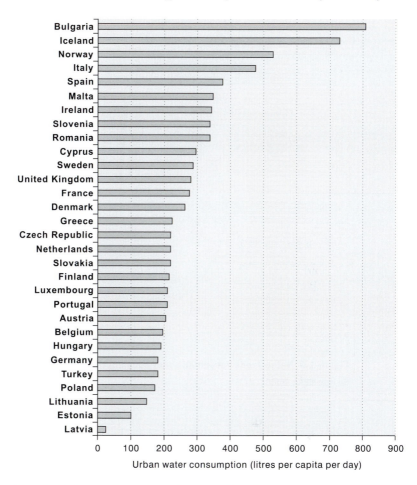

Figure 1.8 Urban water consumption across Europe
Source: EEA, 2009

150 lcd unmetered and 135 lcd for metered households (OFWAT, 2007) which is half that displayed in Figure 1.8 but still higher than the average of 126 lcd for the Netherlands (NEAA, 2003). Water consumption in England and Wales has remained around these magnitudes since the mid-1990s (ONS, 2010). Typical distribution of water in household use is displayed in Figure 1.9 for the Netherlands. Twort et al. (2002) contrast figures from Europe and the USA which vary from 130 to 336 lcd but the proportionate use of water still accords with Figure 1.9. These levels of consumption are much higher than values found in developing countries, primarily due to lack of investment, especially in rural areas.

Rural areas have long lagged behind industrialised cities in the provision of water and in safe sanitation (see Chapter 2). The WHO

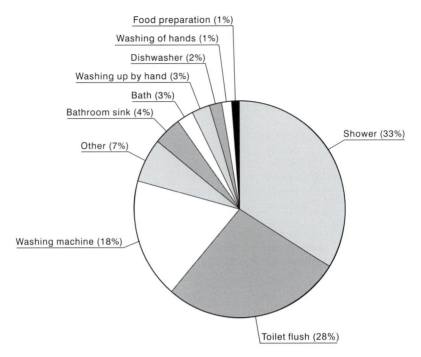

Food preparation (1%)
Washing of hands (1%)
Dishwasher (2%)
Washing up by hand (3%)
Bath (3%)
Bathroom sink (4%)
Other (7%)
Washing machine (18%)
Shower (33%)
Toilet flush (28%)

Figure 1.9 Netherlands household water consumption

Source: NEAA, 2003

guidelines recommend a supply of at least 20 lcd (Cowden et al., 2008), that should be available within 1km and from a source that is free from microbiological contamination, i.e. a safe well or spring. They (WHO, 2009) also note that the minimum adult drinking-water requirement is 2 litres per day (LCD), which is increased to 7.5 litres per capita per day by Howard and Bartram (2003) to include most people in most conditions. Water is also required for personal hygiene and additional household activities including laundry and cooking. Howard and Bartram summarise the health implications for different levels of water access:

> *very high health concern*: water access below 5 lcd ($Lc^{-1}d^{-1}$), water source collection time more than 30 minutes; household consumption and hygiene cannot be assured
> *high health concern*: water access below 20 lcd, water-source collection time 5 to 30 minutes, i.e. no more than 1 km distant; household hand washing and basic food hygiene possible
> *low health concern*: water access around 50 lcd, water-source collection via local tap taking less than 5 minutes; all basic household water requirements are assured including laundry and bathing.

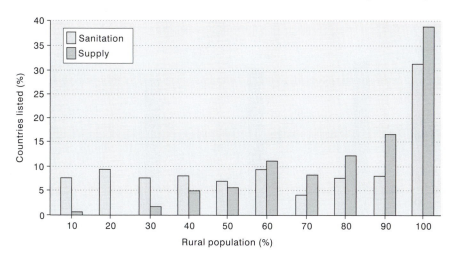

Figure 1.10 Rural water provision (sanitation and potable supply)
Source: World Bank, 2006

very low health concern: water at least 100 lcd, water source typically via multiple taps continuously; all water and hygiene needs are met.

Figure 1.10 illustrates the wide international range of rural water provision, with more than 30 per cent of countries supplying safe water and sanitation to 90 to100 per cent of the rural population, while a significant proportion provide safe sanitation to no more than 50 per cent of the population. The UN estimate that spending of at least $10 billion per year is required to address these problems and to meet international targets for water and sanitation (see the Millennium Development Goals (MDG) in the section below, on Water Markets vs Water Rights, p. 35). Figures 1.11 and 1.12 show that there have been achievements but that rural water provision (supply and sanitation) continues to fall behind that of urban water provision, and that the situation is markedly worse for low income countries.

These continuing failures to assure adequate water supply to poorer and rural populations imply water use must increase. Moreover, aspirations to increase water use beyond the bare minimum will increase both in response to population growth and in response to economic development, thus giving rise to concerns of a future 'global' water crisis. Thus, although water is abundant in aggregate terms, high levels of local demand and inadequate supplies across many parts of the developing world have led many to argue over the past two decades that the world was facing a water crisis (Gleick, 1988, Postel, 1992) and that,

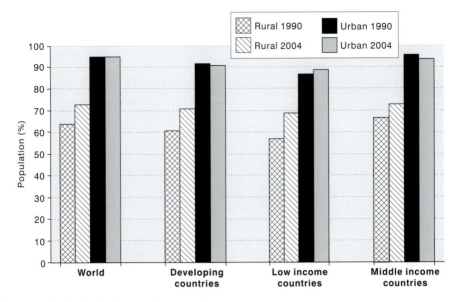

Figure 1.11 Access to improved water source

Source: WRI 2010

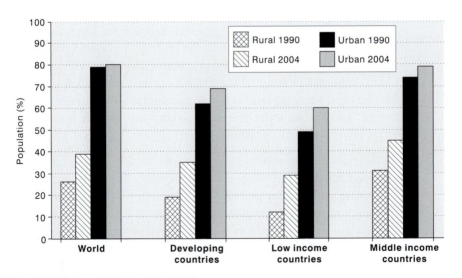

Figure 1.12 Access to improved sanitation

Source: Earth Trends, WRI 2010

'A water problem is really a people problem' (Hagen, 1988). The notion of a global water crisis has often been repeated, most recently by the World Water Council (2010):

> While the world's population tripled in the 20th century, the use of renewable water resources has grown six-fold. Within the next fifty years, the world population will increase by another 40 to 50%. This population growth – coupled with industrialization and urbanization – will result in an increasing demand for water and will have serious consequences on the environment.

The third UN World Water Development Report (WWDR, 2009) presents an equally gloomy future with these opening comments:

> At the beginning of the twenty-first century, the Earth, with its diverse and abundant life forms, including over six billion humans, is facing a serious water crisis. All the signs suggest that it is getting worse and will continue to do so, unless corrective action is taken. This crisis is one of water governance, essentially caused by the ways in which we mismanage water.

Figure 1.13 illustrates the exponential growth of water extraction over at least the last 100 years. It shows the distribution of water use we have noted earlier: agriculture, with 70 per cent of global extraction, dominates use (industry 22 per cent and domestic use 8 per cent), and that much of

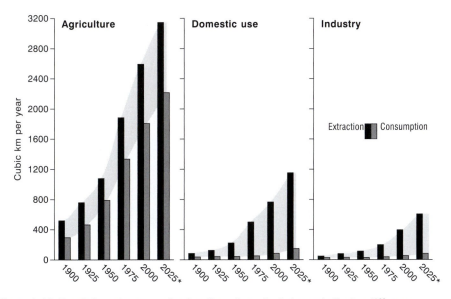

Figure 1.13 Trends in water use and extraction where shaded area indicates difference between extraction and consumption (*predicted for 2025)

Source: source and permission to reproduce from UNEP, 2008 *GRID-Arendal* Trends in global water use

the water extracted for agriculture is consumed and not recycled. In this regard it is important to note that agricultural water use is much less 'efficient' than industrial production. For example, 1 kg of wheat or rice requires around 1000 kg of water, while ratios for beef are much higher, at over 10,000 kg water per kg meat. In contrast, ratios of only 10 s to 100 s are needed for many industrial products. The UNDP (HDR, 2006) provide examples, including: one tonne of steel requires up to 350 m³ (tonnes) of water; a tonne of petrol requires up to 40 m³ and beer between 8 and 25 m³.

As we argued earlier, a global perspective obscures regional trends. Thus, in contrast to a global picture of continuing growth of water consumption, Figure 1.14 shows that in the USA water consumption reached a plateau twenty years ago, after which growth in demand for irrigation and power production ceased.

It is evident that patterns of social and economic development are not dictated by water distribution, although nor are they independent of it. Growth in settlement and economic activity may encroach on floodplains, increasing exposure to flood hazards, or may expand in arid areas, raising the cost and risks of gaining access to water. However, as we have seen above, questions of water scarcity can only be understood in terms of specific types of economic activity and at specific levels of scale.

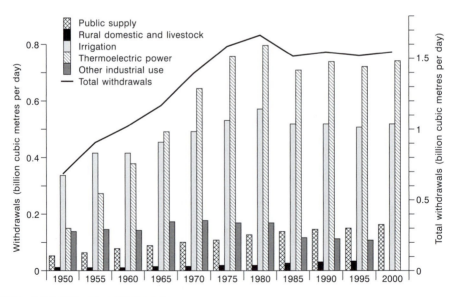

Figure 1.14 USA water withdrawals

Source: adapted from Kenny et al., 2009: 44; permission to reproduce granted by USGS

It is also evident that mismatches between development and 'natural' distribution of water resources may be compounded by changes in hydrological regimes caused by climate change. In particular, IPCC (2007) have identified drought risk as increasing during the twenty-first century, especially in tropical regions, but also with more extremes in precipitation and drier summers in the northern hemisphere, so the incidence of droughts is likely to increase across the globe. Drought is difficult to define (see Box 1.4) and many have called for greater clarity (Wilhite, 2000; Blenkinsop and Fowler, 2007) over this essentially meteorological phenomenon. Most commonly, drought is defined as a short-term water scarcity brought about by an abnormal water shortage and should not be confused with aridity or desiccation which are both longer term. Drought is a natural phenomenon that can occur in all parts of the globe, whereas aridity is a climate norm, defined in terms of water balance (see Box 1.3) based on long-term (30 year) averages according to Warren and Khogali (1992).

Drought can be due to a reduction in supply, such as a decline in rainfall, but can also be caused by an increase in demand, such as a change in land use, or even a combination of both. Since drought is often cited as the world's costliest natural disaster, causing annual damage worth $6 to $8 billion (Wilhite 2000), growing drought risk associated with climate change must be of concern. However, reduction in precipitation is not the sole cause of drought as there are agricultural, hydrological, ecological and socio-economic dimensions to this phenomenon. In thinking about water scarcity and strategies to manage it, therefore, we need to consider both human and environmental causes and also the timescale over which they operate (see Box 1.4).

To summarise, water may be considered 'scarce' in particular local contexts, where cities are constructed in arid areas, for example. Where cities are large, with populations of hundreds of thousands, or, as is increasingly the case, many millions, the water supply immediately available is unlikely to meet demand, and effort is needed to mobilise and invest resources to gain access to additional supply. Alternatively, resources may be employed to reduce the total demand for water, for example by devising more efficient ways of using water to meet human needs (for washing and sanitation, for example). It will be evident that, whichever of these two approaches is pursued, an important role is played by, on the one hand, science and knowledge that can be translated into engineering technology needed to control and direct water flows, and, on the other, the economic and political capacity needed to mobilise resources and to mitigate competition and conflict over water and modify

Box 1.3

The water balance

Assessment of water resources requires an inventory of water inputs and losses: a 'water balance'. The world's water is mostly saline, with 97 per cent found in oceans (Box 1.1). Of freshwater resources, the majority is either frozen (75 per cent) or found subsurface (25 per cent) (Davie, 2003). Only a small amount of fresh water occurs on the surface or in the atmosphere and it is more illuminating to examine the fresh water balance at a catchment scale for a hydrological basin, where inputs and outputs are:

P   Precipitation to include rainfall and sometimes snow, dewfall/occult precipitation and hail.
I   Interception
E   Evaporation
Et   Evapotranspiration (via plants)
Ro   Run-off lost
Rg   Run-off gained
Gi   Groundwater increase
Gp   Groundwater depletion
ΔS   Storage changes

Hydrological models were rapidly developed during the late nineteenth and early twentieth centuries such as the relationship between friction and flow (Manning, 1890), discharge and catchment area (Mulvaney, 1895), infiltration and run-off (Horton, 1919), the unit hydrograph (Sherman, 1932) and evaporation (Penman, 1948) (see Rodda et al., 1976). Water-balance models have since been developed for a number of contexts and purposes, including deforestation in Amazonia (D'Almeida et al., 2006), peatland groundwater (Evans et al., 1999) evaporation from soils (Fowler, (2002) mountainous semi arid water-shed (Flerchinger and Cooley 2000) and to predict climate change impacts in China (Guo et al., 2002).

In practice, for a time-scale of a year, the changes in some elements, such as storage, are negligible (i.e. gains equals losses) and the water balance may simply be represented as:

$$P = Ro + E + Et \qquad \text{i.e. inputs} = \text{output with no storage}$$

This simple model also applies to global water balances, where mean annual rainfall is 857 mm of which only 197 mm (23 per cent) falls on the land and global run-off is 60 mm and the remainder (137 mm) is evaporation, as storage remains constant (Briggs and Smithson, 1992). This simple equation is the basis of hydrology and demonstrates the principle that water is not created but is conserved. It works best for large areas and over long time intervals. Over short time periods, such as a day, and for a small catchment drained by one river system, other elements of the balance will be more significant:

$$Ro = P - I - E - Et + Rg + Gi - Gp + \Delta S$$

or

$$Ro = (P + Rg + Gi) - (I + E + Et + Gp) + \Delta S$$

Run-off = gains minus losses plus change in storage

For a discussion of scale issues in catchment water balances see Bloschl and Sivaplan (1995), Jothityangkoon et al. (2001), Sivaplan et al. (2003) and Xu et al. (2006). Water-balance calculations suffer from difficulties over units: precipitation is typically measured in mm per day, discharge in m^3 per second or litres per second, and groundwater change may be measured in cm depth but this does not easily compute to a rainfall equivalent without knowing the specific yield (the difference between saturation and field capacity, see Ward and Robinson, 1999). However, it is usually possible to express elements of the water balance in various volumetric units, e.g. litres or m^3 of water, or even mm equivalent depths.

Take, for example, a catchment with area 60 km^2 that receives a rainfall of 10 mm, and half of this rainfall goes into storage, with the remainder ('effective rainfall') lost through evaporation or surface run-off. If the mean flow out of the catchment (run-off) for the next 24 hours is 900 litres per second, the rate of evaporation (mm per day) may be calculated as follows:

(NB 1000 litres per second = 1 cumec = 1 m^3s^{-1}).

Convert effective rainfall and surface run-off to a volume in m^3 for the whole day

Effective Rainfall volume ER = 5 mm over 60 km^2/day = 300,000 m^3d^{-1}

Surface run-off Ro = (900/1000) *86,400 m^3 per day = 77,760 m^3d^{-1}

Calculate evaporation in m^3 per day then convert to mm:

Evaporation E (m^3) = 300,000 – 77,760 m^3 per day = 222,240 m^3d^{-1}

Evaporation E (mm) = 222,240 m^3/60 km^2 = 3.7 mmd^{-1}

Box 1.4

Drought

Drought can be defined as a period of below average precipitation (meteorological drought); or a shortage of soil moisture in the root zone of crops that affects productivity (agricultural drought); or a low river discharge that adversely impacts upon the ecosystem (hydrological drought) as illustrated in the following figure.

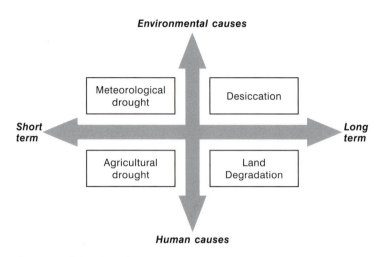

Conceptualising drought

Because of the difficulties of computing water demand and water balances many have opted to measure drought as a statistical index against normal conditions such as standardised rainfall anomalies (SRA) where:

$$SRA = [X_{ik} - X_i]/\sigma_i$$

X_{ik} = Precipitation for ith station and kth observation (often annual precipitation)

X_i = Mean precipitation for ith station

σ_i = standard deviation for ith station

(See, for example, Jones and Hulme, 1996.)

A number of indices have been derived from this basic model and a widely adopted variant is the Standardised Precipitation Index (SPI), defined as the number of standard deviations that observed cumulative precipitation deviates from the climatological average (McKee et al., 1993, 1995, Morid et al., 2007, Cancelliera et al., 2007 and Wu et al., 2007). Keyantash and Dracup (2002) found that, of 18 drought indices compared for robustness, ease of use and versatility in estimating meteorological, hydrological and agricultural drought, a decile precipitation approach was the best method under their criteria, mainly because of its tractability and transparency, but the SPI was a close second. Further details on using the SPI can be found at http://www.drought.unl.edu/monitor/monitor.htm. Hayes et al. (1999) demonstrated that for the USA the SPI proved to be less complicated and more versatile than the previously employed Palmer Drought Severity Index, while Ntale and Gan (2003) found the SPI to be most appropriate for their analysis of drought in East Africa, and Wu et al. (2001) reached the same conclusion for China.

the behaviour of water users. As we shall argue later in this chapter, the management of water resources in development may be understood as a tension between these two requirements, with a tendency for engineering to dominate giving way in recent decades to a greater emphasis on matters of water governance. We will first consider how water resource use has changed at different times of social and economic development.

Water in historical perspective

It is clear that the management of water for irrigated agriculture was the basis of the early civilisations in the Middle East, Asia and America. Karl Wittfogel (1957), one of the first scholars to systematically study what he termed 'hydraulic societies', argued that the particular influence of water on social organisation in such societies was generated by its physical characteristics. More specifically, whereas certain factors in early agricultural production, such as soils and plants, were manipulable by individual cultivators, and others, such as temperature and landscape, were by and large not amenable to manipulation at all, water was manipulable but only at a scale larger than could be managed by an individual cultivator. This was due to water's unique properties of bulkiness and mobility within the landscape, accentuated by the seasonality with which in many cases it is available over time. As a consequence, effective management of water required collaboration and a division of labour in order to achieve the necessary scale of infrastructure to contain and distribute water so as to protect crops from flood and drought. It is worth noting that more recent study of the impact of episodes of climate change in the past has suggested that the desiccation of tropical regions some 4000–6000 years ago caused major disruption of human societies existing at that time (Brooks et al., 2009). This analysis argues that 'civilisation' – and specifically the 'hydraulic societies' established on the floodplains of rivers such as the Nile and Euphrates – may well have been a 'last resort' for human populations decimated by the disruption of food-production systems developed during an earlier, wetter, climatic period.

Wittfogel's study compared a wide range of civilisations he identified as 'hydraulic societies', including those of Egypt, Mesopotamia, China, India, south-east Asia, Aztec Mexico and Inca Peru, as well as the social organisation of smaller-scale water management among, for example, the Chagga people in East Africa. He argued that the demands of constructing and maintaining large water infrastructure, such as dykes, reservoirs and canals, required the mobilisation of a large reserve of labour and this was only achievable under a centralised and coercive authority. In addition to the organisation of labour, the success of 'hydraulic agriculture' required innovations in knowledge, such as astronomy (calendars), mathematics (area measurement) and engineering (water lifting and distribution). Wittfogel argued that the control of such knowledge and organisation by an elite group of administrators was a defining feature of political control in hydraulic societies, enabling the exercise of wider powers, such as the levy of taxes and the construction of other infrastructure, such as roads and aqueducts for drinking water.

While Wittfogel defined hydraulic societies as only arising where neither industry nor rain-fed agriculture were dominant forms of production, it seems possible that other factors have also demanded social organisation to manage water. It is evident, for example, that the much higher productivity of flooded (paddy) rice, relative to other cereals in pre-industrial times, was a major factor prompting the creation of artificially flooded fields through relatively sophisticated water-control systems that required a high degree of social organisation (Bray, 1986). Such systems included a variety of storage works, such as ponds to store rainwater in elevated parts of the Japanese landscape, canals and reservoirs ('tanks') in medieval Sri Lanka, Cambodia and Thailand, and terraces on Philippine hillsides. Stream diversion into contour canals, in which water was distributed over large areas by gravity, were also common in East and south-east Asia. Large-scale versions of such systems might have conformed in certain respects to the centralised political structure of Wittfogel's 'hydraulic societies', such as at Angkor Wat in Cambodia or Guanxian in China. However, many of these water control systems were developed to serve much smaller social units at a village or clan scale, such as the *subaks* in Indonesia, *mahafir* in Jordan or *qanats* in Oman, or the Chagga irrigation systems on the foothills of Mount Kilimanjaro.

What is clear is that all these systems had a predominantly agricultural objective. Political and social organisation was underpinned by the logic of water control in order to improve agricultural output and food security. This remained true of water control developed much later, such as the drainage of low-lying swamps to create agricultural land in seventeenth-century England and Holland. However, following the development of navigation in western Europe in the sixteenth century, and the subsequent expansion of European conquest and power, the growth of investment in manufacturing technology accelerated. This was associated with a rapid diversification in the industrial use of water, for cooling and cleaning, and for navigation canals and power generation. Early manufacturing in the eighteenth century simply scaled up existing water wheel technology traditionally used for grain milling, using ever-larger wheels to supply mechanical power to large-scale manufacturing units. However, the development of steam technology powered by coal enabled greater concentrations of industry and the association of industry with urbanisation. This was exemplified by Manchester, whose population increased from around 95,000 in 1800 to 310,000 in 1840. The rapidly growing scale of industrial activity and its concentration in urban areas meant that local demand for water for industrial use and for human sanitation intensified to the point that local water resources became degraded, as this account from 1840s Manchester makes clear:

> The view from this bridge, mercifully concealed from mortals of small stature by a parapet as high as a man, is characteristic for the whole district. At the bottom flows, or rather stagnates, the Irk, a narrow, coal-black foul-smelling stream, full of debris and refuse, which it deposits on the lower right bank. In dry weather, a long string of the most disgusting blackish-green slime pools are left standing on this bank, from the depths of which bubbles of miasmatic gas constantly arise and give forth a stench unendurable even on the bridge forty or fifty feet above the surface of the stream. ... Above the bridge are tanneries, bonemills, and gasworks, from which all drains and refuse find their way into the Irk, which receives further the contents of all the neighbouring sewers and privies.
> (Engels, 2005 (1845 German original publication): 89)

Apart from documenting the sheer depths to which environmental conditions had sunk in this early industrial heartland, as a result of industrial pollution and lack of human sanitation, Engels's account makes clear that the resulting hazards to human health, notably of typhus and cholera, were endured above all by the poor, who lived in the city, rather than the middle classes in the surrounding suburbs and country areas. It was only in the 1880s that sewage was intercepted from entering the rivers and diverted to a purpose-built treatment plant.

The increased intensity of water use by industrial cities is illustrated by Manchester's widening reach for water supplies: clean water for the city was secured by the municipal authorities initially by constructing reservoirs 12km away, in the nearby uplands of the Peak District in the 1850s, but subsequently (1890s) via an aqueduct from the Lake District, 150km to the north. In this development of urban water infrastructure and services it is possible also to see a historical progression that is replicated in many urban areas in the nineteenth and early twentieth centuries (see Chapters 2 and 5). Swyngedouw (2006) presents these as a series of recognisable steps across Western (OECD) countries:

1 Up to 1850s – small private (commercial) water-supply companies
2 Up to 1920s – municipal bodies investing in water and sanitation measures due to health and environmental concerns
3 Up to 1970s – large-scale, national control of water investment and regulation with goals of economic development and social-welfare improvement
4 Post-1970s – period of privatisation and decline of state-led economic development

The inequality of access to clean water and sanitation that marked industrialisation at its beginnings continues to dominate discussion of water more than 150 years later, translated and projected from the conditions of early industrial towns onto the stage of the global economy of the twenty-first century.

In the intervening period, an unprecedented growth in human population and economic productivity has seen water resources developed on an ever larger scale to meet increasing demands of urban population and industry, both directly through storage and diversion of water to urban areas, and indirectly via the use of water in irrigated agriculture. As Table 1.1 indicates, in the heavily industrialised regions of North America and Europe, industry accounts for about half of all water withdrawals, but a third is used for irrigated agriculture. Elsewhere, agriculture accounts for the overwhelming majority of water use, reaching over 80 per cent of water withdrawals in the least industrialised regions. While such large regions cover great disparities (e.g. 'Asia' includes Japan, as well as India), these summary data underline the concern that demand for water by expanding cities is matched by the demand for water to grow food.

The emergence of an international agenda for water-resource management

Increasing use of water, and consequent depletion, evidenced in reduced river flows, degraded water quality and increasing depths of groundwater, had by the 1970s led some to question whether continued growth of existing patterns of water consumption were sustainable. In many respects this reflected a more widespread groundswell of environmental concern that questioned whether the long economic and demographic boom of the 1950s and 1960s, which saw a doubling of the world population in thirty years, would be sustainable in the future. Books by Carson (S*ilent Spring*, 1962), Ward (*Spaceship Earth*, 1966) and Ehrlich (*The Population Bomb*, 1968) all voiced worries over the future of the planet. This concern, epitomised in the book *Limits to Growth* (Meadows et al., 1972), published by the Club of Rome think tank, and visually reinforced by the first photographs of the Earth taken by astronauts on the Apollo missions to the moon, posited an inevitable contradiction between the finite resources of the planet and the indefinite growth in resource use by an expanding human population.

Just as the original Malthusian predictions of failure of food production to keep pace with population growth failed to materialise in the nineteenth century, so the neo-Malthusian warnings that industrial economies would fail in the twentieth century due to exhaustion of supplies of non-renewable resources (notably metals) proved inaccurate. Nonetheless, the persistent concern over the contradiction between finite resources and continued demographic and economic growth continues to influence many environmental discourses. It has been manifest in increasing formal commitment to international management of natural resources (including water), and gains reinforcement from the evidence of climate change linked to increases in atmosphere concentrations of carbon dioxide and other 'greenhouse gases' generated by industrial societies.

Formal international recognition of this environmental concern was first apparent with the United Nations Conference on the Human Environment, held in Stockholm in June 1972. The conference ended with a declaration of 26 principles, among which a concern with the limited extent of natural resources is evident:

> Principle 2: The natural resources of the earth, including the air, water, land, flora and fauna and especially representative samples of natural ecosystems, must be safeguarded for the benefit of present and future generations through careful planning or management, as appropriate.

> Principle 3: The capacity of the earth to produce vital renewable resources must be maintained and, wherever practicable, restored or improved.

> Principle 4: Man has a special responsibility to safeguard and wisely manage the heritage of wildlife and its habitat, which are now gravely

imperilled by a combination of adverse factors. Nature conservation, including wildlife, must therefore receive importance in planning for economic development.

Principle 5: The non-renewable resources of the earth must be employed in such a way as to guard against the danger of their future exhaustion and to ensure that benefits from such employment are shared by all mankind.

The Stockholm conference led to the establishment of the United Nations Environment Programme (UNEP), and a wave of international and national initiatives as environmental concerns became incorporated explicitly into the mainstream of government responsibility. This growth of an environmental agenda was not unproblematic, however. 'Environmentalists' became quickly identified as opponents to 'developers', not least because early steps by government to give greater consideration to environment, such as the National Environmental Policy Act of 1969 in the United States, specifically sought to entrench account-ability of 'developers' (in that instance US Federal government agencies) for negative environmental impacts of development projects.

In the two decades that followed, the proper balance between environ-ment and development were widely debated, notably through hearings conducted by the World Commission on Environment and Development (WCED – the Brundtland Commission) from 1983 to 1986. The WCED report 'Our Common Future', published in 1987, formulated a reconciliation between 'environmental' and 'development' concerns under the banner 'sustainable development'. In arguing that poor people were led to over-exploit natural resources in order to survive, the report identified poverty as a major driver of environmental damage, and hence more equitable development as a means of resolving both social and environmental ills. This poverty–environment link was strongly manifest in the hydrological link proposed between poor farmers' need to clear forest in upper watersheds and increased flood risks for cities and industries in lower river basins. The 'poverty reduction' message of WCED was, nonetheless, couched within an overall discourse of resource scarcity in the face of growing human population. It should also be noted that while issues of poor sanitation and water pollution make frequent appearances in the WCED report, and ten pages are devoted to marine resources, 'water management' itself is addressed in only one paragraph. The WCED report underpinned the United Nations Conference on Environment and Development in Rio de Janeiro in 1992, and the resultant Declaration, 'Agenda 21', was adopted by leaders of the 178 countries represented at the Conference (UNCED, 1992).

The Brundtland Commission's definition of sustainable development – development that meets the needs of the present without compromising the ability of future generations to meet their own needs – met widespread acceptance, but the means of achieving it were disputed. The measures proposed within the 40 chapters of Agenda 21 were comprehensive in scope but necessarily generic in nature, and provided wide scope for interpretation of priorities. Thus, in relation to water, Chapter 18, 'Protection of the quality and supply of fresh-water resources: application of integrated approaches to the development, management and use of water resources', runs to 90 sections divided among seven 'programme areas'. It identifies as its first objective 'integrated water resources management', which it states (art. 18.8):

> is based on the perception of water as an integral part of the ecosystem, a natural resource and a social and economic good, whose quantity and quality determine the nature of its utilization. To this end water resources have to be protected, … , in order to satisfy and reconcile needs for water in human activities. In developing and using water resources, priority has to be given to the satisfaction of human needs and the safeguarding of ecosystems. Beyond these requirements, however, water users should be charged appropriately.
>
> (http://www.un.org/esa/sustdev/documents/
> agenda21/english/Agenda21.pdf)

This text leaves room for interpretation of the range of 'human needs' for which water is to be prioritised and how the cost of this water is to be financed. In practice, while sharing many of the ideas presented in Agenda 21, international discussions about policy and management of water resources have been characterised by a struggle between competing viewpoints (see, Water Markets vs Water Rights p. 35).

Debates on water-resource management need to be seen in the institutional context of a highly diffused competence on water issues within the United Nations system, with over 20 different UN agencies involved in aspects of water assessment, development or management. Early initiatives focused upon the assessment of water resources and were exemplified by UNESCO's International Hydrological Decade (1965–74) followed by its International Hydrological Programme, the World Meteorological Organisation's Operational Hydrology Programme, and the FAO's programme on irrigation and drainage. The 1977 UN Water Conference at Mar del Plata, Argentina, remains the only United Nations conference ever to have been devoted exclusively to water and demonstrates growing awareness of human impact upon the hydrological cycle. It ended with two resolutions, one concerning the assessment of water resources and the other concerning 'community water supply', and a 'plan of action'. In the light of later developments, it is worth noting from the resolution on water supply that, at that time at least, there was a presumption of a 'human right to water', at least to meet people's 'basic needs'. Moreover, the conference recommendations focused upon mobilising human, technical and financial resources in order to meet those needs (Box 1.5).

The Mar del Plata conference prompted further international water initiatives, such as the International Drinking Water Supply and Sanitation Decade (1981–90), but the 'developmental' emphasis on

---

Box 1.5

Declaration 1977 UN Water Conference at Mar del Plata, Argentina

'In view of the course taken by the discussions and the aspirations of the countries represented at the United Nations Water Conference and in view also of what was proposed at Habitat: United Nations Conference on Human Settlements, and

Considering that:

(a) All peoples, whatever their stage of development and their social and economic conditions, have the right to have access to drinking water in quantities and of a quality equal to their basic needs;
(b) It is universally recognized, that the availability to man of that resource is essential both for life and his full development, both as an individual and as an integral part of society;
(c) To a significant extent similar considerations apply to all that concerns the disposal of waste water, including sewage, industrial and agricultural wastes and other harmful sources, which are the main task of the public sanitation systems of each country;

…

Recommends:

(a) That where human needs have not yet been satisfied, national development policies and plans should give priority to the supplying of drinking water for the entire population and to the final disposal of waste water; and should also actively involve, encourage and support efforts being undertaken by local voluntary organizations;
(b) That Governments reaffirm their commitment made at Habitat to "adopt programmes with realistic standards for quality and quantity to provide water for urban and rural areas by 1990, if possible";

(c) That with a view to achieving these ends, the nations which need to develop their systems for providing drinking water and sanitation should prepare for 1980 programmes and plans to provide coverage for populations and to expand and maintain existing systems; institutional development and human resources utilization; and identification of the resources which are found to be necessary;

(d) That the United Nations agencies should co-ordinate their work efforts to help Member States, when they so request, in the work of preparation referred to in subparagraph (c) above;

(e) That in 1980 the national programmes which have been implemented for that purpose, and the extent to which the countries concerned have succeeded in mobilizing local and national support should be reviewed by an appropriate mechanism to be determined by the Economic and Social Council and based on the use of existing machinery, with a view to attaining co-ordinated action toward agreed targets;

(f) That in accordance with the decisions of the existing structures of the Economic and Social Council, appropriate external assistance should be available in order to assist in building, operating and maintaining these systems;'

meeting human basic needs, which dominated the 1977 conference, was increasingly subsumed within the broader 'sustainable development' agenda, in which 'development' needed to be reconciled to questions of resource conservation.

A recasting of water development priorities emerged from the International Conference on Water and the Environment (ICWE) held in Dublin in 1992. This was conceived as a follow-up to the Mar del Plata conference but also as 'a conference where experts and professionals well versed in the different fields of water would prepare the major input on fresh water to the [1992, Rio de Janeiro] United Nations Conference on Environment and Development' (Young et al., 1994: 32). Convened by the Inter-secretariat Group on Water Resources representing 24 UN agencies, the Dublin meeting was to be for 'government-appointed experts, rather than government delegations. This procedure avoided the danger that one government conference (ICWE) could have had its findings overturned at another government conference (UNCED)' (Young et al., 1994: 33). The ICWE involved some 500 participants from 114 countries, 28 UN agencies and 58 non-government organisations (Young et al., 1994).

While its recommendations were incorporated within Agenda 21, the five-page 'Dublin Statement on water and sustainable development', and particularly its four 'guiding principles' (see Box 1.6), have proved more influential in debates on water-resource management in the subsequent decades. One indicator of the influence of the 'Dublin principles' is their re-working in subsequent reports (e.g. WWC, 2000) to generate three:

> An 'ecological principle' that recognizes the need for holistic management of the water resource, usually interpreted as managing water in an 'integrated' way, according to its hydrological units, such as river basins or subterranean aquifers.

Box 1.6

ICWE 'Dublin' Principles

Principle No. 1: Fresh water is a finite and vulnerable resource, essential to sustain life, development and the environment

Since water sustains life, effective management of water resources demands a holistic approach, linking social and economic development with protection of natural ecosystems. Effective management links land and water uses across the whole of a catchment area or groundwater aquifer.

Principle No. 2: Water development and management should be based on a participatory approach, involving users, planners and policy-makers at all levels

The participatory approach involves raising awareness of the importance of water among policy-makers and the general public. It means that decisions are taken at the lowest appropriate level, with full public consultation and involvement of users in the planning and implementation of water projects.

Principle No. 3: Women play a central part in the provision, management and safeguarding of water

This pivotal role of women as providers and users of water and guardians of the living environment has seldom been reflected in institutional arrangements for the development and management of water resources. Acceptance and implementation of this principle requires positive policies to address women's specific needs and to equip and empower women to participate at all levels in water-resources programmes, including decision-making and implementation, in ways defined by them.

Principle No. 4: Water has an economic value in all its competing uses and should be recognised as an economic good

Within this principle, it is vital to recognise first the basic right of all human beings to have access to clean water and sanitation at an affordable price. Past failure to recognise the economic value of water has led to wasteful and environmentally damaging uses of the resource. Managing water as an economic good is an important way of achieving efficient and equitable use, and of encouraging conservation and protection of water resources.

An 'institutional principle' that combines Dublin principles 2 and 3 in calling for decision-making on water resources to be decentralized to the smallest scale feasible, following criteria of subsidiarity, and to be representative of all water users.

An 'instrument principle' which recognizes water as an economic good whose efficient use and conservation should be promoted by charges payable by users.

These three principles can be found underpinning a large number of reforms of water policy and legislation undertaken in the decade following the Dublin and Rio conferences, with particularly notable examples in Mexico (1992), Brazil (1997) and South Africa (1998), and in the European Union's Water Framework Directive (WFD) (2000).

Although melded into a single framework for water-resource management, the three 'principles' represent quite distinct perspectives on water resources. The 'ecological principle' effectively embodies a perspective that emphasizes scientific management. Quantification of the hydrological cycle within a catchment framework is the basic methodology and this has its roots in a seventeeth-century analysis of flows in the river Seine through Paris (see Box 1.3 for an explanation of the computation of the hydrological cycle). The catchment or river basin has been widely promoted as the ideal framework for water-resource analyses with paired catchment studies being promoted for scientific investigations for much of the twentieth century. By the 1970s scientific doubts were emerging over scale problems and leakage losses. By the 1990s it became evident that such scientific inventories of water were becoming too costly to maintain while attention had started to focus increasingly on water quality and ecosystem functioning. Whereas in the past, from the mid-nineteenth to mid-twentieth century, the scientific perspective characterised hydrology from a primarily engineering standpoint, this had by the latter half of the twentieth century become qualified by environmental considerations, with initiatives to not merely control and utilise water resources, but also to sustain their ecological functions, which included the maintenance of some degree of 'natural flow' in watercourses, and attention to recharge rates of aquifers, for example the assessment of ecological low flows by Petts et al. (1996).

With this modification, however, the conceptualisation of management in terms of hydrological units of 'basin' or 'catchment' effectively frames decisions in terms of the physical and biological properties of the resource. In contrast, the 'institutional principle' and 'instrument principle' do not require that water management be conceptualised in terms of 'natural' hydrological units. Rather, the focus is on processes of water allocation, on the one hand through pricing mechanisms, and in the other through 'deliberative' decision-making in which all water users or 'stakeholders' (i.e. including those whose interests, such as amenity value, may be affected by the water use of others) are represented. As can be seen in the original 'Dublin' formulation (Box 1.6), principles 2 and 3 carried a strongly redistributive message: that those who had been disadvantaged in previous water-management arrangements should gain stronger representation and thus become 'empowered' to take a stronger role in water-management decisions. Women were explicitly identified as needing stronger representation in water management, in recognition of the strongly gendered nature of the labour of providing household water supplies in many societies (see Rathgeber, 1996; Peter, 2006; UNWWD, 2006; discussion in Chapter 6).

These three water-management principles, in turn, represent distinctive, and often opposed, twentieth-century views of how society should be governed. As noted above, the ecological principle embodied 'scientific management', generally invoking the agency of a Weberian state model to ensure that individual, smaller-scale, actions are subordinated to a greater good, defined variously in ecological, political/security, or economic terms. The importance attributed to decentralised and participatory decision-making processes can be traced to the theories of deliberative democracy associated particularly with the writings of Habermas in the 1970s, whereas the emphasis on water pricing as a means of determining allocation priorities is a hallmark of neo-liberal ideas, whose resurgence in the 1980s coincided with the concept of 'sustainable development'. Indeed, the emergence of the sub-discipline 'environmental economics' in the 1980s was the principal means through which both environmental concerns were incorporated into neo-liberal policy and, as a consequence, 'sustainable development' was operationalised as tangible action in the form of 'polluter pays', 'valuation of ecological services', etc. It may be seen, therefore, that the 'Dublin Principles' embedded within water management the major competing visions of social organization in the late twentieth century. The unresolved tensions were to be manifest in the subsequent trajectory of international debate on water-resource development.

Water markets vs water rights

One of a number of networks set up in the aftermath of the 1972 Stockholm conference to link academics and technical professionals working on natural resources was the International Water Resources Association (IWRA), which convened a World Water Congress every three years. At its 8th World Water Congress, in 1994, the IWRA adopted a resolution to create a World Water Council to act as a 'common umbrella' to unite 'the disparate, fragmented and ineffectual efforts in global water management'. Following further preparatory meetings in Montreal and Bari (Italy) in 1995, the IWRA became one of nine founding members of the World Water Council (WWC), established with a headquarters in Marseille, to provide a 'multi-stakeholder platform' to encourage debates and exchange of experience in order to generate a 'common strategic vision on water resources and water services management amongst all stakeholders in the water community' (WWC, 2010).

The WWC embarked on a series of 'World Water Forums' (WWF) held every three years in parallel with the IWRA Water Congress. The Water Forums quickly came to acquire a much higher profile than the Congress. The second WWF, in The Hague, attracted some 5,700 participants, including 114 government ministers. In addition there were 32,500 visitors to a World Water Fair associated with the Forum. The WWF at The Hague also saw the launch of a World Water Vision, a report by the 'World Commission on Water in the 21st Century' convened to be written by the WWC.

The report proved controversial. Much of it adhered to the Dublin Principles. Thus, it advocated 'holistic' management of the water resource using catchments or basins as units of management, rather than 'artificial' political or administrative boundaries (WWV, 2000: 25). It also declared that 'participation is key' with community groups and user associations playing a role at a local scale either in managing services such as sewerage or irrigation or in monitoring providers of such services, and 'user parliaments' working at a larger scale with governments to decide 'what is done, how it is done, and who will pay for it'. Further, it declared 'empowering women's groups, the poor, youth, and community-based groups to have an adequate voice' in decision-making 'is a necessary pillar' of water management. Finally, true to the 'instrument principle' the Water Vision argued that water must be recognised as a scarce commodity, whose price should provide an incentive to transfer use from inefficient to 'higher-valued' uses.

This argument was further extended to claim that the 'gloomy arithmetic' of population growth and rising demand for urban and agricultural water use was so great that investment would be required that was beyond the fund-raising capacity of governments and public-sector water utilities. As a consequence, the report argued, investment for services must come increasingly from the private sector. This would only be possible, however, if water prices covered the 'full cost' of supply and private sector providers of water services were assured of a 'transparent, predictable, regulatory regime to protect consumers and investors'. In effect, the argument had translated a 'scarcity' of water into a 'scarcity' of capital that could be relieved only by recourse to international capital markets through providing the incentive of 'full-cost recovery' through higher water prices.

The report argued that access to water for the poor should be protected by government subsidies, while the private sector got on with the 'business' of providing water services. Candidates to fill this private-sector role were not identified by the World Water Vision, but their importance was underlined by the document's insistence that 'full cost pricing of water' was 'the single most immediate and important measure' (WWV, 2000: 33). Many readers interpreted the WWV as advocating the creation of conditions in developing countries favourable to expansion of the activities of large, mainly French and British-based, water companies that had resulted from privatisation of state water utilities in the 1980s. On this score, the 'Vision' provoked immediate criticism, particularly from NGO groups who rejected it, claiming it emphasised 'a corporate vision of privatisation, large-scale investments and biotechnology as the key answers. ... (and) ... insufficient emphasis and recognition of the rights, knowledge and experience of local people and communities and the need to manage water in ways that protect natural ecosystems' (Morley, 2002: 7).

In the light of this perception, the timing of the World Water Vision was unfortunate since it coincided with growing tension in Bolivia over just the type of 'private-sector' water concession the document advocated. The Cochabamba concession, held by a consortium led by a company owned jointly by Bechtel, from the US, and United Utilities, from the UK, was in the event cancelled by the Bolivian government the month after WWF in The Hague, following two weeks of rioting which left six dead. Nickson and Vargas (2002) have argued that opposition to the concession reflected a coalition of interests which saw the concession as a threat. These included street vendors of water, drillers of wells and wealthier householders who not only saw a rise of 106 per cent in their water tariffs but also found that the wells they had sunk on their own properties were subject to control by the concession-holder. According to Nickson and Vargas, this array of minority interests was able to mobilise the poor, whose water costs had risen relatively little (10 per cent). This was possible because of widespread opposition to the Bolivian government's economic liberalisation programme that had resulted in slow economic growth and widespread unemployment. In this sense, the violence of the struggle over water was symptomatic of wider social discontent. Nonetheless, the cancellation of the Cochabamba concession was widely hailed as a popular victory against global capitalism, and heralded cancellations of concessions held by multinational water companies elsewhere, notably in Dar-es-Salaam (in 2005) and Buenos Aires (in 2007).

More generally, after 2000 the 'water community' which the WWC aimed to unite became increasingly splintered as the World Water Forums became the target of protest and Alternative Water Forums seized opportunities to accuse the WWF of being dominated by commercial interests. One BBC correspondent at the 2003 Forum in Kyoto observed 'international conferences are often described as talking shops. Not this one. It was a giant talking hypermarket' (Hirsch, 2003). Similarly, following violent scuffles at the opening of the 2009 WWF in Istanbul, a member of the Turkish Green Party claimed that the World Water Forum was run by 'a private organisation founded by water companies' and 'not a legitimate organisation. But it is still signing agreements that bind governments. That is what we stand against first and foremost' (*Hurriyet Daily News*, 17 March 2009).

As in Cochabamba, water provided a focus and a means to channel struggle that had much wider origins. However, more formal processes have since 2000 reinforced a retreat from an emphasis on market instruments to allocate water to a renewed emphasis on water rights, more in line with the recommendations of the Mar del Plata Conference. Within six months of The Hague WWF, the United Nations Millennium Summit had declared a series of specific targets to be achieved within 15 years. These targets, later known as the 'Millennium Development Goals', included: 'target 7c: reduce by half the proportion of people without sustainable access to safe drinking water and basic sanitation'. This target was adopted as a national goal by the governments of 189 countries (http://www.undp.org/mdg/basics.shtml). It was further reinforced at the Johannesburg World Summit on Sustainable Development, convened in 2002 to review a decade's progress since the UNCED in 1992, and which declared: 'We welcome the focus of the Johannesburg Summit on the indivisibility of human dignity and are resolved, through decisions on targets, timetables and partnerships, to speedily increase access to such basic requirements as clean water, sanitation, adequate shelter, energy, health care, food security and the protection of biodiversity' (http://www.un.org/esa/sustdev/documents/WSSD_POI_PD/English/POI_PD.htm).

The 'Millennium Development Goals' became a yardstick for progress in reducing poverty, and frequently a reproach to those in power when that progress proved inadequate. The United Nations Development Programme reported in 2008 that, in aggregate, the Millennium Development Goal for access to drinking water was likely to be reached, with 89 per cent of people in developing countries having access to safe drinking water by 2015. This masks an uneven picture, however, with comparatively little progress in some regions, such as sub-Saharan Africa. Moreover progress on improved sanitation was reported to be very much weaker. The impact of these formal international commitments, then, was quite clearly not to bring about achievement of the goals they defined (although there are some years yet to 2015), but their effect in establishing access to water and sanitation as a basic condition of human life and dignity lent weight to a discourse of water that challenged the 'full-cost recovery' and 'private-sector-delivery' emphases of the World Water Vision in 2000.

The shift is evident in the 'Synthesis' of the 4th World Water Forum, in Mexico City (WWC, 2006), which noted 'something new' in that the right to water was 'for the first time debated in a rather consensual spirit' (p. 2), but later observed 'participation of the private sector was a controversial issue during the Forum, but the consensus was that the decision and manner of private participation depends only on the local actors' (p. 29). This caution in relation to the private sector suggests elsewhere in the document the impact of the Cochabamba episode on thinking within the water sector: 'There is a strong need to initiate dialogue within civil society on the involvement of the private sector to choose a model people want to follow for provision and enhancement of their water services. Water service provision is and will remain mainly public. Private sector provision must be considered as only one of the possibilities taking into account the accumulated knowledge and previous experiences' (p. 52). Finally, in its concluding remarks, the 'Synthesis' states that the Forum had shown that the 'right to water' was now better understood, and, despite divergent views, 'the concept is not merely ideological, but has a concrete meaning for all and is endorsed by many, starting with the local authorities and parliamentarians' (p. 108). Immediately following this statement is the conclusion that

> local authorities have an essential role in generating finance for investments in water services. The development of local financing capacity and of local financial markets was emphasized, and a number of ways to achieve this were proposed, but require more debate and testing in real terms. This necessity is founded on the recognition that users and taxpayers are in the

end the main financiers, and on the associated shift from full-cost recovery to a solidarity system of fair tariffs combined with targeted subsidies.

(WWC, 2006: 108)

This view at the conclusion of the 4th WWF is echoed in UNESCO's World Water Development Report for 2009, published at the 5th WWF in Istanbul, which states that in relation to financing: 'while there may appear to be many financing options for water resources development, governments still have only three basic means of financing them: tariffs, taxes and transfers through aid and philanthropy' (WWAP, 2009: xxii). This does not mean that governments will not borrow money from capital markets in order to fund development of water resources. This is often inescapable to meet the large capital investment required by major water infrastructure. However, it makes clear that the cost of borrowed capital, if it is not written off through international aid, must be paid by water users or by taxes. This provides a perspective in which the 'instrument principle' remains – in that access to water is defined as having a cost – but that the social distribution of that cost is governed by social criteria of need and ability to pay.

Summary: water resources and development

In this chapter we have sketched a relationship between water resources and development which is not simply a matter of a 'fixed resource' acting as a constraint to the growth of human society and economy. We have argued instead that human societies have invested in technology and infrastructure that has enabled patterns of settlement that qualitatively and quantitatively do not correspond to natural patterns of water distribution, overcoming in some measure the hazards of water excess or deficit. We have suggested that concepts of best practice in water-resource management that emerged during the course of the twentieth century – most commonly summarised in terms of the 'Dublin principles' – reflected not only 'scientific' understanding in a biochemical or hydrological engineering sense, but also incorporated contrasting assumptions about how society should be governed. Strategies for managing water – and particularly shortages of water – will therefore reflect not only the characteristics of natural water resources and the nature and intensity of human demand (as translated through industry, agriculture, and settlement), but also the outcomes of political contestation over what constitute fair and legitimate processes of allocation of water within society.

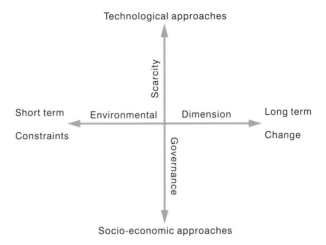

Figure 1.15 Variation in water-resource management strategies

Bearing this in mind, the remainder of the book will explore the role of water resources in development based upon a framework (Figure 1.15) that recognises management strategies as being positioned with respect to two principal axes, one concerned with the environment, and the second with a gradation between 'technological' strategies on the one hand and social or economic (human behaviour) strategies on the other. The environment imposes constraints on human activity, through short-term shocks such as drought and flood, but also longer-term changes, a current focus for debates over global warming impacts. These two axes present four extreme positions which are useful for structuring arguments but are not exclusive. White (2008), for example, presents a vision of human endeavours within a city context to combat environmental constraints and proposes the vision of the 'adaptive city' where responses to water variations include both planned infrastructure and people's behaviours. Nevertheless, technical responses tend to be concerned with water 'scarcity', while human responses are primarily a matter of 'governance'.

Of course there are more than two dimensions to water management and Figure 1.16 depicts the need to take into account both spatial and temporal dimensions. On the temporal scale, or timeframe, it may be supposed the growing importance of issues of sustainability imply the timescales of management strategies are becoming longer-term. On the axis of spatial scale, direction of movement is less clear, as reforms that place more emphasis on social and economic behaviour are overlain with changing spatial scale for intervention. Thus, hydrological analysis is increasingly posited in terms of surface water-catchment units and

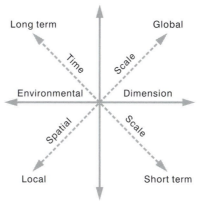

(Technological approaches)

Long term

Global

Time

Scale

Environmental

Dimension

Spatial

Scale

Local

Short term

(Socio-economic approaches)

Figure 1.16 Various dimensions in water-resource management strategies

subunits, or groundwater aquifers. Similarly, but at a larger scale, concerns with climate change increasingly draw attention to water behaviour at the 'global' scale or at the scale of regional climate systems. While taking account of these shifting spatial scales of water management, in structuring this book we have opted to focus on the two axes represented in Figure 1.15. The way the chapters can be located within this framework is illustrated in Figure 1.17.

In this book we seek to step back from existing, frequently prescriptive, 'best practice' consensus and undertake an interdisciplinary review of empirical cases in order to understand the ways in which water management both constrains, and is shaped by, specific development scenarios. Development is understood as an economic or political imperative where access to and utilisation of water is a fundamental requirement for achieving many twenty-first century aspirations and challenges.

In Chapter 2 we explore the role of water resources in underpinning the economic growth of today's wealthiest societies. In examining past trajectories of water development, we can also ask what lessons we can learn for today's 'developing' countries. In Chapter 3 we turn from the management of water as dealing with relatively short-term timescale of economic growth to the longer-term processes of adapting to altered patterns of water availability inherent in a changing climate. Chapters 4 and 5 look at the development of water to enhance supply. In Chapter 4 the focus is on the provision of (more) water to reduce rural poverty through irrigated agriculture. In Chapter 5 we consider the more general

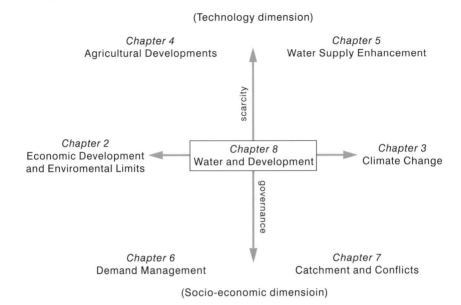

Figure 1.17 Book structure, chapters and water management

problem of enhancing water supply. Both chapters are located at the 'technological' end of the axis for water-management strategies, and both are framed primarily by the logic of combating 'water scarcity'.

The following two chapters, by contrast, are more concerned with social and economic approaches to water management. Chapter 6 explores ways of managing (i.e. reducing) water demand by changing the behaviour of water users. Chapter 7 addresses the management of competing water uses, including within water catchments that cross international boundaries. The final chapter will critically revisit current ideas of 'best practice' in water-resource management and consider their relevance to the development challenges ahead.

Further reading

We recommend the use of the following web-based sources to provide an overview of the global water crisis that can be expected to be updated on a regular basis:

World Bank 2008 *World Development Report 2008: Agriculture for Development*. The World Bank, Washington DC: http://econ.worldbank.org/ WBSITE/EXTERNAL/EXTDEC/EXTRESEARCH/EXTWDRS/EXTWDR2 008/0,,menuPK:2795178~pagePK:64167702~piPK:64167676~theSitePK: 2795143,00.html

WWC 2010 World Water Council WRC: http://www.worldwatercouncil.org/
    index.php?id = 25
WWDR 2006 UN World Water Development Report, 2nd Report: *Water a
    Shared Responsibility*. UNESCO, Paris: http://www.unesco.org/water/wwap/
    wwdr/wwdr2/
WWDR 2009 The UN World Water Development Report, 3rd. Report: *Water in
    a Changing World*. UNESCO, Paris. http://www.unesco.org/water/wwap/
    wwdr/wwdr3/

Data on water resources can be obtained from:

WRI 2010 *Earth Trends* World Resource Institute:http://www.wri.org/
FAO-AQUSTAT (2009) Food and Agricultural Organisation, UN.
    AQUASTAT: http://www.fao.org/nr/water/aquastat/regions/

We also suggest that a historic perspective on key changes in the attitudes
towards water-resource management towards the end of the twentieth century
can be obtained through reading the following:

UNCED 1992 United Nations Conference on Environment and Development,
    The Earth Summit. Rio de Janeiro: http://www.un.org/geninfo/bp/enviro.html
WCED 1987 *Our Common Future*. The World Commission on Environment
    and Development. Oxford: Oxford University Press.
Young, G., Dooge, J. and Rodda, J. 1994 *Global Water Resource Issues*.
    Cambridge: Cambridge University Press.

2 Economic growth, environmental limits and increasing water demand

Introduction

For the past two hundred years economic growth has been driven primarily by social and economic transformation associated with industrialisation and urbanisation. We saw in the previous chapter that from its outset the process of urban and industrial growth profoundly altered patterns of water use, leading to a higher share of total water use by industry, in industrial regions of Europe and North America (Table 2.1). While this may in some instances represent a shift from agricultural to industrial use, in most instances industrial water use has tended to be additional to agricultural water use. Thus, industrial development tends to raise total water use per head of population (Figure 2.1), although comparison between Europe and North America suggests that it is not industrialisation that is the sole reason for the very high water-withdrawal rate in North America. Another factor is undoubtedly the relative aridity of the climate of the western United States where a great deal of economic activity is located, coupled to the widespread adoption of irrigation to overcome this climate constraint. Table 2.1 illustrates agriculture's thirst, withdrawing high volumes of water which may then be transported from the growing region as the products are marketed elsewhere. Thus large amounts of inter-basin water transfers take place through the transport of agricultural produce, evident as an external 'water footprint', or 'virtual water' (see the second section in the previous chapter, Water Scarcity: a Water Crisis p. 7). Hoekstra and Chapagain (2007) calculate the global average water footprint is 1240 $m^3cap^{-1}y^{-1}$ with the highest being the USA at 2480 $m^3cap^{-1}y^{-1}$,

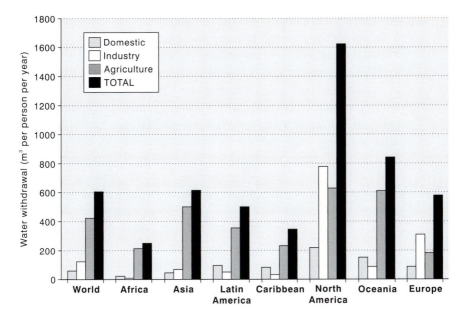

Figure 2.1 Water withdrawal by global region for year 2000
Sources: after WRI, 2010 and FAO-AQUASTAT, 2010

compared to Egypt at 1097 $m^3cap^{-1}y^{-1}$, India at 900 $m^3cap^{-1}y^{-1}$ and China 700 $m^3cap^{-1}y^{-1}$. They relate these figures to particular patterns of development, with differences being explained by gross national income, level of meat consumption, climate conditions and finally water use efficiency in agriculture. Thus demand for water is integral to processes of economic development, whether through irrigation (see third section below, Water and Economic Development p. 54) or industrialisation, and efficient water use requires investment and more highly organised forms of water management.

Table 2.1 *Water withdrawal per person ($m^3c^{-1}y^{-1}$) in 2003*

	Domestic (%)	Industry (%)	Agriculture (%)	Total
World	60 (10)	124 (20)	421 (70)	605
Africa	25 (10)	11 (4)	215 (86)	251
Asia	44 (7)	70 (11)	501 (81)	615
Latin America	94 (19)	52 (10)	353 (71)	499
Caribbean	80 (23)	33 (9)	234 (68)	346
North America	216 (13)	779 (48)	627 (39)	1621
Oceania	148 (18)	85 (10)	610 (72)	842
Europe	87 (15)	307 (53)	182 (32)	576

Source: calculated from FAO-AQUASTAT, 2010

This chapter will examine the role of water use in economic growth in industrial societies during the past two centuries. It will focus on the role of capital investment, and particularly strategic investment by the state, in developing water resources to meet economic goals. It will also chart the evolution of political perceptions about these patterns of investment in water infrastructure particularly. Changes in these perceptions, particularly in the latter half of the twentieth century, resulted in radical shifts in management priorities by the start of the twenty-first century. We will then consider what lessons can be drawn for water-development strategies in today's 'developing countries' – those that remain largely non-industrialised. First, however, we will consider a threat to economic growth caused not by water scarcity but deteriorating water quality caused by rising levels of waste and resultant contamination of water.

Contamination and sanitation

In Chapter 1 we initially presented water availability in terms of quantities per person but the growth of industrial societies has been matched by an increasing volume and diversity of waste products which can impact upon the water environment leading to pollution at times rendering the resource unusable. We observed in the third section, Water in Historical Perspective (p. 23), that deterioration of water quality has historically been a feature of rapid industrialisation. The expansion of early industrial cities was not matched by developments in their sewage and waste water treatment. As a consequence, by the mid-nineteenth century conditions in large urban areas had become intolerable through overcrowded living spaces and the release of untreated industrial effluents. This was not new, as Brimblecombe (1987: 10) records that an investigation in 1307 found the stench from the River Fleet in London was due to waste from the butchers and tanners at Smithfield market. However, five centuries later the scale of the problem had increased to the point that, in 1858, Parliament in London was closed because of the stench from the Thames, conditions captured by Dickens in his novel *Little Dorrit*:

> It was a Sunday evening in London, gloomy, close and stale … Miles of close wells and pits of houses, where the inhabitants gasped for air … In the city it (rain) developed only foul stale smells, and was a sickly, lukewarm, dirt-stained, wretched addition to the gutters.

Although the basic technologies for improving water quality were developed at the time Dickens was writing, the condition of rivers in industrial Europe continued to deteriorate until the last quarter of the twentieth century (see fifth section below, Managing 'Developed' Rivers in Industrialised Societies, p. 66), when major investments began to improve water quality. However, Figure 2.2 shows that even over the last two decades the amount of water contamination globally, as measured by industrial effluents entering water courses, has been rising. To this pollution from industrial 'point sources' has been added waste from 'diffuse sources', particularly that generated by agricultural intensification during the latter half of the twentieth century: irrigation has led to extensive soil and water *salinization* (Jones, 1997); heavy use of fertilisers continues to raise *eutrophication* (nutrient enrichment) levels (Sprague et al., 2009); and *pesticide contamination* of fish remains a problem (see fifth section below, Managing 'Developed' Rivers in Industrialised Societies, p. 66). The history of human development has thus been matched by growing volumes of waste, which at times threaten the very water resources that are the basis for the economic activity and urbanisation, and this problem has continued through the twentieth and into the twenty-first centuries. In Managing 'Developed' Rivers in Industrialised Societies below, we will consider the challenge of reducing waste contamination of the aquatic environment. In this section we will discuss sanitation: the problem of water contamination from human households.

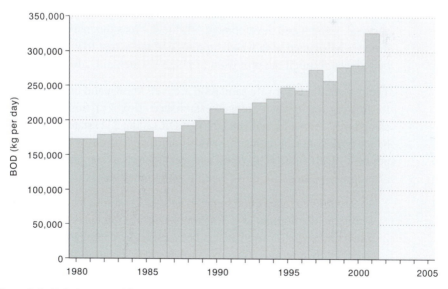

Figure 2.2 Global average biochemical oxygen (BOD) demand due to industrial effluents in water courses

Source: data from WRI Earth Trends 2010

Table 2.2 *Water-related infectious diseases*

Water borne	Risk index	% improvement with potable water and sanitation measures
Cholera	3	90
Diarrhoea	9	50
Typhoid	6	80
Water based		
Guinea worm	4	100
Schistosomiasis	4	60
Vector borne		
Malaria	9	n.a.
Filariasis	4	n.a.
Onchoceriasis	4	20
Trypanosomiasis	1	80
Hygiene based		
Roundworm	3	40
Hookworm	3	n.a.

Notes: Risk index = Frequency * Severity; 1*1 = 1 (lowest); 3*3 = 9 (highest)

Source: after Newsom, 1994: 159

It is the role that potable water plays in human health that creates the most pressing need to secure high-quality fresh water alongside the often deleterious effects of economic development on water quality. As underlined in a recent report by WHO and UNICEF (2006: 2): 'Safe drinking water, sanitation and good hygiene are fundamental to health, survival, growth and development.' Table 2.2 lists the main infectious diseases associated with water supplies and the extent to which they can be reduced through the provision of potable water and safe sanitation.

The requirements for safe sanitation are well known and the technologies for treating waste water are well established (Box 2.1). Some measures can be very simple indeed and UN WWDR (2006: 231) state that 'washing hands with soap can reduce the risk of diarrhoeal diseases by 42 to 47 per cent'. DeBruijne et al. (2007) suggest there is a 'sanitation ladder' of progressive steps for those lacking safe sanitation:

1 *No sanitation* (focus upon hygiene behaviours through education within the community, e.g. washing hands after defecation)
2 *Basic sanitation* (create barriers between waste pathogens and humans through low cost technologies involving safe disposal of wastes, e.g. latrines, composting toilet)
3 *Environmental sanitation* (Box 2.1: reuse of water, preventing pollution of water resources – requires larger scale infrastructure for collection or disposal of wastes)

4 *Ecological sanitation* (recycling of wastes so wastes can be reused – requires separating household urine, faeces and wastewater to be used as a resource after different treatments)

International development agencies have long understood improvement in water supply as a basic precondition for improving the lives of

---

Box 2.1

Basic water-treatment methods

Symons (2006) reviews the history of water treatment, which he notes started four thousand years ago when the use of boiling, sunlight and even charcoal to treat water was recorded, while even two thousand years ago the benefits of filtering were understood, but much of this knowledge was lost until the seventeenth century.

Filtration

The first 'modern' water-treatment plant appears to have been built at the start of the nineteenth century (1804) by John Gibb, for Paisley in Scotland, developed as part of the municipal water supply for Glasgow in 1807 by Thomas Telford, although this suffered clogging from the high water turbidity. This was tackled by Robert Thom in the 1820s using slow sand filtration, which was further developed by James Simpson in 1827 and subsequently adopted throughout England.

Disinfection

A second significant date in nineteenth-century water treatment was the discovery by John Snow in 1854 that the London cholera outbreak could be traced to a sewage-contaminated water pump in Broad Street. Snow is credited with prescribing the use of chlorination to treat such contaminated water. Subsequent scientific discoveries, such as the role of 'microbes' in contaminated water by Louis Pasteur, led to recognition of the need to regulate drinking-water standards (e.g. US 1914 Public Health Standards) and to organise safe sanitation in urban areas (EPA, 2000).

During the twentieth-century sewage treatment technologies were widely adopted in industrialised nations. McGuire, (2006) identifies eight stages in US drinking water disinfection from chlorination at the start of the twentieth-century through to regulation on presence of coliform bacteria in 1914, followed eventually by measures to tackle cryptosporidium by the millennium. Christman (1998) suggests that the use of filtration and chlorination to treat America's drinking water has been responsible for a 50 per cent increase in life expectancy during the twentieth-century. Development of the sewerage infrastructure has been such that the EEA (2007) report 90 per cent of households are connected in Europe.

Although a range of technical solutions exist (Al-Baz et al., 2008, Hlavinek et al., 2006, Tebbutt, 1998 and and Twort et al., 2002) that can be applied to smaller scale rural conditions based upon filtration, coagulation and disinfection, the much higher costs of water and sewerage infrastructure for more dispersed rural populations has meant that even in the 'North' rural populations are at a disadvantage. For example, a survey of the 4 million rural Canadians who rely on private ('non-municipal') supplies mainly through wells (Corkal et al., 2004) showed that more than 65 per cent of waterborne diseases between 1974 and 1996 occurred in rural private or community supplies in non-municipal areas.

poorer people. Effort was focused initially during UNESCO's International Hydrological Decade (1965–74), followed by the 1977 UN Water Conference, which eventually led to the International Drinking Water Supply and Sanitation Decade (1981–90). Reviewing achievements of the latter, Tebbutt (1998: 259) notes that US$ 13.5 billion was invested over the decade, lowering the number of people without access to a safe water supply from 1.826 billion to 1.232 billion by 1990. However, those without access to safe sanitation rose slightly from 1.734 billion to 1.741 billion. At that time it was estimated investment in water supply and sanitation would have to increase five fold to US$50 billion for the next decade. Currently 2.6 billion people lack adequate sanitation services (see Table 2.3 for definitions) with 20 per cent of urban dwellers and 61 per cent of rural populations lacking basic access (World Bank, 2009). Under the current (2015) Millennium Development Goals initiative to reduce by half those lacking access to safe water and sanitation, the UN (2009) report some progress. Since 1990 an additional 1.6 billion people gained access to safe drinking water, which is 'on target' for 2015. Table 2.4 shows that there have been achievements in the supply of water, particularly to urban households.

There is a danger in assuming Table 2.4 presents urban areas as having solved water supply and sanitation problems, and that the problem of

Table 2.3 *Water and sanitation supply terminology used in reports*

	Improved	Unimproved
Water supply	Household piped water or public standpipe, borehole, protected well or spring, rainwater collection	Unprotected well or spring, bottled water, tanker-truck, direct use of surface water
Sanitation	Flush to sewer system, septic tank, pit latrine with slab or ventilated, composting toilet	Latrine without slab or ventilation, flushing without treatment e.g. into street, bucket, use of bush or field

Source: after WHO-UNICEF, 2006

Table 2.4 *Proportion (%) of world population with basic access to water and sanitation*

	Urban	Rural	Total
Water	95	73	83
Sanitation	80	39	59

Source: World Bank, 2004

water supply in developing countries is a matter of the greater proportion of the population that lived in rural areas until 2008. As we saw in Chapter 1, provision of water and sanitation infrastructure is more costly in rural areas (see also Box 2.1). However, the lack of safe sanitation is not just a problem for rural areas. The rapid growth of urban populations in the 'South' and growing numbers of urban poor means that many urban dwellers now live in squalor without appropriate disposal of wastes. De Bruijne et al. (2007) note that in many respects the situation facing urban areas in the South is far worse than for Europe 150 years ago. Poverty is deeper, there are fewer opportunities to access land and water resources, and existing resources are being degraded through pollution and over abstraction. We will see in Chapter 7 that future population growth is forecast to take place in the urban areas of Africa and Asia, where currently some of the highest levels of poverty are to be found. WHO–UNICEF (2006) note that in 2004 this meant over 600 million people in urban areas lacked sanitation. Living in urban slums these people are the least able to help themselves due to lack of tenure, poverty, high population density and poor infrastructure.

These circumstances are well exemplified by the Indian city of Mumbai, projected to have a population of 22 million, the second largest in the world, by 2015. Gandy's (2008) account observes that, despite major investment in infrastructure to increase water supply to the city, water is available for only a few hours each day to many of the less than 70 per cent of the population served by piped water connections in their houses. The remaining 30 per cent or so must obtain their water from even less reliable standpipe supplies or informal water vendors operating water tanker trucks. Gandy (2008) argues that these shortcomings in water supply are not primarily technical, but result from the politics of inequality within the city. In this analysis, not only are there strong vested interests (water tanker operators) who actively lobby for the delay of water infrastructure improvement, but the middle-class beneficiaries of the existing municipal water system have little interest in extending it to improve supply to poorer slum areas. Instead they favour slum clearance to make way for commercial real-estate development. Equally, the shortcomings of the existing system are addressed through water users' individual acquisition of wells, pumps or water purification equipment, and the use of bottled water. Observing that an 'array of technological, scientific and architectural innovations … enable wealthy households to insulate themselves from the environmental conditions of the poor', he concludes that 'The recent history of Mumbai has militated against the kind of progressive political movements that galvanized processes in European cities during the second half of the 19th century … ' (Gandy, 2008: 122).

The uphill struggle to secure safe drinking water in developing countries is illustrated by data from East Africa showing mean water use had *declined* from 61.4 lcd in the 1960s to 39.6 lcd at the end of the 1990s (Thompson et al., 2001). Much of this decline was due to the drop by almost 50 per cent in average water consumption by households in areas with piped water supply (Table 2.5). This was mainly due to the inability of municipal water services to maintain infrastructure, so that water was supplied for fewer hours per day.

A further development indicated by the same study is the greater heterogeneity of supply within a given area. Whereas in the 1960s population in a given location could be classified as all obtaining water by 'piped' or 'unpiped' supply, by the 1990s, this was no longer the case, with 'unpiped' water use by some households within areas in which others had piped water supply. This suggests increasing social inequality in these areas. Greater water use from unpiped sources involved use of surface sources (seepage zones, rivers, lakes) in rural areas, and increasing purchase of water from private water sellers in urban areas. Despite the deteriorating reliability, those with piped supplies still used more than three times as much water as those reliant on unpiped supply. This difference has critical implications for health, since, while water used for cooking and drinking was more or less the same (4 lcd) for all households, the amount used for washing and bathing averaged 33 lcd for those with piped supplies, against only 14 lcd for those using unpiped sources. This is reflected in the much higher rates of diarrhoea (during the week previous to survey) reported within households

Table 2.5 Changes in average water use (litres per capita per day – lcd) over 30 years in East Africa

Year	Region		Urban		Rural
	'piped'	'unpiped'	'piped'	'unpiped'	'unpiped'
1966–8	128	11	128	15	10
1997–9	66	20	66	24	18

Source: data from Thompson et al., 2001

reliant on unpiped water (23 per cent) compared to those using piped water (10 per cent). For households using surface sources of unpiped water, rates of diarrhoea were particularly high, at 33 per cent (Thompson et al., 2001).

A recent and comprehensive summary of the nature of the sanitation problem is provided by De Bruijne et al. (2007). They also present a critique of the various approaches, from collection through to treatment. This reduces the need for a lengthy discussion here, but it is worth noting their figures on the economics of sanitation. They report that the MDG for water and sanitation require $11.3 billion a year, but if implemented this would yield $84 billion savings in health care and productivity (less labour lost through illness). Whittington et al. (2009) note, however, that the incremental benefits of supplying water and sanitation may be rather small, and insufficient to cover the high investment costs, leading to less willingness to pay. Both sets of authors agree that there is no 'magic bullet', nor consensus over the best approach, while other policy priorities such as 'alleviating poverty' may also require water-resources development. As the international funding institutions are unlikely to fund expensive infrastructure water projects, the solutions for DCs may lie in community-level changes through boreholes, sand filters and dams.

Finally, it needs to be remembered, that sewage waste is a resource, not just as 'night soil' fertiliser to increase soil fertility, but also because it is mostly water. Recycling of wastewater is addressed in Chapter 6 but, as an illustration of its significance, Cook (1998) notes that in England and Wales around 30 per cent of public water supply used recycled effluent, a figure that rises to 70 per cent in the high demand south-east where during drought conditions river flows can be mostly recycled effluent. Although it is reasonably simple to treat the 'point sources' of sewage contamination in water systems (compared to 'diffuse sources' of agricultural pollution (Merrington et al., 2002)), Hunt (2004) cites very low values of wastewater treatment with just 14 per cent in Latin America and the Caribbean and 35 per cent in Asia. This raises questions about how to compare 'withdrawal rates' of water in different regions (see Chapter 1). In particular it alerts us to the need to consider rates of water recycling and re-use, which may be much higher in industrialised economies. We shall consider this in more detail in Chapter 6.

Water and economic development

The steam technology that drove the growth of the first industrial centres such as Manchester in the early nineteenth century also provided new means of transportation that greatly expanded the reach of markets and accelerated processes of human migration. The construction of railways across the western United States was the key factor in increasing the scale of European settlement that had hitherto been limited to restricted enclaves, such as San Francisco, whose population had grown from eight hundred to fifty thousand during the five years following the discovery of gold in 1849. The problem for many of the settlers was that, in much of the vast region west of the 100-degree meridian, rainfall was insufficiently reliable to support settlement based on arable agriculture (Reisner, 1993). Agricultural settlement was therefore premised on irrigation. Among the most favourable irrigation sites was the arid basin between the Pacific Ocean and the San Bernadino mountains in southern California, which combined low humidity, mild maritime winters and shallow groundwater maintained by intermittent flooding of the rivers fed by melting snow from the mountain ranges to the east. The mild winters were of particular significance since they allowed year-round crop production whereas in much of the western United States aridity was compounded by an altitude of more than 1500m above sea level, so that frost limited the growing season to no more than four to six months.

First developed during the 1850s by religious communities (Mormons, Quakers and others) who supplied the transport routes generated by the goldfields to the north, the year-round productive potential of irrigated agriculture in southern California began to be exploited on a much larger scale with the construction of a railway spur from San Francisco in 1867 and a further line to Kansas in 1885. The latter brought a flood of settlers, including many leaving failed farms in the mid-west after a severe drought in 1888–90 that saw the population of Kansas and Nebraska decline by 25–30 per cent (Reisner, 1993: 107). This pattern of migration was to be repeated a generation later: high cereal prices during World War I coupled with a series of years of exceptionally high rainfall encouraged a new expansion of arable farming in the mid-western US during the early decades of the twentieth century. This was followed by seven years of below-average rainfall in 1928–35 that resulted in the destruction of much of this agriculture in the 'dust bowls' that saw '35 million acres destroyed', '125 million acres degraded' and 750 thousand bankrupt farmers migrating to California, Oregon and Washington (Reisner, 1993: 150). Some have questioned the claims of such widespread apocalyptic land degradation, pointing out that data may have been inflated by the incentive to claim compensation from the US government (Warren and Agnew, 1988).

It is nonetheless the case that successive waves of migration fuelled a rapid growth of population in southern California, centred on the city of

Los Angeles whose population grew from about 320,000 in 1910 to over 1.2 million in 1930 and 1.5 million by 1940 (population of the city in 2008 was 3.8 million, and 9.86 million in Los Angeles County – US census, 2009). As in the case of expanding manufacturing towns in nineteenth-century Britain (Chapter 1), the water demand of the expanding city exceeded water available in the immediate vicinity, and required new supplies from ever more distant sources (Box 2.2). This trajectory also parallels a common path identified for industrialised countries,

---

Box 2.2

The increasing reach of water demand in southern California (after Reisner, 1993)

As in the case of the growing industrial cities of Europe, the huge influx of people into southern California, and Los Angeles County in particular, had by the end of the nineteenth century exhausted local supplies of water. However, in the case of Los Angeles the quest for more water was intensified by its arid climate and by the fact that much of its economy was based on irrigated agriculture. The development of more efficient centrifugal pumps at the end of the First World War enabled irrigation to expand more than tenfold (from 6000 to 75,000 acres) between 1913 and 1918. This resulted in rapid depletion of groundwater, leading Los Angeles to develop aqueducts to bring $0.31 \text{ km}^3/\text{year}$ from Owens Valley, some 250 km distant, in 1913, and a further $6.6 \text{ km}^3/\text{year}$ from the Colorado River, 350 km away, in the 1930s.

Further water supplies were developed for irrigation of California's Central Valley, and in particular the San Joaquin Valley where newly arrived migrants had expanded irrigation to about 600,000 ha by the early 1930s. The Central Valley Project involved not only construction of the Friant Dam on the San Joaquin River, draining the southern Sierra Nevada, but also Shasta Dam and Trinity Dam on tributaries of the Sacramento River draining the northern Sierra Nevada. A canal system was to allow transfer of water from northern to southern California. The Central Valley Project was expected to deliver an annual total of 8.6 km^3 of additional water, principally for irrigation in southern California.

The history of water supply to Los Angeles was marked by long-running struggles over water rights. These involved legal subterfuge and sometimes violent conflict, such as during the construction of the Owens Valley aqueduct. It also witnessed epic engineering feats, of which the Hoover Dam on the Colorado River, built in 1931–6 and 220 m high, was perhaps emblematic.

This history also registered an increasing scale of water management, involving a transition in the institutions responsible. An initial phase of private enterprise – the Los Angeles City Water Company, formed in 1878 – gave way in 1904 to a tax-funded municipal corporation – the Los Angeles Department of Water and Power. By the 1920s, an increasing role was taken by the US Federal government, notably in funding the Hoover Dam and the associated Colorado aqueduct. In 1935 the Federal government's Bureau of Land Reclamation took over implementation of California's Central Valley Project when the project's original proposer, the State of California, failed to raise the necessary funding due to the 1929 stock-market crash.

Agriculture accounts for 70–80 per cent of water consumption in California and some 60 per cent of water used in agriculture is supplied through surface water infrastructure, of which the (federal-run) Central Valley Project and (state-run) State Water Project are major components. Water supplied through federal projects has historically been subsidised to enable a settler family to make a living on a holding of 320 acres (128ha) in the 1920s, a limit raised to 1260 acres (504 ha) in 1979. Writing a decade later, Marc Reisner (1993: 324) commented: "In California, with 1260 acres and subsidized water costing between $3.50 and $9 per acre-foot, a halfway ambitious farmer could become a millionaire – which was not exactly the intention of the Reclamation Act." In practice, 45 per cent of California's farmland is owned by farms whose outputs individually are worth more than $0.5 million (California Department of Food and Agriculture, 2009), and have historically included land holdings of many thousands of hectares owned by some of the largest corporations in the United States (Reisner, 1993: 338). Although these wealthy farms may be subsidised, Sumner and Brunke (2003) calculated subsidies for water costs (estimated at $10–$40 per acre-foot) amounted to only 3 per cent of the Producer Support Estimate (PSE – a money measure of explicit or implicit income transfers to agriculture) for California in 1998, which is insignificant when compared to import tariff barriers worth 40 per cent of PSE.

where private water companies were replaced by municipal authorities in the mid-nineteenth century and water supply was thereafter funded from local taxation, but with an increasing role of national government in investment and regulation of water allocation. In the case of the western United States the government agency most involved in developing water resources has been the Bureau of Land Reclamation, formed in 1923 specifically with the aim of supporting agricultural settlement. From the 1920s to the end of the 1970s the role of this and other agencies of the US federal government in developing water resources may be seen as classic examples of the role of the state in leading development.

The developmental state

The centrality of water use to economic activity means that development of water resources and water infrastructure will always feature as part of economic development. In some cases, such as that of Spain, the development of water infrastructure may be a leading idea in the construction of a national identity (Swyngedouw, 1999). However, few countries have matched in scale the state-led development based on investment in water infrastructure that was undertaken in the United States. Early federal activity focused on supporting settlement in the areas of unreliable rainfall of the west. Spurred by the widespread failure of private irrigation schemes in the nineteenth century, the 1902 Reclamation Act mandated government 'reclamation' water projects to be funded from the proceeds of sales of federal government land to settlers. The projects' costs were to be recouped over time (initially 10–20 years) from the sale of water to the settlers. The pace and scale of this federal government activity rapidly increased from the late 1920s, however, when large water infrastructure – particularly dams – were seen as a means of both creating jobs on construction sites themselves and also underpinning the viability of agricultural employment for the hundreds of thousands moving westwards from failed farms in the mid-west.

The initial focus of much of the infrastructure was simply to secure water supply for agriculture, such as the Central Valley irrigation project (see Box 2.2). The dominance of agriculture in the way water was conceptualised in the US is manifest in the unit of volume adopted to measure water: the 'acre-foot' (1235 m^3) – the volume of water equivalent to a foot depth of water on an acre of land. However, the realisation that agriculture alone would be unable to repay the costs of the dams meant that their construction could only be justified through recourse to a wider cost–benefit accounting. In addition to their role in flood control and regulation of river flow to enable navigation, the capacity of dams to generate electric power was the key factor in enabling projects to promise a positive return on government investment. In particular, the sale of electricity from hydroelectric turbines incorporated in the dams enabled cheap water to be supplied to agriculture from 'multi-purpose' dams.

This integration of hydropower with irrigation water supply in development planning objectives thus radically changed the economics of dam construction, and the early 1930s saw the simultaneous construction of four major dams: Hoover (35.2 km^3 water storage and 2080 MW electricity generating capacity) on the Colorado river; Shasta (5.5 km^3 and 1042 MW) on the Sacramento river; and Grand Coulee (11.5 km^3 and 6809 MW) and Bonneville (167 km^3 and 1104 MW) both on the Columbia River. The dam-building boom ran from the 1930s until the early 1970s, leaving a legacy of some 79 thousand dams on the rivers of the United States, of which over 8000 are 'major dams' (> 50 m in height) (National Inventory of Dams, 2009). California alone has some 1200 dams, and the Columbia river has 13 dams on its main stem, with a further 36 on its tributaries, all constructed within 40 years (Reisner, 1993: 164). On the other side of the country, the Tennessee Valley Authority similarly constructed more than thirty dams within thirty years.

The developmental impacts of this enormous government investment in water infrastructure were first and foremost to foster growth in private sector production in industry and agriculture. The federal agency construction contracts provided a platform for the growth of US civil engineering expertise in companies such as Bechtel (one of the members of the original consortium that tendered to build the Hoover Dam) and Parsons Engineering. The huge increase in hydroelectricity generating capacity provided a cheap source of energy for industry and effectively enabled industrialisation in previously undeveloped areas. This was particularly exemplified by sparsely populated Washington State in the north-west of the country. Here, the abundance of cheap hydroelectricity from the Columbia River dams was exploited to rapidly expand capacity in energy-intensive industries. Two of these, aluminium smelting and plutonium production, and their associated manufacturing industries, were critical to the development of US military capacity from the 1940s onwards. The provision of cheap irrigation water from government-funded dams enabled Californian agriculture to dominate the North American food market, generating US$36.6 billion in sales in 2007, or 13 per cent of all US farm sales on only 1.6 per cent of the farm acreage, accounting for half of all US-grown fruit, nuts and vegetables and 22 per cent of all US-produced milk (California Department of Food and Agriculture (CDFA), 2010).

In addition to its effects on US industry and agriculture, the development of water-resource infrastructure was associated with an expansion in government planning capacity. The two main federal agencies charged with water-development infrastructure, the Bureau of Reclamation (irrigation and hydropower) and the US Army Corps of Engineers (flood control, navigation and hydropower) were joined in the 1930s by the Tennessee Valley Authority (TVA) (see Box 2.3).The TVA marked an important shift in conceptualisation of water management. The perception of water as connecting and integrating different uses in different parts of the river valley gave rise to concepts of integrated water management that influenced water policy for more than sixty years.

Apart from the sheer growth in scale, the operation of the US public agencies charged with water-resource development has been marked by changing social priorities. To some extent the emphasis on constructing

Box 2.3

The Tennessee Valley Authority

Established in 1933, the purpose of the Tennessee Valley Authority (TVA) was to 'improve the navigability and to provide for the flood control of the Tennessee river, to provide for reforestation and the proper use of marginal lands in the Tennessee Valley, to provide for the agricultural and industrial development of said valley ...' (Tennessee Valley Authority Act, 1933).

The TVA was also charged with developing nitrate-fertiliser production at the site of an existing government nitrate-explosives factory, using hydropower from the Wilson Dam at Muscle Shoals, Alabama.

The TVA was set up as a response to a number of different problems identified by President Roosevelt's 'New Deal':

- to rehabilitate a region subject to destructive soil erosion arising from nearly a century of mining, logging and agriculture;
- to prevent damaging floods;
- to reduce energy costs, and particularly the high prices then being charged by private electricity utilities.

The solutions interconnected through the construction of water infrastructure:

- hydroelectricity from TVA-built dams provided publicly owned power supplies;
- fertiliser produced with cheap hydroelectricity assisted programmes of reforestation and rehabilitation of farmland;
- reduced soil erosion from newly terraced hillsides diminished the threat of silting damaging the operation of the dams.

As a consequence, the TVA epitomised a perception of 'the river valley as a whole' and integrated water-resource management as essential to its effective development.

The imperatives of increasing demands for electricity, initially for the war effort in the 1940s, saw the TVA reach beyond its original hydroelectricity brief and develop coal-powered electricity using Appalachian coal resources, and by the 1960s TVA was also building nuclear-power plants. By 2005, more than 60 per cent of TVA's total power generation of about 23 MW was from 11 coal-fired plants, and another 30 per cent was from six nuclear reactors (TVA, 2009).

dams simply exhausted the availability of suitable sites, with those remaining increasingly costly or hazardous – the latter underlined by the failure of the Teton dam, in Idaho, as it was being filled for the first time in 1973. Whereas the Hoover Dam had cost $50 million with a payback period of 50 years (which ended in 1984), new dams under consideration for the Colorado River in the 1970s were budgeted to cost on the order of $1 billion (Reisner, 1993: 290). By the 1980s, this was a receding prospect. Expansion of federal government budgets was subject to increasing political opposition and an expectation that local or commercial funding should be found for water infrastructure.

This meshed with another major change that increasingly shifted the focus of state-led water development. Growing environmental concern during the 1960s resulted in the passage of the Environmental Protection Act in 1969 which required federal agencies to undertake an environmental impact assessment of any proposed development. This established a platform for re-thinking goals for water-resource development, and in particular a new focus on ensuring the ecological functioning of water resources. This re-thinking was partly a reflection of increased public concern with environmental quality in general, but also of the increasing economic value of commercial activity that required high-quality water resources to meet growing recreational demands – for fishing, boating, rafting, etc. – of an increasingly affluent society. Operators of hydro-electric dams or coal-fired power stations were now required to explain how their operation avoided environmental damage, or, more frequently, repaired harm already caused. The TVA, for example, has developed a programme for monitoring pollutants in its 31 reservoirs from which it publishes a series of water-quality ratings with health advice to recreational water users (TVA 2009). Nonetheless, it is also clear that the scale of past development has left environmental hazards that have yet to bring forth an adequate response. For example, wet ash landfill dumps from some TVA coal-fired power stations are considered so hazardous that their location is classified (*Guardian*, 18 June 2009). In December 2008, two of these landfills, at TVA's Kingston and Widows Creek power stations, leaked 1.1 billion and 10,000 gallons respectively of heavy-metal contaminated ash deposits into rivers and onto land, resulting in heavy clean-up costs and questions raised about the accountability and priorities of the TVA (*Chattanooga Times Free Press*, 15 January 2009).

These shifts have led many to call into question the US model of state-led investment in major water infrastructure as a strategy to promote economic growth (see also section below, Sustainability and 'Post-industrial' Water-resource Management, p. 75). However, it is a model that has been widely pursued elsewhere, and at an increasing scale. Thus the size of individual dams, such as the Hoover and Grand Coulee, constructed in the western US in the 1930s – and for many years the largest in the world – has now been surpassed by the South American

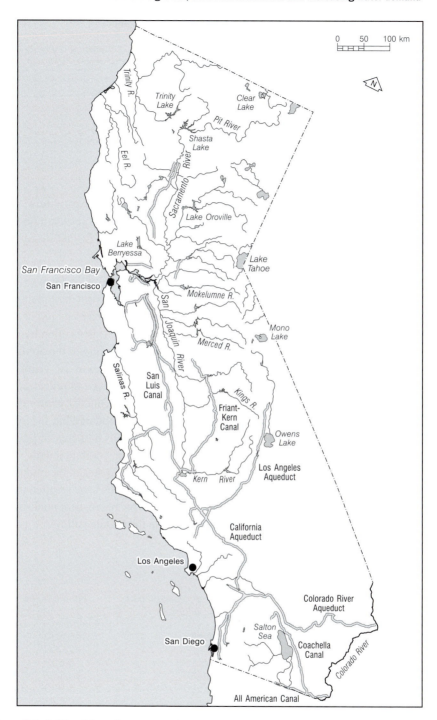

Figure 2.3 California's major water projects

Source: CDWR, 2009: 12; permission to reproduce granted by CDWR

Table 2.6 Hydroelectricity-generating capacity of large dams compared

Dam (year constructed)	Country	Hydropower generating capacity (MW)
Hoover (1934)	US	2,080
Grand Coulee (1935)	US	6,809
Akasombo (1965)	Ghana	1,020 (in 2006)
Kariba (1959)	Zambia/Zimbabwe	1,266
Aswan (1971)	Egypt	2,100
Guri (1968)	Venezuela	10,000
Itaipu (1984)	Brazil/Paraguay	12,500
Three Gorges (2004)	China	18,200

dams built at Guri (Venezuela) and Itaipu (Brazil/Paraguay). More recently these have also been dwarfed by the Three Gorges Dam on the Yangtze River in China (Table 2.6). See Chapter 5 for further discussion on the social and environmental impacts of large dams.

As with the TVA, the primary purpose of the Three Gorges Dam is flood control and hydroelectricity generation. The Yangtze (or Changjiang) river, flowing some 6300 km from south-western China to its mouth at Shanghai on the east coast, is China's longest river. It is swollen by the monsoon rains in July–September giving rise to flooding in the autumn months, with serious floods recurring on average cycles of 2.5 years and 16.4 years (Yu et al., 2009). The shorter cycle is believed to reflect greater monsoon rainfall associated with El Nino events. Twentieth-century trends suggest increasing height of floods, smaller area inundated but for longer duration, which Yu et al. (2009) attribute to the effects of increasing run-off associated with deforestation of the upper catchment, combined with greater containment of the river by some 2000km of dykes in the middle and lower river valley. Historically, the Yangtze floods have resulted in the loss of millions of hectares of crops, millions of people made homeless and thousands of deaths (Table 2.7), principally in the middle valley.

Table 2.7 Floods on the Yangtze River

Year	Farmland flooded (million ha)	People displaced (million)	People killed
1931	3.39	28.55	145,000
1935	1.51	?	142,000
1949	1.81	8.1	5,699
1954	3.18	18.88	33,169
1998	4.4	13.80	1,526–4,100
2004*	–		120

Source: CTGCP (2009)
*post dam construction

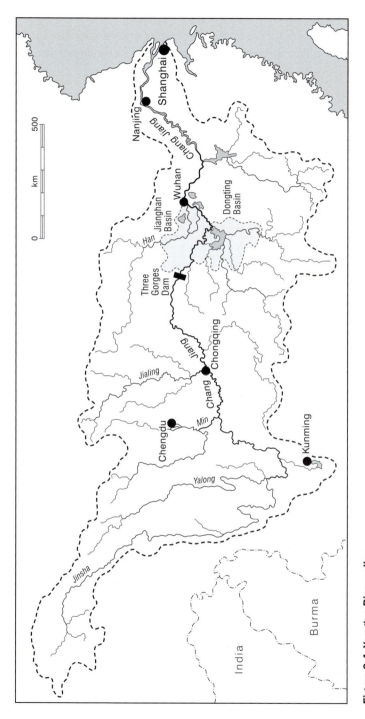

Figure 2.4 Yangtze River valley

The Three Gorges Dam, constructed at the turn of the century (Box 2.4) aimed to eliminate this threat. In 2004, a year after the dam first filled, the Yangtze experienced one of its highest ever flows (60500 m³/s) and the dam was reported as successfully retaining 0.5 km³ of additional floodwater (CTGPC, 2009).

Despite this achievement the Three Gorges Dam has been highly controversial. International criticism has emphasised the 1.13 million people (critics quote figures as high as 1.5 million) relocated from towns and villages submerged by the reservoir, the submersion of archaeological sites, and the threat to water quality posed by accumulation in the reservoir of untreated sewage and toxic contamination from submerged industrial plants. It is also argued that the very high sediment loads due to soil erosion in the deforested upper catchment of the Yangtze will

Box 2.4

The Three Gorges Construction Project

The site of the Three Gorges Dam, just where the Yangtze emerges from its mountainous upper catchment into the flat floodplains of the middle valley, had been identified early in the twentieth century, and preliminary designs for a scheme were undertaken in the 1950s. A pilot hydroelectric scheme on the site, the Gezhouba project, was approved in 1970 and began generating electricity in 1981. In 1982 the Chinese leadership instructed that a project for a full-scale dam at the Three Gorges site be included in the national plan, and feasibility studies for a 175 m high dam were completed in 1989. Legislative approval of the Three Gorges Project was given in the form of a resolution adopted by the fifth meeting of the Seventh People's Congress in April 1992, and within a year the State Council Three Gorges Project Construction Committee had been established and had set up funding arrangements and elements of the project-management system, including the China Yangtze Development Corporation (CTGPC) to build and operate the dam and the ship navigation locks integrated with it, and the 'migrant relocation development bureau' (CTGPC, 2009). The project design was undertaken by the Changjiang Design Institute, the engineering design enterprise of the Changjiang Water Resources Commission, the state agency charged with oversight of development in the Yangtze Valley. Work on the site started in 1994 and in 2003 the dam began producing hydroelectricity and the locks for ship navigation were opened for traffic. The project echoes in ambition and execution the US federal dams of the 1930s but surpasses them in scale: a dam 181 m in height and 2.3 km in width, storing 39 km³ of water and installed hydroelectricity generating capacity of 18,200 MW (to be increased by a further 4200 MW).

cause the dam to silt up. Critics say that the dam's sluice system, designed to prevent silt accumulation behind the dam, is 'untried' (see, for example, PBS, 2009). Yu et al. (2009) have also cautioned that the dam will not necessarily always protect the valley downstream of the dam from flooding. They suggest flood hazards in the lower valley have increased: first, due to drainage and cultivation of areas once occupied by large lakes (the Dongting lake region) that served to receive flood water; second, due to a tendency to increasing precipitation in the lower valley which they suggest may be a feature of climate change. It is, however, too early to judge how effectively the Three Gorges project will meet the challenges of managing water quality or what impacts the dam will have on China's economic and social development.

To some extent the critics of the Three Gorges Dam can justifiably point to water infrastructure that has failed to benefit the majority of people living in the locations where it has been constructed, for example in Africa and India (see Chapter 5). In particular, large dams have been conspicuously negative in their impacts on people displaced from areas submerged by their reservoirs. Criticism of large dams also encapsulates the wider criticism of state-led development: that it subordinates complex social realities to simplified visions of progress, producing an uneven distribution of social benefits and costs. However, what the controversy also reveals, perhaps, is how the role of water resources in development changes as society itself changes. In the US, concern with the impact of water infrastructure on long-run ecology of rivers has begun to significantly modify planning decisions that were previously dominated by the simple delivery of water and electricity to industry, agriculture or urban areas. To an extent this reflects greater affluence, but also a shift in political perceptions of threats to human wellbeing and of the importance in managing those threats of maintaining 'ecological services' provided by 'natural' rivers. These are not always fully understood, but concerns to improve water quality in lakes and rivers – a basic social benefit well understood by society in general – have led to a growing acceptance that water management needs to be conceptualised in ecological, as well as hydrological, terms. To explore the implications of this, we now examine the management of a river that flows through one of the world's industrial heartlands.

Managing 'developed' rivers in industrialised societies

The Rhine is the most densely populated and industrialised river basin of Western Europe, with more than 54 million inhabitants and about 10 per cent of the world's chemical industry (van Dijk et al., 1995).

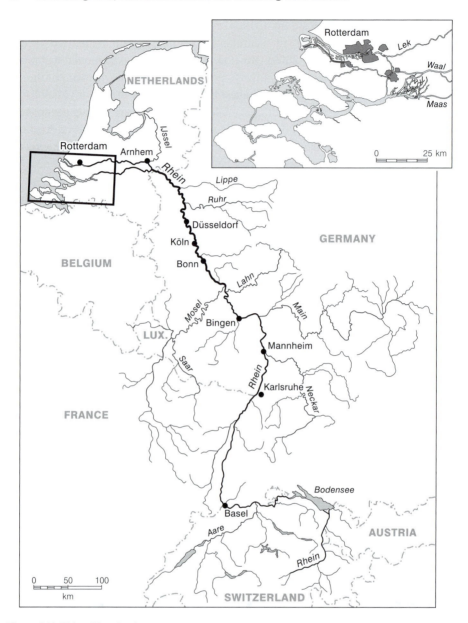

Figure 2.5 Rhine River basin

The Rhine and its principal tributaries (the Main, Moselle and Neckar) also constitute a key navigation artery linking the port of Rotterdam at its estuary to the principal manufacturing centres of France, Germany, Switzerland and the Netherlands. The river's management presents formidable challenges of international cooperation, since the basin covers parts of Switzerland, Germany, France, Netherlands and Luxembourg, and smaller areas of Austria, Belgium and Italy.

The importance of the Rhine as a transportation route meant that the river channel has been subject to modification at least since Roman times. However, this remained relatively small-scale until the nineteenth century, when the multi-channel reaches were simplified into fewer, deeper and straighter, channels (see Figure 2.6). As a consequence river flow velocity and channel scouring increased. In addition to channel straightening, the hydrology of the Rhine valley has been modified by flood-protection works wherever the river flows through a floodplain (see Box 2.5). Two important consequences of these dykes were: first, that large floodplain areas that historically were subject to flooding at intervals of several years were no longer flooded; and second, many of these protected floodplains were used for development as residential or industrial areas (see Figure 2.7). According to van Dijk et al. (1995), 90 per cent of Rhine floodplains between Basel and Karlsruhe have 'vanished'. Neumann (2002) observes that dykes separate about 85 per cent of the former floodplain from the river.

As a consequence of the flood prevention measures taken in the nineteenth and twentieth centuries, flooding was experienced in the

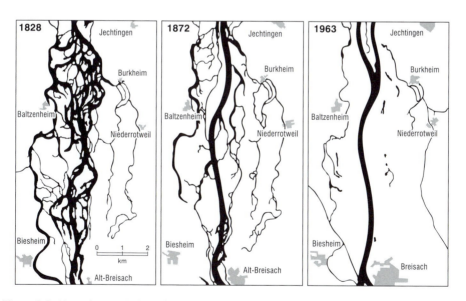

Figure 2.6 Alterations to Rhine channels from nineteenth to twentieth century

Source: redrawn from van Dijk et al., 1995: 380; permission to reproduce granted by J. Wiley (Regulated Rivers: Research and management)

Box 2.5

The Rhine: a 'developed' river

The River Rhine flows from its source in the Swiss Alps some 1320 km to its outlet into the North Sea. The 190,000 km^2 basin of the Rhine is conventionally divided into six sections along its main stem and three tributary sub-basins (see Figure 2.5). The main-stem divisions (ICPR, 2009) are: the 'Alpine Rhine' including Lake Constance and the headwaters draining into it; the 'High Rhine' from Lake Constance to Basel and including the tributary river Aare; the 'Upper Rhine' between Basel and Bingen forms the frontier between France and Germany and then flows through Germany to its confluence with the Main; the 'Middle Rhine' between Bingen and Bonn flows through a deep and narrow valley cut into the Slate mountains; the 'Lower Rhine' from Bonn to the Dutch border at Arnhem; the 'Rhine Delta' consists of three main channels flowing through Holland, the Waal carrying 66 per cent of the flow, the Nederijn-Lek (22 per cent) and the Ijssel (12 per cent). The three main tributary systems are the Neckar and the Main, both draining from the east into the Upper Rhine, and the Saar-Moselle, draining into the Middle Rhine from the west. The Rhine is supplied by a combination of winter rainfall in the lower parts of the catchment and · melting snow in the upper catchment in the spring and summer, providing a strong year-round flow, but with seasonal flooding: average discharge (at Lobith on the Dutch border) is about 2040 m^3/s but this may rise to around 10,000 m^3/s (van der Lee et al., 2004).

Major modifications to river flow and floodplain hydrology date from the early nineteenth century. In particular, Johan Gottfried Tulla's design, implemented from 1817 and 1890, straightened and rechannelled the braided river spread across a 10 km wide floodplain upstream of Worms in the Upper Rhine (see Figure 2.6), effectively reducing the transport distance between Basel and Mannheim by 81 km, or 20 per cent (Pinter et al., 2006). The Lower Rhine was similarly increasingly 'channelised' from the later nineteenth century, with a major surge in hydraulic engineering in the 1940s, including the construction of weirs and navigation locks, extending to the Neckar and Main, and in the post-war period to the Moselle (de Groot, 1992).

In the Delta (the Netherlands) the elevations below sea level of much of the land meant that dykes have been built since the Middle Ages to protect land from flooding. By the late twentieth century the dykes constructed on either side of the river effectively enclosed an 'embanked floodplain' some 0.5–1 km wide. The lower-lying parts of this embanked floodplain would be flooded when river flows reached 3500 m^3/s, and the higher parts would normally be flooded by flows of 6500 m^3/s. All floodplain dykes in the Netherlands are required by law to be designed to withstand a flood with a recurrence probability of 1250 years. Elsewhere, on the Lower Rhine floodplains, flood-protection dykes are generally designed to withstand floods with a 200-year recurrence probability (Hooijer et al., 2004). On the floodplain of the Upper Rhine, Bohm et al. (2004) suggest flood-protection design may vary: to withstand floods of between 100 and 1000 years recurrence probability.

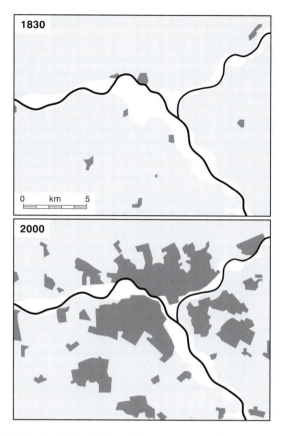

Figure 2.7 Urbanisation in the Rhine valley: comparison of Arnhem in nineteenth and twentieth century

Source: redrawn from Hooijer et al., 2004: 348; permission to use granted by J. Wiley (River Research and Applications)

Rhine valley much less frequently. By the end of the twentieth century, however, this strategy of flood management had been called into question. In December 1993 record flooding caused nearly US $33 billion in damages and forced the evacuation of a quarter of a million people (Pinter et al., 2006). A little over a year later, in January 1995, a further record flow of 12,000 m³/s recorded just south of the German–Dutch border resulted in similar damage, but additionally threatened to breach the Dutch dyke system for the first time since 1926.

The events of 1993 and 1995 revealed the seriousness of the threat of catastrophic flooding of areas with very high population densities and major centres of agricultural and industrial production. A re-evaluation of flood management was undertaken, including a European Union-funded programme of research cooperation between Belgium, Germany,

France, Netherlands, Luxembourg and Switzerland that had as its goal: 'development of methods and tools to assess the impact of flood risk reduction measures and of land use and climate change scenarios in order to support the spatial planning process of the Rhine and Meuse River Basins' (Hooijer et al., 2004). Along with a number of other studies (Pinter et al., 2006, Engel, 1997), this work, summarised by Hooijer et al. (2004), concluded that flood risk on the Rhine is increasing. This is in part due to a trend of increasing rainfall, attributed to climate change, resulting in more extreme flow events. These would result partly from increasing winter rainfall in the Middle and Lower Rhine, and also in the tributary sub-catchments, each of which can independently generate substantial peak flows that may – as in 1995 – coincide when these tributary flows reach the Lower Rhine. In addition, increasing temperature is expected to cause less of the Alpine precipitation to be stored as snow until the summer and more to fall as rain, so that the flows arising in the upper basin will be more synchronised with those of the lower catchment than in the past. As a consequence, data presented by Hooijer et al. (2004) suggest that future flood management would need to accommodate peak flows (typically measured at Lobith on the lower Rhine/ Delta boundary) on the order of 18,000 m^3/s, some 50 per cent higher than the peak flow in 1995. This demands a reconsideration of the presumed safety against flooding provided by the existing protective dykes.

The research on flood hazards in the Rhine basin also makes clear, however, that the risk of flood damage is now much greater than in the past because of the intensive urban and industrial encroachment onto the river floodplains. The economic and social consequences of flooding have thus become far greater (Bohm et al., 2004; Geilen et al., 2004; Pinter et al., 2006), and potential damage from floods is estimated to have doubled every thirty years (Hooijer et al., 2004). Strategies to manage the flood hazard generally agree that planning authorities need to incorporate flood risks into land-use decisions, rather than assuming that dykes can be raised to maintain 'full control' of the river flood. Hooijer et al. (2004), in particular, argue that reliance on dykes for protection generates a false sense of security and that, rather than assumptions of 'zero-risk' implicit in the construction by government of flood-protection dykes, decisions that raise risks of flood damage (by investing in vulnerable construction in flood-hazard zones) should carry an obligation of compulsory insurance against flood damage. Strategies for future flood management on the Rhine generally also point to a radical overhaul to the management of floodwaters that emphasises flexibility and adaptability, providing more 'room for rivers'. The main elements of such an approach have been set out by Geilen et al. (2004) and also by Hooijer et al. (2004). These include:

- Changing land use, especially in upper parts of the catchment and tributaries to slow the run-off of precipitation into the main river channels and thus lowering the potential peak flows in the river.
- Retention of peak discharges in the floodplain, by recreating 'retention basins': floodable ponds, reed-beds and 'backwaters' that connect to the main channel only during flood events. This principle is extended through the identification of areas of floodplain, usually farmland, that can be sacrificed to inundation in order to protect residential areas.
- Increase in the flow capacity of the river by widening the river bed or lowering the floodplain.

Despite general agreement on these elements of flood-management strategies, there is disagreement on the emphasis to be given to each element in managing floods on the Rhine. Pinter et al. (2006) argue that the main emphasis should be on changing land use in the upper catchment so as to slow run-off and prevent the generation of high river flows, pointing to changing farming practices in the twentieth century as a major cause of increased run-off to rivers. In contrast, Hooijer et al. (2004) argue that, while upper-catchment land use change can lower the frequency of floods in small sub-catchments, especially 'flashfloods' arising from localised convective rainstorms, it cannot prevent the extreme flood events recently experienced in the Lower Rhine and Delta. They argue that these extreme floods are associated with prolonged winter rainfall when soil saturation, often associated with frozen ground, provides little scope for increased rainfall infiltration. As a consequence, they argue, the trends towards increased rainfall arising from climate change have much greater influence on peak river flows than changes in land use. They conclude that the upper catchment of the Rhine cannot be relied upon to reduce peak flows by retaining run-off. Instead they argue that the floodplain of the Lower Rhine needs to be managed to create more flood retention areas.

To this basically hydrological argument they add an ecological one: that a floodplain of reconstituted flood retention basins of varying kinds provides an opportunity to use differential flooding to maintain habitat and landscape diversity. This is consistent with concepts of ecological rehabilitation outlined earlier by van Dijk et al. (1995), in which flood events were perceived as moments in which normally separate bodies of water are connected, providing a key function in allowing the movement and dispersal of aquatic species between these backwaters, and between them and the main river channel. The significance of floods in linking fragmented ecological zones along the Rhine is also suggested by Geilen et al. (2004). At this point we need to link our discussion of flood management with that of another major challenge to water-resource management under conditions of economic growth, noted earlier in this chapter: the problem of deteriorating water quality.

As noted at the start of this chapter, quality of river water was an early casualty of the industrialisation of towns in nineteenth century England. If industrialisation affected the Rhine a little later, its effect on the quality of the river water was no less devastating, as the data for salmon catches from the Rhine show (Figure 2.8). Concern at decline of salmon stocks in the Rhine had already arisen in the latter half of the nineteenth century and the Salmon Convention drafted by the Rhine riparian states promoted re-stocking efforts that included releasing some 160 million young salmon into the Rhine between 1879 and 1912 (de Groot, 1992). However, the combination of declining water quality and more efficient fishing using new designs of 'seine' nets prompted a rapid decline in stocks. The Dutch salmon fishery was exhausted by the 1930s and the German fishery by 1950 (Figure 2.8, de Groot, 1992; van Dijk et al., 1995). In addition to the chemical pollution by urban and industrial waste, which in the early twentieth century many believed would simply be washed out to sea by the Rhine's year-round fast-flowing waters, salmon breeding and migration were physically disrupted by the hydraulic engineering of weirs and locks for ship navigation, hydroelectric schemes (in the High Rhine), extraction of gravel from salmon spawning

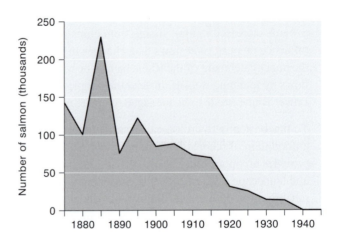

Figure 2.8 Decline of salmon population in the Rhine

Source: redrawn from van Dijk et al., 1995: 386; permission to reproduce granted by J. Wiley (Regulated Rivers: Research and Management)

sites in the upper reaches of the Rhine and its main tributaries, and by the constant heavy traffic of ships and barges on the river, particularly once the construction of locks at Haringvleit and Zuidersee had restricted direct open access between river and sea to only one channel (de Groot, 1992).

The quality of the Rhine River water continued to decline during the post-war European economic boom as the river was used to dispose of an ever-growing range of industrial chemicals, as well as urban sewage and growing quantities of agricultural chemicals. By the 1970s the water quality in the river was so poor that coordinated action was begun to reduce pollution, especially by upgrading sewage treatment. However, it was only in the wake of the catastrophic pollution caused by the Sandoz fire in Basel in 1986 that clear ecological goals were incorporated as part of water-quality management (Box 2.6). Of these, the one with most immediate public recognition was the commitment to make the Rhine fit again for salmon.

The scale of this task cannot be exaggerated. The accumulation of contaminated sediment in the Rhine Delta was such that in 1987 the port

---

Box 2.6

Rehabilitating Rhine River water quality

The Rhine's water reached its worst condition in the 1970s when researchers found that 'fish died instantly when exposed to the water of the Lower Rhine' (van Dijk et al., 1995: 381). Growing environmental concern prompted the riparian states to seek international cooperation to reduce the growing health hazards posed by the 'largest sewer in Europe'. The Convention of the International Commission for the Protection of the Rhine (ICPR) brokered agreements on emission reductions focusing on improving oxygen concentrations and reducing organic compounds and heavy metals in the river, particularly through improvements in treatment of sewage before its release into the river. By the 1980s the Lower Rhine was being recolonised by molluscs, amphipods (shrimps) and insects. The first clear signs of this recovery were visible in dense swarms of caddis fly (*Hydropsyche contubernalis*) along the river banks in summer (Neumann, 2002). Then a chemical spill on 1 November 1986 following a fire at the Sandoz pesticide factory in Basel caused 30 tons of insecticide, fungicide and herbicide to be released into the Upper Rhine, killing fish on a stretch of the river extending hundreds of kilometres downstream from Basel (van Dijk et al., 1995). The scale and highly visible consequences of the disaster demonstrated the high vulnerability of the Rhine to catastrophic pollution, and gave new impetus to efforts towards its ecological rehabilitation. The 1987 Rhine Action Plan agreed by the riparian states had as its goals:

1. the creation of conditions in the river that would enable the return of 'higher trophic species' of which the most emblematic is the salmon;
2. to safeguard the Rhine as a source of water for treatment and supply as drinking water;
3. to stop the contamination of sediments with toxic compounds;
4. to meet the water quality requirements of the North Sea Action Plan.

By 2004, the ICPR reported, target water-quality standards for the Rhine had been achieved for 37 pollutants, including DDT and related pesticides, and a series of hydrocarbons. In addition, levels of a further 21 pollutants, including arsenic, chromium, lead, nickel and mercury, and agricultural nutrients phosphate and ammonium were close to their target levels. For other heavy metals (cadmium, copper and zinc) and for the dangerous organic compounds polychlorinated biphenyls (PCBs), concentrations in the Rhine remain above target levels (ICPR, 2009). This has important consequences. For example, zinc has specific inhibitory effects on microbial organisms that break down organic pollutants, so that zinc pollution retards natural biodegration processes (van Dijk et al., 1995). PCBs are particularly problematic because they accumulate through the food chain. Levels of PCBs in eels in the lower Rhine were no different in the 1990s than in the 1960s and were four times the level considered safe for human consumption (van Dijk, 1995). Three 'priority pollutants' had also not been reduced to target levels by 2004: the insecticide lindane, the herbicide diurone and the anti-fouling agent tributyltin (applied to boats) (ICPR, 2009).

of Rotterdam was obliged to dredge some 10 million m^3 of sludge from the harbour basin for disposal as hazardous waste. Since 1999, the annual amount of hazardous sludge removal from the harbour has been running at 5 million m^3 (ICPR, 2009). Improving water quality also confronts challenges posed by less obvious forms of pollution. Thus, despite reducing heavy-metal pollution from industrial 'point sources' by between 72 and 95 per cent, the reduction of pollution of the water of the Rhine has not been commensurate (Vink and Behrendt, 2002). Remaining point sources of pollution are municipal waste-water-treatment plants, but 'diffuse sources' (soil erosion, groundwater drainage and urban surface runoff) account for about half of heavy-metal emissions overall, but as much as 70 per cent of copper and mercury and 76 per cent of cadmium emissions into the Rhine (Vink and Behrendt, 2002). Such diffuse sources of contamination present much more intractable problems of heavy-metal emissions reduction. Moreover, the levels of contamination need to take account of a component of heavy metals in the river which is contained in sediment carried by the river flow. For the Rhine, Vink and Behrendt (2002) found this sediment component varied from 12–15 per cent for cadmium and zinc, to 33–46 per cent for mercury and lead. Retention of pollutants in sediments is cumulative and poses questions about the possible release of heavy metals back into river water from contaminated sediments in the future.

This is not a problem solely for the Rhine basin. Rothwell et al. (2008, 2009) have found the uplands of North West England to be severely contaminated by heavy metals, in particular lead, released over the last 150 years from the industrial centres of Manchester and Sheffield and then subject to atmospheric deposition. These contaminants have remained locked to the surface 10 cm of peat, which is now being eroded into the local water courses that drain into nearby reservoirs. For the time being, it appears the adsorption by the peat remains strong, so that the lead remains largely trapped in the sediment of the water courses.

In addition to problems arising from accumulated 'historic' pollution, the achievement of safe-water quality confronts the challenge of newer synthetic organic substances, generally referred to as 'micro-pollutants', that arise from the use of modern pharmaceuticals, cosmetics and cleaning products. These are often not removed in conventional waste-water-treatment processes and, although present in low concentrations in river water, may have significant impacts on specific species. Van Dijk et al. (1995), for example, refer to an inhibitory effect of micro-pollutants on phytoplankton.

Sustainability and 'post-industrial' water resource management

The Rhine exemplifies the challenges and priorities of managing water resources in industrialised economies at the start of the twenty-first century. It suggests that two key priorities, flood control and water-quality improvement, lead in similar directions. In particular the need to provide more 'room for the river'. This approach has led to efforts to reverse the confinement of the river to a single channel with floods contained by dykes and embankments. Instead, efforts are made to provide greater flood retention by the floodplain, re-opening secondary channels and meanders as 'retention basins' into which floodwater can be diverted and retained during high flow periods. As well as serving a hydrological function of physically managing river flow, these retention basins are also seen as a means of re-establishing floodplain ecology and landscape for amenity (leisure and tourism) purposes, and as means of reducing flow rates and hence increasing precipitation rates of sediment. The latter has a significant impact on removing certain pollutants, notably phosphates derived largely from agricultural fertilisers and farm effluents, as these are retained in sediments (van der Lee et al., 2004).

The problems of water quality in the Rhine typify a key aspect of degradation of water resources over the past two hundred years. Point sources of pollution are most associated with urban and industrial sources. As we noted in the second section above, Contamination and Sanitation (p. 46), in developing countries there is a focus upon pathogens and potential human health impacts because of the limited treatment of sewage wastes. In industrialised contexts, as in the case of the Rhine, concern has focused on heavy metals but today there is an increasingly complex array of harmful substances potentially released at catastrophic scale. Examples noted above were the 1986 Sandoz case on the Rhine, and the spillages from TVA power station ash landfills. Further examples identified by WWDR (2006: 283) include incidents at Tisza, Romania, in 2000 (breach of a tailings dam released 100,000 m^3 of waste contaminated with cyanide and heavy metal which disrupted drinking water extraction and caused massive fish kills in the Danube and Black sea); Kentucky, USA, 2000 (coal tailings dam failed, dumping 950,000 m^3 of slurry causing fish kills in some 120km of streams and disrupting drinking water abstraction); and in Brazil (iron mine waste) and the Philippines (copper and silver tailings waste). The case of the Rhine indicates that major investment can substantially reduce pollution from such point sources.

More extensive impacts have arisen through agricultural intensification. Increases in crop yields associated with the green revolution have led to greater leaching of fertilisers, herbicides and pesticides into water courses. Many of these are toxic to human health as well accelerating eutrophication in ecosystems where phosphates and nitrates are limiting nutrients. Irrigation, because of its hugely inefficient use of water (see Chapter 4) and massive twentieth-century expansion, has become one of the major causes of environmental degradation through salinisation. Jones (1997), for example, notes that salinity of the Rio Ebro, Spain increased from below 600 TDS mgl^{-1} before irrigation in 1960, to sustained values over 800 TDS mgl^{-1} within 25 years. Similar problems have arisen with groundwater becoming contaminated as in Pakistan and India, while the water returned to the Colorado river by irrigated farms in the western US has proved so saline that the US government had to invest in a desalination plant in order to honour its 1944 agreement to supply 1.8 km^3 of the river's annual flow to Mexico.

Reducing pollution from agricultural sources is not straightforward. Nienhuis (2008) charts conditions in the lower Rhine (Lobith) and Meuse (Eijsden) and notes significant reductions in total phosphorous load over the last 50 years with a peak of P 2.0 kgs[-1] falling rapidly to 0.5 kgs[-1]. Nitrogen load tends to plateau over the same period at N 17 kgs[-1] declining to 10 kgs[-1] by the end of the last century. Comparison with nineteenth-century data (1870) however suggests that conditions have not yet returned to pre-industrial levels.

Box 2.7

Defining and measuring water quality

Establishing a sufficiently good quality of water is not straightforward. There is general agreement on harmful substances but the method of quantifying and setting acceptable levels varies. Consideration needs to be given to both the level of exposure and the duration of exposure to establish the dose–response relationship but there will also be variations due to body mass (size, age). The UK Environment Agency (Environment Agency, 2010a General Quality Assessment GQA) for example, uses a combination of biological indicators (invertebrates); chemical indicators (ammonia, biochemical oxygen demand and dissolved oxygen) and nutrient status (Phosphates and Nitrates). This provides a comprehensive assessment for the ecosystem functioning of the water courses but the same criteria are not necessarily appropriate for drinking water supplies. UK drinking-water standards (Thames water, 2009) are set around microbiological conditions (absence per 100 ml of coliforms, *E.Coli*, *Enterococci* and *Clostridium perfringens*) and chemical conditions for lifetime consumption based upon 43 criteria including taste, colour and turbidity, and concentrations of chemicals ranging from aluminium to vinyl chloride.

The WHO (2009) has published guidelines for drinking water that establish microbial and radiological aspects for water. The report contains over 160 pages on possible chemical effects for over 120 substances. For example, compared with the UK–GQA method, nitrate (and nitrite) is noted to have a potential health impact with a WHO guideline value of 50 mg/l for short-term exposure. WHO notes that most natural occurrences will be below 10 mg/l except in contaminated groundwater, for example. The GQA similarly uses a value of 40 mg/l as being high and possibly harmful. The GQA level for ammonia is set at 9 mg/l as being classified as indicative of a poor environment while the WHO note that ammonia normally occurs at values less than 0.2 mg/l and as there is no immediate health relevance no health-based guideline is provided. Assessing water quality is both complex and costly and in many areas there are insufficient data to establish trends.

Across the USA, Sprague et al. (2009) found no significant changes in total loads of phosphorous, total nitrogen or nitrate from 1993 to 2003. After adjusting data for changes in flow arising from drought conditions and other hydrological impacts over this period, concentrations revealed increasing trends at 33 per cent of sites for phosphorous, 21 per cent for nitrogen and 12 per cent for nitrate, compared to decreases at only 16 per cent of sites for phosphorous and 25 per cent for nitrate The majority of sites showed no change. They conclude (p. 101) somewhat bleakly:

> While modest gains have been made in nutrient quality in some heavily impacted streams, nutrient enrichment has increased in many streams that were least impacted in 1993. The Nation's least-impacted streams are increasingly being affected by population growth, and new strategies designed to reduce these effects may need to be integrated with ongoing actions to improve nutrient quality in heavily impacted streams.

Similarly, Wolff et al. (2007) report that for the USA between 1992 and 2001 pesticides were detected in 65 per cent of stream samples, with almost all fish examined exhibiting some organochlorine contamination even though these pesticides had not been used for several years prior to sampling.

The European Environment Agency (2007) presents generally a more favourable picture of water quality. The UK (EA, 2010b) announced in 2008 the eighteenth consecutive improvement in surface water quality with 76 per cent of English and 95 per cent of Welsh rivers being good or very good quality based on their chemical characteristics. This is explained by investments by the water industry over the last two decades in waste treatment. The same pattern is observed in West and Central Europe. The EEA's fourth Environmental Assessment of Europe (2007) reports water-quality improvements within the EU-25 region where 90 per cent of the population are now connected to sewerage networks and waste treatment is established. Yet Merrington et al. (2002) noted that in the UK since 1979 the number of agricultural pollution incidents rose from below 2000 a year to around 4000 two decades later. There are problems with data availability for some European areas but in general West-Central Europe demonstrates decreases in biochemical oxygen demand and ammonium from 1992 to 2004 in keeping with the UK–EA assessment. Nitrate levels have not decreased as rapidly overall although some countries such as Germany, Denmark and the Czech Republic report significant improve-ments, compared to deterioration for Slovenia, and little change between 1994 and 2004 for Italy, Luxembourg and the Netherlands. Remaining areas of high water pollution through municipal and industrial wastes include rivers in Russia, Albania and Kazakhstan. The Volga is particularly noted for its poor condition while watercourses in the Ukraine are noted to have high nitrates and, in places, pesticides due to agricultural waste contamination. There is, then, a mixed picture of improvements where investments have been made in waste management and water infrastructure, but also areas of continuing contamination through urbanisation and industrialisation. Of more long-term concern, however, is the rapid increase in new synthetic organic pollutants – so-called micro pollutants – that include biologically active substances such as pharmaceuticals for which standard waste-treatment approaches were not designed. The range of compounds involved, their potential interaction and complexity of ecological effects arising from this type of pollution suggests a need for a major review of waste treatment, and perhaps a re-appraisal of industrial societies' dependence on water as a principal waste disposal route.

The Rhine experience shows convergence of flood control and water quality objectives in an 'ecological' approach to water management. Hooijer et al. (2004) have pointed out that this demands a different type of decision-making from that of the past. In particular, planners are now confronted with decisions about multiple objectives to be achieved with river management: if flood management is to be integrated with restoring floodplain ecology, what kind of 'nature' do we want? Particular species of fish (e.g. salmon), or birds, or mammals (e.g. beavers)? Active sedimentation processes to reduce pollutant concentrations? Ecological linkage of fragmented riverine and floodplain habitats? Above all, however, the concept of restoring 'room for rivers' implies reconciling agricultural, urban and industrial development with riverine ecology in ways radically different from past approaches to water management. In particular, this will mean re-conceptualising river floodplains as an environment that will be (*should* be) flooded from time to time. In many cases 'reduction in flood damage' will involve planning authorities preventing development that is not 'flood-tolerant'. In other instances, Hooijer et al. (2004) have argued, flood risks need to be made explicit to floodplain residents and businesses, and the costs shared through requirements for mandatory flood insurance appropriate to the likelihood of floods occurring.

Similar shifts in water-management thinking are evident even in California, where the pre-eminence of agricultural water use no longer goes unchallenged. The role of dams in preventing migratory fish species reaching their upstream spawning grounds – and the impact on economically significant salmon fisheries of the north-west – has become more publicly appreciated, and a five-year drought in 1987–92 in California brought the environmental stress in the system to a head. In particular the transfer of water from wetter northern California via the State Water Project to irrigate the drier south of the state was seen increasingly as threatening the survival of fish species in the San Francisco Bay and Sacramento River delta. In 1992 increasing resistance to the water transfers resulted in an explicit re-allocation of 1.23 km^3 water from agriculture to protect salmon runs and wildlife refuges, and in 1994 the Monterrey agreement made a further permanent transfer of 0.13 km^3 from agriculture to urban water use. A further consequence of the 1987–92 drought was the organisation of a government-supervised water market to re-balance supply and demand using transfers through the state's water infrastructure. Farmers could sell their irrigation water by fallowing all or part of their land. According to Howitt and Sunding (2003), during the 1990s tight restrictions meant that agricultural water users could not sell their water to neighbouring urban areas, where water prices were three times higher, but only to other agricultural districts. Even at the height of the drought, overall water sales amounted to only 3 per cent of total water use. Most buyers of water were district authorities seeking to increase stream flows in order to meet environmental regulations.

By early 2009 California was again confronting the need to choose between ecological and agricultural priorities as it entered a third consecutive year of low rainfall. Following rainfall of only a third of average in January, State Governor Schwarzenegger proclaimed a state of emergency in February 2009 due to state-wide water shortages. Legal requirements to protect fish species in the Sacramento/San Joaquin Delta meant reduced pumping of water from north to south via the state's water projects (SWP and CVP). This resulted in steep reductions in deliveries of water to irrigated agriculture, notably in the San Joaquin valley, where allocations, already reduced to 50 and 40 per cent respectively in 2007 and 2008, were further reduced to 10 per cent in 2009. Most irrigation districts were less badly affected, and some degree of water reallocation (about 10 Mm^3 in total) was effected through a state-operated 'water bank' (whereby the state paid $275 per acre-foot to irrigators for water they did not need). Nonetheless, it was estimated in June that lost agricultural production in 2009 would amount to almost $1 billion (CDWR, 2009) and in some areas, notably the San Joaquin valley where groundwater levels had been severely depleted by many years of over-draft (Figure 2.9), local losses would be much more severe. Loss of irrigation quickly translated into social consequences, with some 35,000 people unemployed in Fresno County by April and more than 3000 people seeking food aid at distribution centres in the City of Mendota the same month (CDWR, 2009). In the event, rainfall improved and state-wide run-off was estimated at 70 per cent of average for 2009

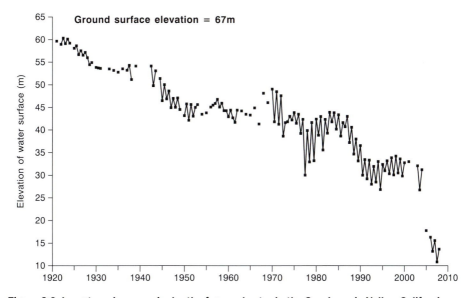

Figure 2.9 Long-term increase in depth of groundwater in the San Joaquin Valley, California

Source: California Department of Water Resources, 2009: 18; permission to reproduce granted by CDWR

compared with a little over 50 per cent of average in the previous two years. Yet, as earlier in 1987–92, the 2007–9 drought exposed the vulnerability to water shortage of one of the richest economies in the world, and raised questions about the sustainability of the Californian model of water-resource management. It represents in extreme form the dilemma confronting future water-resource development: the economic challenge of increasing costs of expanding supply from ever more distant and/or capital-intensive sources, or the political challenge of reducing demand through changes in consumption behaviour. We will return to these issues in Chapters 7 and 8.

Summary and conclusions

In this chapter we have focused on the role water has played in the development of those economies that have proved most successful in generating economic growth. Over the past two hundred years these have been largely based on urban and industrial development. The development of water resources has been integral to this pattern of economic growth through providing transport links, hydroelectricity for industry, irrigation for agricultural intensification and drinking water for urban populations. Investment in water infrastructure has also been critical to enabling economic activity to grow by protecting it from the threats of floods or drought.

We have seen, however, that approaches to water-resource development have changed. Whereas the nineteenth and much of the twentieth centuries were dominated by engineering infrastructure to move water to where it was needed (or move it from where it was not needed), there has in recent decades been recognition of the limitations of such approaches. Instead, there is growing emphasis on managing water as a 'multifunctional' ecological resource, providing a number of ecological services (such as clean water and biodiverse environments). Priorities among such services are not clear cut, however. As a consequence, definition of the goals of water-resource development implies the need for discussion and negotiation, calling into question both the existing planning procedures of government agencies and the existing behaviour of water users. This is associated with a greater emphasis on the means by which water-resource use is governed. As we outlined at the end of Chapter 1, water-resource development may be seen as constituting a tension between approaches that rely on technology to improve water supply and those that rely on governance to modify the amount of water demand.

What can 'less-developed' (less industrialised) countries learn from the experience outlined in this chapter? It is clear that many have embarked on patterns of development modelled on that of the US administrations from the 1930s to 1970s. As a consequence, many encounter the criticisms earlier levelled at 'top–down' state-led development of water infrastructure, as exemplified by international controversy over China's Three Gorges Dam project. Yet, one lesson that seems clear is that large-scale water infrastructure projects, such as the Hoover Dam, have in the past been associated with major shifts in economic activity. The lessons for development planners are surely that: first, such shifts are not necessarily definitive nor sustainable, as exemplified by the changes in policy on flood control on the Rhine and on water transfers for irrigation in California; and second, the effects of major infrastructure investments must be seen as highly dependent upon economic and political context. It was observed, for example, that the trajectory of water supply development in rapidly expanding cities such as Mumbai does not appear to be following that of cities such as Los Angeles a century earlier or Manchester 150 years earlier. Would the experience of arid California be relevant to developing the West African Sahel?

In Chapters 4 and 5 we will consider approaches that emphasise the use of technology to mitigate water 'scarcity' through increasing water supply. Chapters 6 and 7 will consider governance approaches to both total demand and competing demand. In the next chapter, however, we will first consider the longer-term relationship of water resources and development in the light of projections of climate change.

Further reading

This chapter has discussed how economic development has been reliant upon the development of water resources leading to increases in demand while also causing problems of contamination requiring treatment. Additional reading on the water crisis and demand following on from Chapter 1 can be found in:

Lomborg, B. ed. 2009 *Global Crises, Global Solutions* 2nd edn. Cambridge University Press, Cambridge.
Lundqvist, J. 2000 A global perspective on water and the environment. *Physics and Chemistry of the Earth* 25(3): 259–64.

Reading on water treatment and improved sanitation can be found in:

Al-Baz, I. Otterphohl, R. and Wendland, C. eds. *Efficient Management of Wastewater*. Springer-Verlag, Berlin.
De Bruijne, G., Geurts, M and Appleton, B. 2007 Sanitation for all ? IRC International Water and Sanitation Centre (www.irc.nl).
Symons, G.E. 2006 Water treatment through the ages. *Journal of American Water Works Association* 98(3): 87–98.

Water standards are summarised for human health and ecological environments by:

WHO 2009 *World Health Organisation*: WHO Guidelines for *Drinking-water Quality* 3rd edition World Health Organization Geneva http://www.who.int/water_sanitation_health/dwq/gdwq3rev/en/index.html
Environment Agency (UK) 2010 Water Quality Assessment http://www.environment-agency.gov.uk/

The need to exploit water for economic development is epitomised by the nineteenth- and twentieth-century western expansion into the arid American West by:

Reisner, C. 1993 *Cadillac Desert. The American West and its Disappearing Water*. London: Pimlico.

Such development comes at a price, both capital costs and environmental costs, over which attitudes have changed and discussion of these changing positions for water management can be found in:

Gandy, M. (2008) Landscapes of disaster: water, modernity, and urban fragmentation in Mumbai *Environment and Planning A* 40: 108–30.
Swyngedouw, E. (2006) Power, water and money: exploring the nexus. Human Development Report 2006 Occasional Paper. New York: United Nations Development Programme.

3 Climate change and fresh water resources

Changing climates

The relationship between climate and natural resource use has long been the subject of analysis and speculation and has formed the basis of teaching environmental systems and geography into the twenty-first century (Huggett et al., 2004; Strahler and Strahler, 2006; Holden, 2008). The relationship is often considered deterministic, exemplified in the notion that global vegetation distribution is dependent upon climate leading to an 'equilibrium' distribution, widely used to map and explain the world's major biomes. There are several well-used global classifications of climate, (e.g. by Koppen, 1931; Thornthwaite, 1948; UNESCO, 1977) using rainfall, temperature, evaporation and water balances. Time and stability are key factors in such classifications but more detailed analysis reveals that rarely are ecosystems in such a steady state or equilibrium. Climate has undergone change at both global and local levels, and on both geological and historic (human) timescales. Early twentieth-century discussions of climate too often failed to take into account the dynamic nature of the atmosphere and, further, the need to distinguish between climate, weather and meteorology (see Box 3.1) but today we are only too well aware of the need to view climate as a temporal as well as a spatial environmental variable.

Climate change, then, is not a new phenomenon. Sources of change in the Earth's climate include solar activity, changing composition of the atmosphere, continental drift and associated tectonic activity, plus geomorphological processes (Harvey, 2000). Continental drift, tectonic

Box 3.1

Climate, weather and meteorology

The atmosphere can be examined at a variety of scales with four usually identified:

Micro-scale
> Microclimate phenomenon that change over intervals of a few minutes to milliseconds and that occur at a spatial scale of less than 1km and usually sub-metre scale. Such investigations are dominated by vertical transfers (rather than horizontal) of heat and water involving turbulent transfers through eddies in the lower layers of the atmosphere.

Meso-scale
> Atmospheric motions that take minutes or hours to complete, over tens of kilometres (horizontal scale). Studies of individual clouds have helped understanding of precipitation but conversely it is still difficult to predict cloud behaviour, as cloud seeding attempts have proved. Usually meso-scale studies are at a larger scale than individual clouds, examining land and see breezes, mountain and valley winds such as the **chinook** or **fohn**, and urban–rural climate differences.

Synoptic-scale
> Phenomena lasting from one to several days and extending horizontally over 100s of kilometres, i.e. the view accorded by most weather predictions. This has been the most prevalent scale of study conveyed to the public until the advent of global warming.

Global-scale
> Global circulations taking weeks to months to complete at horizontal scales of thousands of kilometres. Such planetary-wide phenomena as the jet streams, Rossby waves, monsoonal circulations and trade winds are included. There is growing appreciation of the feedbacks within the climate system and the teleconnections that appear to exist between ocean temperatures and atmospheric circulations, for example the association between EL Niño (the occasional upwelling of warm rather than cold waters off the Pacific coast of South America) and Asian monsoon.

Meteorology examines the physical nature of the lower atmosphere, i.e. the weather-forming processes such as condensation, evaporation and energy fluxes, but in recent years the chemical composition of the atmosphere has received increasing attention as concerns over global warming and ozone depletion have come to the fore. Weather is the daily state of the atmosphere (wet, humid, anticyclonic) but this tends to be descriptive and used mainly as a non-scientific term. Climate describes the average conditions experienced by an area, based upon long-term estimates of what is normal, e.g. arithmetic means using 30 years of observations. Given the dynamic nature the use of thirty-year averages to determine climate can be problematic. The figure shows the departures of annual rainfalls for West Africa based upon the 30-year period 1931–1960. In 1991 the averaging period changed to 1961–1990. It is evident from the figure that identification of wet and dry periods is influenced by the averaging period selected, and where records are incomplete this can introduce sources of error, a point that will be elaborated upon in the section Climate Change Uncertainties, p. 101. Much of this chapter will try to explain the dangers of simply accepting the broad generalisations from IPCC reports and extrapolating these to local conditions, thus uncertainty and complexity will be common themes.

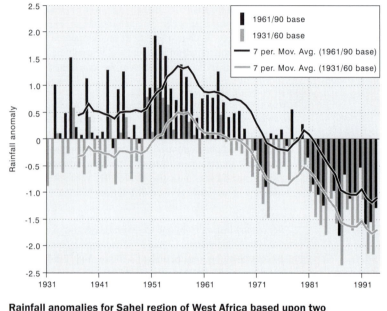

Rainfall anomalies for Sahel region of West Africa based upon two sampling periods (7 year period moving average)

Source: after Agnew and Chappell, 1999; permission to use granted by Springer (*GeoJournal*)

activity and rock weathering, for example, have atmospheric impacts over geological timescales of millions of years with weathering, for example, explaining Cenozoic cooling. On a shorter timescale climate change is linked to alterations of the energy received at the earth's surface forced by orbital eccentricities, as explained by the Milankovitch theory which creates cycles of 20 thousand to 100 thousand years. Variations in solar activity can be mapped to much shorter cycles – an 11-year period often being cited. Despite a number of cool periods in recorded history such as the 'Little Ice Age', which reached its minimum temperatures during the seventeenth century in Western Europe, most current interest is focused on the accelerated warming over the last century and its relationship to atmospheric carbon dioxide (CO_2) emitted in increasing quantities since the onset of the Industrial Revolution. Global average atmospheric concentrations of CO_2 have risen from 270 ppm (parts per million) at the end of the eighteenth century to 355 ppm in the 1990s and to 377 ppm by 2004 (CDIAC, 2008). In 1995, the IPCC confirmed its belief that human activity was responsible for these increases and that they had an influence on the global climate (IPCC, 1995). This was a much stronger statement than in previous reports and by the Third Assessment Report (TAR) of the IPCC (2001) human (anthropogenic) activity was held to be the cause of global warming leading to climate change, a view subsequently confirmed in the Fourth Assessment Report (FAR) in 2007 which states: 'Warming of the climate system is unequivocal, as is now evident from observations of increases in global average air and ocean temperatures, widespread melting of snow and ice and rising global average sea level' (IPCC 2007: 30). The rate of warming of 0.13°C per decade from 1956 to 2005 was nearly twice that observed over the whole of the twentieth century, for which the total increase was 0.74°C. Over this period precipitation is reported to have increased in eastern areas of the Americas, northern Europe and central Asia, while it has decreased in southern Asia, Southern Africa and the Mediterranean. There is also high confidence that for glacier- and snow-fed areas run-off has increased while rivers have warmed. The IPCC–TAR (2001) reported 'high confidence' that ecosystems were being affected. The IPCC–FAR went further, stating that there is now ecological evidence globally, but especially from studies in North America and Europe, based on observations 1970–2004, with 89 per cent of observed changes in biological systems consistent with warming.

In the future the IPCC forecasts a warming of 0.4°C until 2025, after which temperature predictions vary depending upon the different energy scenarios modelled, leading to a 'best estimate' of a temperature increase of between 1.8 and 4.0°C during the twenty-first century. In terms of water resources, IPCC (2007) have high confidence that at higher latitudes and in East and South East Asia river flows will increase (10 to 40 per cent by 2050), while in dry-subtropics decreases in rainfall will mean a drop of 10 to 30 per cent in run-off. Drought is expected to increase and cause water shortages in semi-arid areas, including Southern Africa, North East Brazil, Western USA and the Mediterranean. Even where increased precipitation is predicted to cause higher run-off this water-resource benefit is likely to be moderated by increasing variability and seasonal changes. IPCC (2007) also calculate that it is likely that up to 20 per cent of the world's population will inhabit areas where flood risk has increased due to climate change. There are, then, a number of threats identified as arising from climate-change effects on water resources, including more floods, more droughts and increased variability. These will present fresh challenges for those seeking to develop water resources (Kerr, 2007a).

Much has now been written on climate change, from the scientific understanding through to the social ramifications of adaptation, and most generally now accept climate forcing through anthropogenic induced global warming (Harvey 2000; Houghton, 2004; Pittock, 2005; Dessler and Parson, 2006; Dow and Downing, 2006; Walker and King, 2008). This was not always the case and it is worth noting that following a Northern Hemisphere cooling trend in the 1960s, some wrote about the onset of the next ice age, for example Bryson (1974) and, later, Ponte (1976) in an aptly titled book *Global Cooling*. There have also been suggestions (Rahmstorf, 1997) that, as a consequence of global warming, changes to circulation in the North Atlantic could bring much colder conditions to Europe. Despite the recent (2008) respite in the warming trend which has been linked to ocean circulation (Keenlyside et al., 2008), as we write, the scientific and political consensus is that human-induced emissions of greenhouse gases has in the recent past, and will in the near future, bring about climate change (Kerr, 2007b; Anderson and Bows, 2008; Meteorological Office, 2010; Hulme, 2009a). On the other hand, during preparations for the Copenhagen Climate Change Conference (2009) there was greater acknowledgement that there remain climate-change sceptics. Hulme, (2009b: 1), reflecting upon the reasons why science rather than politics has become the 'battleground' for supporters and opponents of the global warming thesis, notes that in the UK 'only 41 per cent believed humans are causing climate change, 32 per cent remained unsure and 15 per cent were convinced we aren't'. Critics of global warming were arguably more evident in the twentieth century than today but still this is a controversial topic. Oldfield (2005) lists the following topics raised by sceptics of climate change, some of which have been labelled as myths and dismissed by The Royal Society (2008) but recently revisited by Schiermeier (2010),

- Historical interpretations of temperature record are incorrect
- Model limitations
- Feedbacks not taken sufficiently into account
- Interrannual variation greater than observed warming

- Solar variability and other natural causes
- Resilience of the climate system.

Conversely, some, such as Kevin Anderson, Director of the Tyndall Climate Change Centre, argue (Anderson and Bows, 2008) that influential policy analysis, such as the Stern report (2006), has underestimated global temperature rises based upon what we already know. This has potentially critical implications. IPCC (2007) states that observed global warming is 0.6°C, and that a further warming of 1.1°C is inevitable, given past greenhouse emissions. IPCC poses a question of whether a warming of 2°C is the 'tipping point' for many natural and agricultural ecosystems. However, by taking account of accumulation of greenhouse gas emissions, rather than annual emission rates, Anderson and Bows (2008) predict that we are already past the 2°C 'tipping point' and that we will most likely exceed a 4°C warming during this century.

We note here an interesting complementary discussion on the evidence of our (in)ability to artificially change weather patterns, which we review in our analysis of the results of cloud seeding, presented in Chapter 5. However, it is clear that, despite sceptics' questioning of the evidence, there is overwhelming scientific and political support for the view that the global climate has been undergoing accelerated warming during the twentieth century and that this is likely to continue in the current century. Despite the apocalyptic future some writers associate with such a scenario, linkages between human activity, greenhouse emissions, climate change, hydrological impacts and water-resource management are complex, and many have yet to be fully quantified. The main topics discussed in this chapter, that is, climate change and hydrological impacts in the context of natural variability, are highlighted in Figure 3.1.

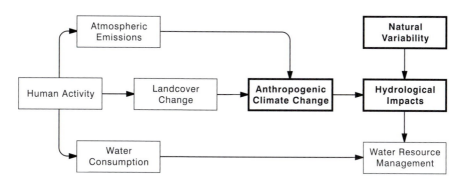

Figure 3.1 Links between climate change and water-resource management

Source: after Kundzewicz et al., 2007

After four IPCC (1990, 1995, 2001, 2007) reports and numerous other publications on the possible impacts of climate change upon fresh water resources it could be argued that there is little more to be written but we hope to show that there is still much uncertainty. This not to call into question the view that climate change is inevitable and that observed global warming is commensurate with physical models of greenhouse-gas behaviour as explained by Houghton (2009). However, a chain of uncertainty, that commences with assumptions over future energy production and ends with extrapolations from regional climate models to local hydrological impacts, means that there is still discussion over the future impacts on water resources. This chapter will explore these uncertainties and will conclude that the best approach is to ensure water-resource management systems are resilient and have the flexibility to respond to a range of extreme conditions that cannot yet be predicted with high confidence at local resolutions. We will discuss the predictions for changes in temperature, precipitation and evaporation as these are three key climate variables that influence the hydrological cycle. The hydrological impacts of these climatological changes will then be investigated through river flows, impacts upon crop growth and drought.

Climate-change predictions

Temperature and climate change

Figure 3.2 shows the global temperature record with steady temperature rise over the last two decades and strong warming evident since 1975. However, the graph also shows this was not always the case. Air temperatures in the northern hemisphere have not risen continually, and have at times declined, most noticeably in the 1960s (Pearce, 1997, Gordon et al., 1998). Temperatures have not increased uniformly across the world since the 1970s, as is so often assumed, with differences observed between northern and southern hemispheres. Moreover, despite the consensus on temperature rise, the year 2008 saw a marked decline in temperatures, which has been linked to La Niña circulation, which began to develop in 2007 and is now the strongest since 1988/89.

Global temperatures were expected by the IPCC to increase by 1.4°C by the year 2030 (Houghton et al., 1991) within a range of 0.7–2.0°C. During the 1990s these figures were modified in light of the effect of sulphate aerosols and the UK Countryside Commission (1995) reported at the time that we could expect a rapid rate of global mean warming of

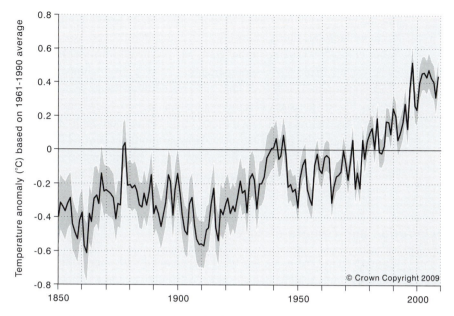

Figure 3.2 Global average temperatures based on HadCruT3 globally gridded data; shaded area indicates sampling uncertainty

Source: adapted from MO, 2010 data; permission to reproduce date granted by UK meteorological office who retain the copyright

0.2°C per decade with the greatest warming expected in higher latitudes. Early climate models failed to accurately reproduce past temperature changes due to weak ocean-atmospheric coupling, some bias in the climate record, sunspot activity and atmospheric turbidity. The Meteorological office reported (2010) that, since the 1990s, model predictions have vastly improved but uncertainty remains because of the need to forecast future atmospheric conditions based upon assumptions such as economic activity, deforestation and energy consumption. Houghton (1997), summarising the IPCC (1995) report, reconfirmed earlier predictions that a doubling of CO_2 levels from 280 ppmv pre-industrial levels by around 2030 would lead to a rise of 1.6°C increasing to 2.5°C (within a 1.5 to 4.5°C extreme range) by 2100 under a 'business as usual' scenario IS92a.

Houghton (1997) also noted the greater difficulties of predicting regional changes and the need to understand alterations in terms of extreme events not just averages. It is extreme events that have the greatest impacts. Unfortunately, as one increases cartographic scale (i.e. 1:1000 > 1:1million) model prediction uncertainty increases, but in general it is possible to observe that in winter temperature rises will be greater over land than over oceans with corresponding reduction in the diurnal range, and that raised temperatures will be found in higher northern latitudes due to reduced sea ice and snow cover. More recently Johns et al. (2003) using model HadCM3 predicted warming by 2100 of between 2.0 and 5.0°C (compared to 1.5 to 6°C from comparable models) but noted rises over land surfaces could be twice that over the oceans. These increases in temperature (see section The Chain of Uncertainty, below p. 101) will have further effects on the hydrological cycle, in particular precipitation and evaporation, but beyond 2020 the details are far from certain, while many argue 'heroic' action is required now to keep twenty-first-century warming to below a 2°C threshold (Tollefson, 2010).

Precipitation and climate change

Climate-change research has been preoccupied with energy fluxes and has only reached a consensus on modelling average temperature changes. Precipitation is arguably a more significant factor for hydrological changes but attempts to predict precipitation have been beset with difficulties. In the 1980s this was reported through generalised statements that a warming earth would intensify the hydrological cycle, i.e. more winter precipitation in temperate lands but in some parts (e.g. semi-arid regions) this would mean reductions in annual precipitations. Forecasts at a more local scale were often contradictory and complex. For example, Hulme (1996) reported predictions by different models of rainfall changes for Southern Africa varied between a 2.5 to 7.5 per cent increase, a 20 per cent increase, or, at the other extreme, a decline of 7.5 per cent. This uncertainty is reflected in empirical records of rainfall. Kruger (2006) noted previous studies on South Africa rainfall had reported a range of trends, from more days with heavy rain through to no trend at all for annual totals. Kruger found that while for most of the region there was no strong evidence of changes in precipitation, in some parts there were changes in the number of consecutive rain days (both positive and negative). Given this level of historical variability it would be surprising if Global Circulation Models (GCMs) of climate did provide consistent predictions at the local scale, and yet this is what Hewitson and Crane (2006) report. They used the results from three GCMs for South Africa precipitation and found that, while the magnitudes varied, there was agreement in the spatial patterns with a prediction of increased summer rainfall over central and eastern plateau (Drakensberg mountains) but little change in the Western Cape. Offering a similar approach, the accumulation of flow and rainfall data for the Nile River basin over the course of the twentieth century has enabled the identification of a relationship between rainfall intensity in East Africa and atmospheric circulation patterns resulting from changes in ocean surface temperature differentials in the Indian Ocean (Conway et al., 2007).

Although there is disagreement in detailed locations most models find there is an increase in higher-latitude precipitation as temperatures rise (Johns et al., 2003) with a complementary decline in precipitation in the subtropics. There is no clear trend in the frequency of tropical storms but their power does appear to have increased since the 1970s with a doubling of the most severe storms (Emanuel, 2005; Webster et al., 2005). Moving towards a more local scale of prediction demands an understanding of local precipitation patterns. Studies in Europe demonstrate a high level of consistency, with a consensus predicting future drier summers and wetter winters (Buonomo et al., 2007). At a more detailed level there are, of course, variations. UKCIP (2002) note drier summers in the UK would have a north–south gradient with desiccation greatest in the south, while increases in winter rainfall would have a slight west–east gradient with highest rises to the east. The test of the reliability of these predictions is generally sought in models' ability to reproduce the patterns of the past. Fowler et al. (2005) illustrate differences between regional climate model predictions and past observations, notably for upland and coastal areas, but at regional and local scale there is greater uncertainty. Not only is precipitation expected to increase in some parts, but a possible shift to convective processes should lead to higher intensities of rainfall. Osborn et al. (2000) examined trends in the UK, finding that, after filtering out the influence of the number of rain days, over the period 1961–95 there was evidence of more intense precipitation in winter and less intense precipitation in summer, possibly linked to the North Atlantic Oscillation (NAO). As winter rainfalls increase, the winter:summer ratio in southern England, which was at unity in the nineteenth century (Agnew et al., 2000) is now approaching a value of 1.3. In contrast, examining rainfalls in northern England (Durham) Burt and Horton (2007) found after a very wet period in the 1870s a less marked increase in this ratio with a narrowing in the differences between the seasons.

A problem with these predictions, then, is that they are most consistent at only global and continental spatial scales. The coarse spatial resolution of GCMs presents challenges for the prediction of local spatial variation. Over two decades ago Kim et al. (1984) proposed a statistical approach they labelled 'climate inversion' to link local and large-scale climate predictions, in this case successfully for monthly precipitation and temperature in Oregon. More recently Hundecha and Bardossy (2008) have reviewed developments in downscaling methods to address this problem, noting the popularity of the faster statistical approach compared to the dynamic approach based upon a meso-scale climate model. Downscaling has come to the fore based on the assumption that synoptic (regional) conditions remain homogeneous so that statistical means can be used to infer local conditions from the larger-scale synoptic changes contained in GCM prediction. The IPCC (2001) promoted this approach to explore regional and local change. Using this technique, Hundecha and Bardossy (2008), in their study of precipitation over Germany based upon GCM predictors from the Hadley Centre climate model, found that while winter precipitation indices were effectively captured, this was not the case for summer conditions, possibly because of the prevalence of convective process for the latter. They concluded that statistical downscaling models still produced uncertain predictions for extremes of both precipitation and temperature. In a similar manner, working on South Africa rainfalls, Hewitson and Crane (2006) argue that there has been little evaluation of the plethora of empirical downscaling techniques deployed since IPCC promoted this approach nearly two decades ago, and yet Salathe et al. (2007) report success in using a statistical downscaling approach to predict precipitation in the Northwest USA. They note that important local differences are found through this method but improvements through use of a meso-scale dynamical approach would require the resolution to change from 50 km to 15 km, with consequent increase in data requirements. Prudhomme et al. (2002) also believe that at present there is little confidence in precipitation predictions at timescales of less than a month and they state: 'There exists no universal downscaling method for all situations' (p. 1139). There remains much to be done on precipitation forecasting and the use of downscaling, especially the need to assess not just mean changes but those at the extremes of probability i.e. storms and droughts. The development of downscaling methods over the last decade does promise greatly improved predictions in the near future.

Evaporation and climate change

Evaporation is a key hydrological variable but is beset with a confusing terminology and debate over the most reliable method for measurement. Unlike precipitation and run-off it is often overlooked in discussions of climate change, (Calder's 'Blue Revolution' (2006) being a notable exception). The term evaporation is taken here to embody all water vapour that diffuses into the atmosphere, whether it be from a bare soil, water or vegetated surfaces (Monteith and Unsworth, 1990). The term 'evapotranspiration', which is often used to only apply to vegetated surfaces, will be avoided as it adds unnecessary complications under partial vegetation cover, although we note that, in predicting biomass or crop yield, transpiration is far more significant than total evaporation.

Knowledge of evaporation is important in calculating soil–water balances, vegetation–water demand, or drought assessment. It is also a key element transferring energy within the atmospheric system such that moist air produced by high evaporation rates over the oceans contains high potential energy, which can be released as latent heat through condensation leading to large spatial transfers of energy around the globe. Thus, understanding evaporation rates is important to our ability to analyse the redistribution of energy during global warming.

Any change to the earth's energy balance (such as through global warming) will have consequences for evaporation but this has proved difficult to assess. The amount of solar energy received at the outer edge of the Earth's atmosphere is a function of latitude and season, it averages to around 160 wm^{-2} increasing to 300 to 450 wm^{-2} in equatorial regions. Much of this is lost as terrestrial re-radiation but some 100 wm^{-2} is surplus at the surface of the Earth and powers the major atmospheric and oceanic circulations (McIlveen, 1998). Because of global warming the energy balance is frequently presented as:

$$Q^* = Si{\downarrow} - So{\uparrow} + Li{\downarrow} - Lo{\uparrow} \text{ (wm}^{-2}) \tag{1}$$

$Si{\downarrow}$ incoming shortwave radiation

$So{\uparrow}$ outgoing shortwave radiation

$Li{\downarrow}$ received longwave radiation

$Lo{\uparrow}$ outgoing longwave radiation

Thus enhanced atmospheric absorption of longwave radiation is presented as the basic process of global warming, although the Meteorological Office (MO, 2005) point out that increased cooling from the troposphere is a more exact explanation. The earth's surface energy balance can also be presented as (Oke, 1987):

$$Q^* = QH + QE + \Delta QS \text{ (wm}^{-2}) \tag{2}$$

Q^* = Net energy balance at the surface of the Earth

QH = energy expended on sensible heat and turbulent transfer

QE = energy expended on evaporation

QS = stored energy (significant for oceans and/or small time periods)

Bowen ratio β = QH/QE   (3)

Where the Bowen ratio of QH/QE demonstrates the relative amounts devoted to sensible heat and evaporation. The Bowen ratio has been used in global climate studies as it can be useful where desiccation is taking place β > 1, compared to wetter phases β < 1.

The problem with equation 2 is that it is difficult to solve for extensive areas and for short time periods without significant meteorological instrumentation. As a consequence, more approximate empirical methods have been used. Many of the earliest twentieth-century attempts to estimate evaporation relied upon simply relating change in temperature to change in evaporation, often at a monthly timescale through equations such as those devised by Thornthwaite or Blaney Criddle as listed by Jones (1997). Despite their empirical nature these methods have recently proved to be highly useful where climate-change data is either limited or unreliable. Another problem facing the hydrologist is that evaporation is often a point estimate whereas areal or volumetric values are needed spatial extrapolation has proved difficult (Morton 1983; 1994).

Given these difficulties, calculating evaporation rates is not straight-forward. Simple mass balance equations based upon vapour pressure gradients (Dalton equation) have recently been used by Singh and Xu (1997) for total evaporation from Canadian lakes but this was only successful where equations could be calibrated, while daily values are required for many prediction models which we noted above is problematic. A possible alternative is an equilibrium model based on the Priestly–Taylor approach, which only requires energy as an input (Gavin and Agnew, 2004). There is also a plethora of models to assess evaporation alongside direct measurements through hydrological (water balance) or meteorological (eddy correlation or bowen ratio) approaches. Of these, the Penman–Monteith equation has gained widespread accept-ance over the last two decades (see Box 3.2 for a detailed explanation of the approach and terms used below).

What changes have been predicted for evaporation as a consequence of climate change? In the 1990s Arnell's (1992) analysis of the impact of climate change upon UK river flows typically used monthly estimates of potential evaporation (PE) given that GCM predictions were not then sufficiently robust for shorter time periods. Ten years later Arnell (2003) was able to examine streamflows using daily PE values derived from the

Box 3.2

Penman–Monteith approach to evaporation

The FAO (Smith 2008) recommends a meteorological approach for the determination of evaporation through the Penman–Monteith equation. A full review of this method is available at http://www.fao.org/docrep/X0490E/x0490e00 – hence the equations are not reproduced here. There are two versions for computing actual evaporation (AE), a one-step model and a two-step model (note that we are avoiding the term evapotranspiration hence we will not use AEt or PEt). In the one-step model AE is calculated directly from standard meteorological variables: i.e. temperature, wind speed, relative humidity and net radiation. Note this approach is only suitable for the determination of AE for periods of one day or less if the full energy balance is computed. Sources of error may arise if observations from a climatological station are not collected over an appropriate surface (this should be short wet grass, not gravel or asphalt). An additional problem is the need for GCMs to provide daily values of wind speed and humidity. Perhaps most difficult for the one-step approach is the assessment of surface (canopy) and aerodynamic resistances. Providing vegetation height and wind speed are known, it is fairly simple to compute aerodynamic resistances but surface (canopy) resistances (as shown in the following figure) are more difficult to obtain and require some understanding of the relationship between resistance and available water capacity. Where such data are not available the two-step approach may be used and AE is inferred from potential evaporation (PE).

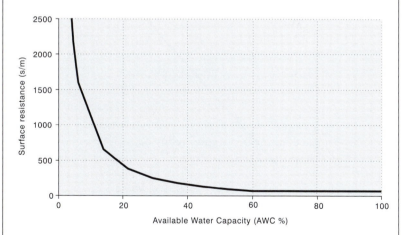

Changes in Penman–Monteith surface resistance (rs) for grass with an initial value of 70 sm⁻¹ as soil water available capacity (AWC) decreases beyond a readily available value of 60% AWC

Source: based on MORECS; see Gavin and Agnew, 2003 and Hough and Jones, 1997

For the two-step Penman–Monteith Evaporation Model, step 1 is to calculate the potential evaporation for the surface using a crop coefficient (KC) which predicts vegetation leaf area cover through a plant's phonological stages.

$$PE_{surf} = KC * PE_{ref} \tag{i}$$

Where,

PE_{surf} is the loss through evaporation from that surface, which may be a crop type, when water availability is not limiting.

PE_{ref} uses the notion of evaporation of water from an extended surface of a short green crop, actively growing, shading the ground, of uniform height, and not short of water, later revised to an extensive surface of 12 cm high green grass with a surface albedo of 0.23 and a canopy resistance of 70 s/m (Smith, 2008).

Step 2 is to calculate actual evaporation,

$$AE = R * PE_{surf} \tag{ii}$$

R   is the ratio of actual to potential evaporation. It is determined by the extent to which soil moisture may limit the availability of water for evaporation. The following figure indicates a typical relationship whereby AE/PE = 1 until 50 per cent of available water is consumed and then reduces to zero when wilting point is reached, i.e. available water capacity (AWC) is zero.

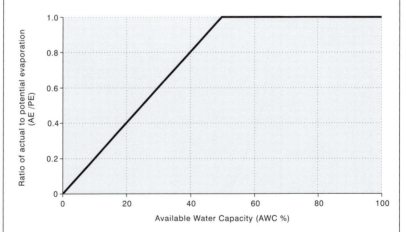

Relationship between available water capacity (AWC) and ratio of actual to potential evaporation based upon 50% of AWC being freely available.

The amount of soil water available to plants for evaporation (AWC) is normally taken to lie between field capacity (FC the upper limit of soil-water storage when free drainage has ceased) and crop wilting point (WP is generally around a soil water potential of −1.5 MPa), i.e.

AWC = FC − WP (mm of water)　　　　　　　　　　　　　(iii)

Soil moisture deficits (SMD) occur when soil moisture content (SMC) falls below field capacity, with the maximum SMD being AWC. Thus,

SMD = FC − SMC (mm of water)　　　　　　　　　　　　　(iv)

Soil moisture deficits can be used to measure drought conditions, often more effectively than simply computing the differences between precipitation and PE_{ref} because SMDs can take into account soil water storage and plant rooting depths (Mpelasoka et al., 2007).There have been criticisms of the highly empirical nature of the two-step model while Morton (1983) raises a much more fundamental objection about the ability of these point-based models to estimate areal evaporation as illustrated in the following figure. Here, PE is calculated as explained above based on point meteorological observations whereas AE is an areal or catchment assessment of the total volume of water lost through evaporation. It can be seen that PE is at a maximum while AE is at a minimum when water is limiting hence using a ratio of AE/PE can be erroneous as they appear to be complementary. Morton's CRAE model is used for extensive areal averages and does not take account of local variations in slope, vegetation and moisture. Xu and Singh (2005) reviewed three evaporation models based upon this complementary relationship for a range of climate regions. Compared to observed monthly water balances, reasonable results were found for the wet conditions experienced in Sweden, but the results become worse in China and Cyprus as the climate became more semi-arid and soil moisture influence was greater. The Penman − Monteith approach remains then the recommended model for determining evaporation.

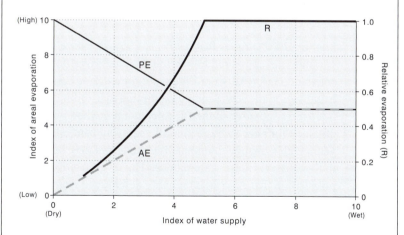

Morton's Complementary Regional Areal Evaporation (CRAE) model predictions based upon actual evaporation (AE), potential evaporation (PE) and ratio (R) of AE to PE.

UK Climate Impacts Programme (UKCIP02) which provides sufficient data for daily PE to be computed for three thirty-year scenarios during the twenty-first century but this still leaves a question about actual evaporation (AE). Arnell (2003) computes that for UK PE rates there will be a reduction by 2050 up to 50 per cent in winter months but as the absolute values are so small this is not significant. For the rest of the year in the UK PE increases especially in summer and early autumn. In northern England this increase is around 10 per cent by 2050, but in the south increases above 40 per cent are predicted which could have significant impacts on soil moisture deficits (see Box 3.2), and hence groundwater recharge and river flows.

A more recent analysis by Ekstrom et al. (2007) examined PE predictions under climate-change scenarios for Europe (i.e. 1960–90 compared to 2070–2100). They found increases of around 1 mm/day in winter (contradicting Arnell), and very large increases of 4 to 8 mm/day in summer for Europe, with the exception of the UK and the most northerly areas. Validation of the calculations against past meteorological and hydrological records raised important questions, suggesting under-estimation of UK values compared to historic observations. This led to a revised approach, whereby PE was calculated using a Blaney–Criddle-derived equation using only monthly temperatures which is similar to the strategy used by Burt and Shahgedanova (1998). A parallel validation based on the Rhine basin found better agreement between observations and predictions but led Ekstrom et al. to conclude that predictions of PE values were problematic and should not be used for hydrological simulations in these regions.

Fowler (2002) argues that, as PE is a 'conservative' hydrological variable (i.e. compared to precipitation and river flow), it is not unreasonable to infer AE from PE except for extreme conditions. An examination of New Zealand confirmed that soil moisture balances were not overly sensitive to the PE calculation method with best results obtained for near average rainfall years, but cautioned that in drought years with a change in climate regime errors may be introduced, though in this analysis these were modest. However, we disagree with the proposition that AE can be reliably determined from PE without additional consideration of hydrological conditions. For example, the largest water loss from many wetlands is not run-off but evaporation, and under conditions of high transpiration the water losses may greatly exceed that of open water evaporation – hardly a 'conservative' variable. However, although it is widely understood that predicted climate changes will lead to higher temperatures and resultant lower relative humidity, it cannot be conclusively stated that evaporation rates will be higher. Variations in cloud cover and sometimes higher precipitation may mitigate some warming effects, although it is likely that stronger wind speeds and other changes will enhance evaporative demand and lead to higher potential evaporation rates. For these elevated rates to lead to higher actual evaporation will require higher precipitation and for large parts of the world this is not predicted, hence actual evaporation will not increase as much as predicted unless irrigation is provided. In summary, therefore, while temperature is reasonably well predicted under different climate-change scenarios, there is greater uncertainty – particularly at the local level – for key hydrological variables such as precipitation, evaporation and soil moisture.

Climate-change uncertainties

The predicted changes of the three key climate variables discussed above are subject to different levels of uncertainty. This has led sceptics to seek to undermine GCM predictions through listing the shortcomings of the models and inconsistencies in their predictions. While this was more convincing when the GCMs were still rather crude and were unable to accurately reproduce historic observations, by the twenty-first century a scientific consensus supports climate change – but only within a range of predictions. There are three main areas of concern over climate-change predictions for water resources:

a) The chain of uncertainty
b) Unforeseen impacts of other climate-forcing mechanisms.
c) Model and data inadequacies including poor spatial resolution.

We will consider each of these in turn.

The chain of uncertainty

We can depict the origins of uncertainty as flowing from sets of assumptions about future human activity (see Figure 3.3). Forecasts estimate the rates of gas emissions, their atmospheric residence, computed climate change and finally regional and local impacts. If the forecasts of energy production and greenhouse gas emission are incorrect then predictions of impacts will be unreliable. Harvey (2000) reports the results of computer simulations that suggest atmospheric levels of CO_2 in the past may have

Changes in human activities and controls (mitigation)
Future emissions of greenhouse gases
Residence and concentration in the atmosphere
Global climate change
Regional and local changes in climate
Impacts upon water resources
Water resource management

Figure 3.3 Climate-change impacts chain of uncertainty

Source: Agnew and Fennessey, 2001

been 4 to 16 times higher than pre-industrial concentrations of 280 ppmv, but that over the last 160,000 years current levels (377 ppmv) are unprecedented. Early climate models were based on the assumption that there would a doubling of pre-industrial CO_2 concentrations (the equilibrium models), but found that that atmospheric CO_2 concentrations were below expected values. This was explained by greater absorption of CO_2 by 'carbon sinks', in particular the world's oceans, than had been expected by the IPCC at that time (1991). The degree of 'atmospheric leaching' of CO_2 is still an area of discussion today as it depends not only on rainfall but also on the temperature of the oceans and the functioning of other natural carbon sinks such as forests, peat uplands and other wetlands (Evans and Warburton, 2007; Worrall et al., 2007; Laffoley and Grimsditch, 2009).

There is also uncertainty over the future level of emissions from combustion of fossil fuels, which is related to economic activity. In 1990, the IPCC's 'business as usual' scenario (SA90), assuming that no steps were taken to curb greenhouse-gas emissions, predicted carbon releases would rise from 6 Gty^{-1} to over 20 Gty^{-1} by 2100. Under this scenario global temperatures were predicted to rise by 0.3°C per decade, or 3.0°C by the year 2100 (a rise of 4°C from pre-industrial times). After 1992, six alternative scenarios were used, ranging from the highest greenhouse gas forecast IS92e, based on the phase-out of nuclear energy, to the lowest scenario (IS92c), assuming fossil fuel shortages. In the latest forecasts IPCC (2007) uses the following scenarios:

A1   Rapid economic growth with population peaking around 2050
A1F1 High GHG emissions for a scenario of fossil energy intensive
     consumption
A1T Non-fossil intensive scenario (second lowest GHG emissions)
A1B Balance across A1 with mid range predictions
A2   Heterogeneous patterns with slow technological change and high
     GHG emissions
B1   The lowest GHG emissions with a service and information
     economy

Scenarios A1F1 and A2 produce GHG emissions of around 120 Gt CO_2 eq.y^{-1} by 2100, while B1 and A1T predict emissions that would be below the current level of 40 Gt CO_2 eq.y^{-1}. These extremes are associated with temperature increases of 4.0°C (range 2.4 to 6.4) and 1.8 (range 1.1 to 2.9)°C, respectively. It may be that actual conditions lie outside of even these six scenarios, and predictions diverge at timescales beyond the next 20 years. However, the notion of a 'tipping point' has gained momentum. In the past, doubling of CO_2 concentrations was used as a threshold, or reference point, for climate modellers, although this had little policy impact, probably because the point at which this would be reached could not be agreed and was too far off in the future. More recently, the notion of a 2°C temperature rise has become accepted as a 'tipping' point for climate change. It is argued that a rise of 2°C will have significant impacts on human activities and the environment. Moreover, it is highly likely to occur during the twenty-first century. This threshold has therefore been widely adopted by policy makers, even though its significance in terms of impact is still debated. As we noted above, Anderson and Bows (2008) argue forcibly that such emission reduction scenarios, which were the basis of the highly influential Stern (2006) report are far too optimistic. Their paper presents the following scenarios (p. 3879):

- Assuming emissions peak 2015, 450 ppmv CO_2eq. (equivalent) stabilisation requires subsequent 4 per cent annual reductions of CO_2eq. and 6.5 per cent reductions in energy and process emissions
- Assuming emissions peak 2010, 550 ppmv CO_2eq. stabilisation requires subsequent 6 per cent annual reductions of CO_2eq. and 9 per cent reductions in energy and process emissions
- Assuming emissions peak 2010, 650 ppmv CO_2eq. stabilisation requires subsequent 3 per cent annual reductions of CO_2eq. and 3.5 per cent reductions in energy and process emissions.

They conclude that it is unlikely that stabilisation will be possible below 650 ppmv CO_2eq., stating: 'In the absence of an almost immediate step change in mitigation (away from the current trend of 3 per cent annual emission growth), adaptation would be much better guided by stabilization at around 650 ppmv CO_2eq.' They note Meinshausen's (2006) calculation that this is likely to result in a 4°C rise in global temperatures.

Other climate-forcing mechanisms

It has long been reported that changes in rainfall are coupled to sea-surface temperature anomalies (Robertson and Frankignoul, 1990;

Diaz and Markgraf 2000). Glantz (1987) suggested a link between droughts in Africa and Atlantic temperatures, and these were successfully modelled (Folland, 1987) some time ago. Mason (1995) identified sea-surface temperature changes in the South Atlantic and South-West Indian Oceans that might be significant for rainfall in southern Africa. The most widely discussed coupling is that of the El Niño Southern Oscillation (ENSO) with monsoonal rainfalls (El-Niño is the oceanic component and Southern Oscillation is the atmospheric) (Clarke, 2008; Fedorov and Philander, 2000; Glantz et al., 1991; Marshall and Plumb, 2007). The El Niño, or cessation of cold water upwelling in the eastern Pacific, is associated with large-scale weather oscillations. Prior to an El Niño there are powerful south-easterly winds, which produce a build-up of water in the western Pacific. This is later advected eastwards and interferes with the normal upwelling of cold water off the South American coast (Caviedes, 1988). The normal or 'Walker' circulation is reversed and there is a surface westerly wind and persistent high-pressure subsidence in the south-western Pacific. This in turn promotes changes in the monsoonal circulation leading to extreme weather conditions in India and drought in Australia (Mpelasoka et al., 2007). The impact is wider than Asian rainfalls, and recently (2007/8) the Northern Hemisphere experienced a notable period of cooling coincident with opposite circulation to El Niño, that of La Niña.

The ENSO has also been shown to influence flows in major river basins in tropical regions such as the Nile and the Congo. Schongart and Junk (2007) demonstrate that flood levels in the Amazon are lower during El Niño but there are much higher flood flows and inundation with La Niña. A positive correlation between ENSO and higher summer monsoonal precipitation across India has received much attention. Revadekar and Kulkarni (2008) found that large daily rainfalls could be predicted by an ENSO index but also noted that less intense periods of either wet or dry spells were not clearly associated with ENSO. In the same region, Chandimala and Zubair (2007) link flows of the Kelani (Sri Lanka) with ENSO variability, although Zubair et al. (2008) found only 'modest' reliability in predictions of Sri Lanka precipitation and needed to take account of season and geographical location to improve prediction. Taking a longer time perspective based upon a proxy (tree-ring record) of past climate from Java (Indonesia), and computing the Palmer Drought Severity Index (see Box 1.2), D'Arrigo et al. (2008) found close associations of this to Indian Ocean sea surface temperatures and, finally, to ENSO events from the 1950s.

There is, then, overwhelming evidence linking sea-surface temperature changes and weather systems in the tropics in ways that may enhance, or even oppose, changes due to global warming. It has also been possible to establish patterns between ENSO and precipitation in Europe (Zanchettin et al., 2008) and North America (Sobolowski and Frei, 2007) but both studies also found influences from the North Atlantic Oscillation (NAO) and the Pacific Decadal Oscillation (PDO), indicating that such global mechanisms are complex. Zhao and Moore (2008), however, sound a cautionary note that the ENSO–monsoon relationship has weakened in recent decades and prediction may be becoming more difficult. They note that while Amazonian rainfall has been increasing over the past four decades, no trend is evident in Indian summer monsoon rainfall. Vecchi and Soden (2007) suggest that under global warming there is a weakening of the Walker circulation leading to an overall weakening of tropical circulation thus creating greater uncertainty for the prediction of climate changes. Under global warming, therefore, there may be alteration of important climate-forming processes. Henson (2005), for example, reported that by 2005 temperatures had increased to rival the previous hottest year on record, that of 1998, but, whereas 1998 coincided with an El Niño event, the 2005 temperatures had increased without being augmented by El Niño.

Another natural forcing mechanism is a change in solar emission and, possibly, cosmic rays although the Meteorological Office (MO, 2010) strongly refute that warming over the last 50 years can be explained by anything other than greenhouse gas concentrations. It is known that changes in atmospheric turbidity due to volcanic eruptions or air pollution – especially sulphate aerosols – can offset warming through raising the planetary albedo and reflecting more solar radiation (Gribben and Gribben, 1996; Andronova and Schlesinger, 2004; Kim et al., 2007). This may be enhanced by a possible secondary effect of inducing more cloud growth through higher concentrations of hygroscopic nuclei. Global temperatures were expected by the IPCC to increase by 1.4°C by the year 2030 (1991) within a range of 0.7–2.0°C. During the 1990s these were modified in light of the effect of sulphate aerosols but by the end of the twentieth century it was estimated (Stott et al., 2000) that natural forcings had been negligible since the 1960s. Taking into account the possible impacts of atmospheric sulphates and likely volcanic activity they concluded that, in the second half of the twenty-first century, perhaps one third of warming may be offset by these atmospheric changes.

Model and data inadequacies

Predictions of global climate change have become more consistent over the past three decades. They were subject to considerable variability during the development of coupled ocean-atmosphere models in the 1980s and 1990s. Earlier models gave a warming trend roughly double that actually observed over the last 100 years and it proved necessary to take into account energy exchanges with, and circulations of, the world's oceans, as well as sunspot activity and atmospheric turbidity. Developments over the last two decades have addressed a number of these concerns which has promoted greater confidence in the predictions of temperature changes. An account of the development of climate models can be found in McGuffie and Henderson-Sellers (1997) where it is explained that simpler energy balance models were superseded by General Circulation Models (GCMs). These were initially based upon assumptions of doubling of CO_2 concentrations but were developed into models that could take account of gradual changes in atmospheric composition. It is now familiar to see predictions based upon a range of future energy scenarios, showing quite high consistency in their identification of conditions for a temperature rise of 2°C (Kerr, 2007b). The exact timing is more difficult to predict, especially as after 2025 model predictions tend to diverge. Kerr (2007a) notes that ice sheets take centuries to melt, while atmospheric response lags by decades to changes in green house gas emissions.

The most important areas of uncertainty within the current climate change consensus are potentially important feedbacks and variations that could influence when changes will be detectable. Oppenheimer et al. (2007) list these as:

- Increased contribution from melting glaciers to raise sea level
- An *amplifying* effect between climate and carbon-cycle feedbacks

  Permafrost stores
  Marine ecosystems structure
  Land use change
  Air pollution

- Processes coupling SSTs and tropical circulation
- Socio-economic changes

  Costs of mitigation
  Confined emission scenarios

Meier et al. (2007) believe that more rapid than expected melting of glaciers will generate a higher than expected rise in sea levels, while

Heimann and Reichstien (2008) argue that terrestrial ecosystems interact in a more complex manner than expected. As we noted above, ENSO and atmospheric CO_2 are linked due to the impact of drought and heat upon plant growth and respiration in the Amazon and S.E. Asia. A negative feedback mechanism (reducing warming) is provided by photosynthesis that, in a warming world, will increase and thus absorb CO_2 at faster rates – at least until higher temperatures start to suppress plant respiration. In fact, the processes are more complicated, depending in part upon above-ground and subsurface respiration rates, such that experiments produce ambiguous results and large uncertainties remain. Heimann and Reichstien (2008) conclude that at a global scale terrestrial ecosystems will provide a greater than expected *positive* feedback (i.e. accentuating CO_2 emission) as temperatures rise.

Models are verified both through their underlying physics and through comparison with the historical observations. Climate data networks are not consistent over time, while natural fluctuations in data mean that identifying the anthropogenic impact needed for GCMs is often a challenge. Much attention has been paid to temporal change: i.e. the date at which CO_2 will double; the comparison of observed and predicted annual global temperatures since 1850; and likely changes in emissions of greenhouse gases over time. By contrast, although spatial variability has always been understood to be important, it has required the development of geostatistical tools (Oliver, 1987; Chappell and Agnew, 2001) for this element to be more fully investigated. The climate record can be criticised for being spatially biased. There are more land-based than ocean-based stations and more in the northern than in the southern Hemisphere. Globally, the distribution of climate stations is uneven and in parts of Africa the database has deteriorated over the last few decades, (Chappell and Agnew, 2004). Thus predictions of change, especially over the oceans and developing countries, are super-imposed upon uncertain data for local conditions.

A further key element of uncertainty surrounds the spatial scale at which predictions are made. GCMs have typically modelled the earth with a resolution in excess of 100 km and around 300 km, i.e. 3° latitude and 4° longitude. The resolution of GCMs has been greatly improved. For example, the replacement of HadCM3 by HadGEM1 led to a doubling of vertical atmospheric (30 km deep) and oceanic (5km deep) levels in the model, and a 1.25° lat and 1.875° long spatial resolution for terrain (1° ocean). Nevertheless, there remains a huge challenge in converting global temperatures to hydrological changes, especially at regional and local scales where regional climate models need to operate at scales of 25 to 50 km resolution.

We noted above that downscaling GCM has become an important goal in climate change prediction, but it is dangerous to assume that hydrological processes operate in a similar fashion across all scales. It should also be understood that scale is not only represented as a Euclidian distance, but may be a social construct (for a discussion of scale issues and social science see Sheppard and McMaster, 1994). Bloschl and Sivaplan (1995) have reviewed and summarised the major scale issues. Of particular interest here is heterogeneity and variability in catchments as issues when linking processes across scales – or 'downscaling' – as discussed further by Sivaplan et al. (2003). They identify three types of variability, in order of decreasing predictability: discontinuity, periodicity and randomness. Alongside discussions of downscaling GCM predictions is research on upscaling from point observations and surface processes. Ahrens and Beck (2008) examine problems of upscaling rain-gauge data, Crosbie et al. (2008) consider transpiration rates from individual trees, and Intsiful and Kunstmann (2008) present work on energy fluxes and report some successes in dealing with heterogeneity. Wood (2009: 735) reviewed the role of scaling laws in upscaling and states that 'Hydrological models are *quasi-scalable* … such models are subject to substantial uncertainty'.

Thus, we cannot assume a continuum of hydrological processes operate across scales. This presents problems for assessing the impacts of climate changes upon water resources. Woods et al. (1995) were similarly concerned that variability changed with catchment size such that (p. 310) 'variance between catchments decreases as catchment area increases'. They introduced the idea of the Representative Elementary Area (REA) where spatial averages were meaningful and caution against extrapolating from areas of less than approximately 1 km^2.

A complication for climate analyses is that the actual networks of stations have changed during the observation period, creating the possibility of an added artefact in the climate signal. Agnew and Chappell (2000) demonstrated local variability in West African rainfall trends, e.g. marked desiccation is evident in Burkina Faso but not in neighbouring Niger. Changes in the rain gauge network have been reported several times (Chappell and Agnew, 2004). For example, in West Africa the number of stations has decreased in the wetter coastal area, and increased in the drier continental area, while the total number of stations with complete records decreased dramatically after the late 1970s. The same observation was recently made for Africa as a whole, and Kenya in particular, at the 2nd Africa Water Week in Johannesburg: 'You can only get reliable rainfall data in Kenya up to 1980 as the existing infrastructure, like rain gauges and stream gauges, are no longer working,' (SDN, 2009). Dai et al. (2004) challenged concerns raised over changing networks, arguing that the conventional approach of area-weighted normalised/standardised anomalies would deal with this issue. In a more recent review of network changes in West Africa, Chappell and Agnew (2008) argue that there is an artifact in the climate signal that needs to be taken into account when assessing climate changes. It should not be assumed, however, that spatial heterogeneity is evident for all locations. Croxton et al. (2006) examined the Central England Temperature and the England and Wales Precipitation series to see if they were consistent with observed major changes across the UK climate. Significant correlation was a strong feature of the temperature record suggesting that regional GCM predictions for the UK can generally be assumed to hold true, and while the precipitation record was not quite as strong, it was still significant.

Climate-change impacts

The IPCC has stated in both the Third (2001) and Fourth Assessment Report (2007) that water resources are being affected by climate change. By 2050, river run-off is expected to increase 10–40 per cent in higher latitudes and wet tropics, and 10–30 per cent in dry tropics fuelled by intensity changes. More extreme precipitation patterns will lead to more droughts *and* more floods; while water resources stored in glaciers will have declined. Regionally, the greatest increases in water stress are likely to occur in Africa, exacerbating impacts upon agriculture due to reductions in the length of growing season, and possibly reducing rain-fed crop yields by 50 per cent by 2020. Australia is also likely to experience water security problems through reduced precipitation. In Asia, crop yields may increase in south east Asia, but any gains will possibly be offset by decreases in central and south Asia due to decreased glacial meltwater flows into the region. IPCC working group II concludes that hundreds of millions will be exposed to increased water stress, but the actual impact will depend upon human resilience and adaptation. A summary of regional impacts for the twenty-first century from IPCC (2007: 50) is presented in Figure 3.4.

Fresh water resources are threatened by sea-level rises, both by surface flooding in coastal areas and saline intrusion into groundwater.

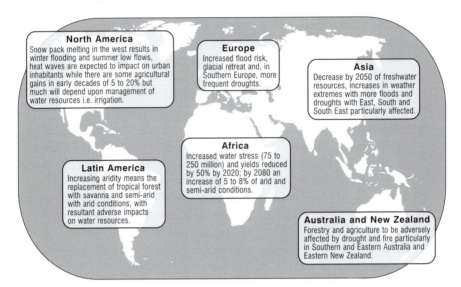

North America
Snow pack melting in the west results in winter flooding and summer low flows, heat waves are expected to impact on urban inhabitants while there are some agricultural gains in early decades of 5 to 20% but much will depend upon management of water resources i.e. irrigation.

Europe
Increased flood risk, glacial retreat and, in Southern Europe, more frequent droughts.

Asia
Decrease by 2050 of freshwater resources, increases in weather extremes with more floods and droughts with East, South and South East particularly affected.

Africa
Increased water stress (75 to 250 million) and yields reduced by 50% by 2020; by 2080 an increase of 5 to 8% of arid and semi-arid conditions.

Latin America
Increasing aridity means the replacement of tropical forest with savanna and semi-arid with arid conditions, with resultant adverse impacts on water resources.

Australia and New Zealand
Forestry and agriculture to be adversely affected by drought and fire particularly in Southern and Eastern Australia and Eastern New Zealand.

Figure 3.4 World map of predicted climate change impacts
Source: based upon IPCC, 2007

Concerns over sea-level rise due to global warming were expressed two decades ago, with the expectation of a 20 cm rise by today and a 50 cm by the year 2045, mainly through thermal expansion but augmented by ice melt (Wigley and Raper, 1987; Warrick et al., 1993). More recently Gregory and Lowe (2000) forecast a rise of around 40 cm from 1990 to 2100 with 15 cm already observed during the twentieth century (Dessler and Parson, 2006), although Cabanes et al. (2001) note that uneven geographical distribution of tide gauges may mean this is an overestimate. More controversially, Meir et al. (2007) suggest that previously unaccounted contributions through glacial melting will add a further 10 to 25 cm sea level rise by 2100, or 50 per cent more than the IPCC (2007) overall projections of 20 to 50 cm (Oppenheimer et al., 2007; Solomon et al., 2008). As 100 million people live less than 1m above sea level, and 40 per cent of the world's population live within 60 miles of the coast this is a serious threat, especially in areas of south east Asia (Dow and Downing, 2006). As set out in the first chapter, our focus in this book is on fresh-water resources, Therefore, while not wishing to diminish the significance of this threat – both of flooding from coastal storms and salt water intrusion – to coastal communities, the remainder of this chapter will not discuss this further, but will focus on fresh-water impacts of climate change: surface flows, plant–soil–water interactions and drought. These three are the most fundamental hydrological impacts, as they determine secondary changes to both surface (lakes and reservoirs) and sub-surface (groundwater) water storage.

River flows and flooding

We noted above that global warming is likely to cause more weather extremes: more drought, more storms. Both can lead to increased surface run-off thus increasing flood risk. Conventional understanding of surface run-off based upon temperate areas (Ward and Robinson, 1999) is that higher rainfall intensity can lead to Hortonian overland flow in the rare occasions that surface infiltration rate is exceeded. Saturation overland flow may occur where surface soil is bare and/or shallow – as is common in tropical and semi-arid climates – or where surface soil quickly becomes saturated (e.g. moist areas close to rivers). In humid areas, subsurface throughflow is more common (Bevan, 2003). Although one might expect that drier conditions will lead to less run-off, desiccation of the soil surface can produce hydrophobic conditions where water is repelled, and surface crusts can also reduce infiltration rates (Evans et al., 1999; Patrick, 2000). Thus, even where drought increases, higher storm flows can be expected to result from global warming, as summarised by Arnell et al. (2001) for the IPCC. It should be noted that in urban areas intense downpours can cause flooding as impermeable surfaces enhance rapid runoff and urban drainage infrastructure may not be able to cope (Marsalek et al., 2006). These are commonly referred to as pluvial floods, as distinct to fluvial flooding where inundation arises because the river channel capacity is exceeded. The rest of this section will focus on the latter as much of the past analysis has had this same focus, but recent events in the UK have shifted attention to flooding that also occurs away from the river channel, while sea level has already been addressed above.

Methods for the prediction of river-flow range from simple models based upon annual areal catchment characteristics, requiring few data, through to more detailed storm-based distributed models using topography and surface wetness, where data demands are often excessive. Guo et al. (2002) used less demanding monthly data to investigate impacts of possible climate changes to precipitation and temperature upon river flows in China, but found that outcomes were dependent upon the model used. Bevan (2003) confirms that flood prediction can take place using monthly or even annual data. However, detailed forecasting of changes to flood magnitude and frequency requires hourly, or at least daily, rainfalls. This places great data demands upon the outputs of GCMs, as shown by Prudhomme et al. (2002) who obtained daily rainfalls through linking regional climate models (RCM) to GCMs. Downscaling can provide daily rainfalls for resolutions of 50 km or less, but a key assumption is that the spatial variability and intensity underlying the extreme hydrological events does not change. Although the most common downscaling methods are statistical and dynamic, Bell et al. (2007) suggest a perturbation approach has the advantage of using observed baseline data rather than a modelled rainfall series, but this approach cannot be used for future simulation without making assumptions about future variability. Arnell et al. (2001) also noted the need to move beyond the consideration of mean climate changes and take account of extreme variability. This is reflected in recent attention upon predicted increases in daily rainfall totals or the incidence of high intensity rainfalls leading to increased run-off.

The UK can be used as a useful exemplar of the types of changes that can be expected as Fowler et al. (2005) note that forecast increases in rainfalls generally match those expected elsewhere. Osborn and Hulme (2002) report that observed changes for the period 1961–95 reveal winter precipitation has generally increased, in some places by as much as 50 per cent, while summer precipitation has typically declined by 10 to 40 per cent. However, it is the change in extreme events that is of paramount interest, especially over the risk for increased winter flooding (Marsh, 2001). It was noted above, following the paper by Osborn et al. (2000) that rainfall intensities are expected to increase in the UK, possibly through the influence of the NAO. This could be expected to generate more surface flows leading to an increase in flood events. Indeed Prudhomme et al.'s (2002) analysis of flows in the River Severn found, irrespective of the downscaling approach, that there was a significant increase in flood frequency, such that a 50-year return period event may in future occur every five years. They caution, however, that impacts on flood frequency can be different to those on flood magnitude. For example, a milder winter with more days of rain can be expected to increase the magnitude of floods by 6 per cent, whereas warmer conditions and more convective storms (more intensive rainfall) may increase magnitude of the 50 year flood by 30 per cent. Bell et al. (2007) used 25 km resolution inputs of hourly precipitation and daily PE to predict 1km resolution run-off in 15-minute time steps for 1961–90 and 2071–2100. They found that it was possible to have great confidence for predictions in upland catchments, where topography was an important control and there was a general increase in flood magnitude and frequency. It was more difficult to predict flows for low lying rivers, and earlier work by Arnell (1992) also pointed to upland catchments being most sensitive to climate changes. Bell et al. (2007) caution that the variability observed in the results, and the dependence of predicted impacts upon few but extreme rainfall events, requires several climate change scenarios to be modelled before future flood risk can be fully assessed. Wilby et al. (2007) queried whether there were clear trends in UK river flows, noting that the 1990s was an abnormal period for the NAO and concluded that there was need for greater understanding of anthropogenic and catchment change effects. Furthermore, Arnell (2003) notes that previous studies were too reliant upon mean climate forecasts and identified the need to superimpose predictions upon the natural multi-decadal variability of rainfall and run-off. They concluded that climate-change-induced run-off values were outside the range of natural variability, but that, when these are combined, the range increases substantially so there is even greater uncertainty than that implied by climate change alone.

A further challenge for those wishing to link climate forecasts to river flow predictions is that the impacts of climate change suggested above take place in a world that is increasingly under human influence as noted in Wilby et al. in their concluding comments (2007).

Any assessment of impact through climate change needs to take human modification into account. 'Mitigation and Adaptation' has become a topical phrase indicating the need to both curb greenhouse-gas emissions and to plan for inevitable change, but confronts the difficulty that predictions of climate change are superimposed upon an environment that, as we saw in Chapter 2, has undergone significant alterations during the twentieth century. The creation of impermeable paved areas has had profound impacts on hydrology (Marsalek et al., 2006), and much was previously written on the urban heat island effect, recently reviewed by Roth (2007). There are noticeable differences between urban and rural climates with respect to higher incidences of storms, and higher nocturnal temperatures but these differences are not consistent. Rural land-use change for example can also present flashy run-off responses similar to urban areas because of the need for enhanced drainage in both systems. In West Africa the River Niger has been modified by the Office du Mali irrigation system upstream of Niamey, leading to changes in the flood regime, and barrages on the River Senegal have similarly modified flows, while the Grand Anatolian Programme in Turkey with 22 dams on the headwaters of the Tigris and Euphrates has modified the annual flood wave (Agnew and Anderson, 1992). It is then problematic to use historic records of flow as examples of 'natural' conditions without taking account of catchment anthropogenic changes, while superimposing the effects of global warming-induced run-off adds even further challenges for flood prediction.

Drought and agro-climatological impacts

Box 1.4 identified differences between drought and desiccation, the former being a short-lived and exceptional phenomenon, whereas the latter is a long-term change in climate towards greater aridity. Drought may be further differentiated between 'meteorological' drought, defined in terms of variables such as rainfall, and 'agricultural' drought, defined in terms of impacts upon plant growth and crop yields. It is the latter with which we are concerned when discussing climate-change impacts, because agriculture, through irrigation, consumes globally more water than industry and domestic demand combined. Climate change impacts upon plant growth and food production may therefore be as significant as the more obvious increases in flood events. The science of agro-climatology is complex due to the need to take account of climate–plant–soil water interactions over a wide range of environments (see Box 3.3 and 3.4 for details). The key conclusions are that 'C3' crops such as wheat, soyabean and rice, and trees are relatively well placed to respond positively to a world that is warmer and richer in CO_2, if there is suffi-cient water to offset the impacts of higher temperatures. This may be valid only within certain limits, while temperate crops may be more productive within the expected warming, even higher temperature increases could induce stress. Predictions for the tropics are unambigu-ously negative, and by 2030, southern Africa could lose almost a third of maize production while losses of rice and millet could exceed 10 per cent in south Asia. The key factor here is an increase in drought impacts, as the opportunities for enhanced plant growth are not expected to be matched by water supply.

There are various meteorological (rainfall) indexes for drought determi-nation such as the SPI (see Box 1.4) but agricultural drought cannot be understood merely in terms of a reduction in rainfall. The complex interaction of meteorological and other factors in determining drought is exemplified in Africa, and particularly the Sahel, where a history of reported natural disasters arising from droughts in the 1910s, 1940s and 1970s was followed in the 1990s by predictions of long-term desiccation (D'Amato and Lebel, 1998; Hulme, 2001), and increased vulnerability to water shortages due to climate change (Thornton et al., 2006, Washington et al., 2006). Some have questioned an interpretation of widespread and persistent drought in the sub-Saharan region based upon

Box 3.3

Predicting crop yield under changing water supply

In order to assess crop yield and thus agricultural production the first step is to establish the relationship between economic yield, vegetative growth and water supply. The effect of climate change upon evaporation and hence crop yield is often captured by the following relationship

$$[1 - (Ya/Ym) = Ky\,[1 - (AE/PE)] \tag{i}$$

where,

Ya	= Actual yield
Ym	= Maximum yield
Ky	= Yield response factor
AE	= Actual evaporation
PE	= Maximum (potential) evaporation

Box 3.2 demonstrates that evaporation is related to AWC either through surface resistance in the one step Penman–Monteith model, or via an Actual:Potential evaporation ratio.

As an alternative, Water Use Efficiency (WUE) is widely used to quantify the relationship between crop production and water consumption in an empirical form,

$$WUE = Yield\ per\ unit\ area/Water\ used\ to\ produce\ yield \tag{ii}$$

or

$$WUE = Y/T$$

where,

Y	= Yield
T	= Crop transpiration (inferred from AE estimates)

The following table shows examples of WUE for some tropical crops (adapted from Jones, 1992).

	$(Y/T) * 1000$
C3 plants	
Wheat	1.93–2.20
Rice	1.47
Alfalfa	1.09–1.60
Pulses	1.33–1.76
C3 Range	0.88–2.65
C4 plants	
Millet	2.72–3.88
Sorghum	2.63–3.65
Maize	2.67–3.34
C4 Range	2.41–3.88

This table shows the greater WUE of C4 plants, but determination of transpiration is difficult for operational estimates.

Box 3.4

Agroclimatological processes and plant growth

Photosynthesis is vital for plant growth whereby hydrogen derived from water is combined with carbon dioxide derived from the atmosphere to produce carbohydrates using energy derived from the sun. Three mechanisms can be identified by which plants assimilate CO_2 during photosynthesis (after Agnew, 1998):

- C3 crops such as wheat, soyabean and rice and trees (CO_2 fixed by a 3 carbon compound phosphoglyceric acid)
- C4 crops such as maize, sugarcane, millet and the majority of savanna grasses (CO_2 fixed by a 4 carbon compound oxaloacetic acid)
- CAM plants such as cacti and other succulents (CO_2 fixed by crassulacean acid)

C4 plants found in tropical climates where prevailing conditions are high temperatures and high levels of sunlight have developed much higher rates of photosynthesis which in turn results in higher water-use efficiencies (WUE, see Box 3.3) as stomata are not open for as long (Ehlers and Goss, 2003). Where atmospheric CO_2 concentrations rise, there is a potential for increases in growth rate, biomass accumulation and yield, providing water stress does not occur. Bell and Treshaw (2002) note, however, that the existing high rates of C4 plant photosynthesis mean that gains under higher CO_2 are likely to be greater for C3 plants. Thus, higher temperatures and higher concentrations of CO_2 will favour the C3 temperate plants, as long as water can be provided in response to higher growth rates and water demand, and temperatures do not rise above optimum levels.

As temperature increases so does the rate of chemical reactions in plants although growth by cell elongation is most rapid at temperatures around 20°C, while above 40°C enzyme reactions are reduced. Higher temperatures produce increased rates of both photosynthesis and respiration with the exception of the extremely hot conditions found in tropical deserts. The negative effect of the highest temperatures on assimilation rates causing an increase in respiration can be offset by increases in sunlight absorption. The reduction in net assimilation rates due to high temperatures (below 40°C) is, however, seldom significant except in dry conditions when water supply inhibits photosynthesis. In such arid conditions leaf temperature cannot be maintained at optimum temperatures through evaporative cooling, hence drought and then desiccation will have negative impacts on growth.

Concern has been raised in particular for the future of maize production in Africa under climate change and there are several research initiatives, e.g. CIMMYT (2009) following from plant-breeding advances under the 'green revolution'. Monneveux et al. (2005) observe that in Africa 17 per cent of maize losses were due to drought, hence the need for 'more crop per drop' but this is challenging as maize is phenologically a complex plant compared to the indigenous (in Africa) crops of millet and sorghum, that is maize may be much more sensitive to climate change. There are then many factors under climate change scenarios that may impact plant growth: changes to CO_2 concentrations, day length and temperature, but it is rainfall that has received most attention, i.e. agricultural drought.

precipitation analysis (Agnew, 2002, Glantz, 1994, Mortimore, 1989). At the heart of the debate about drought occurrence is disagreement on the exact nature of this phenomenon and its impacts, and, specifically, whether drought should be defined by outcomes, not inputs. It is clear that rainfall is highly variable in space and time so there will be occasions when rainfall is below average. Whether or not this has an impact on human activities or the ecosystem, and is regarded as an agricultural drought, will depend upon the resilience of that system, and thus on the assets available to enable a response or adaptation to drought. This will also depend on the spatial and temporal scales being considered.

With respect to spatial scale, for example, Africa is highly heterogeneous, making it difficult to predict and manage drought. Raynaut et al. (1997) divided West Africa into a mosaic of 21 different classes according to ecological conditions and population trends. If one superimposes climate conditions onto this there are numerous additional permutations. At the macro level drought may be quite easily understood and presented. At the local level, where intervention often needs to take place, much less is known. Vulnerability to drought will also change over time. It may be argued that in developing countries communities have become less resilient as a consequence of greater exposure to world market forces during the latter half of the twentieth century. The example of the expansion of arable cultivation in the Sahel, driven by the European market for vegetable oil, during the relatively wet years of the 1960s has been identified as one of the main reasons the 1970s Sahel drought had a major impact on rural populations (Franke and Chasin, 1980). This echoes the nineteenth-century expansion of arable cultivation by European settlers in the mid-western United States, facilitated by unusually wet years during the 1860s and 1870s, leaving the settlers vulnerable to subsequent years of lower rainfall, and the 1888–90 drought saw the population of Kansas and Nebraska decline by between 25 and 50 per cent (Reisner, 1993: 107). More generally, intensification of agriculture and greater water demands for irrigation and urban expansion create an environment that is more sensitive to rainfall variations (Glantz, 1994), and thus more dependent on investment in infrastructure (water storage, flood defences) or adaptive responses. A reduction in rainfall, then, is often seen as a trigger mechanism for drought but the root causes are more usefully explained in terms of social and economic change. Such an interpretation is not merely the province of pioneers at the agricultural frontier. The UK has recently encountered a number of droughts, and blame for river low flows has been apportioned to rainfall and climate change alongside increasing consumer demand and over-abstractions as well as other catchment changes (Agnew et al., 2000).

As we noted in Chapter 1, drought is a naturally occurring phenomenon, with many parts of the world regularly experiencing droughts 1 or 2 years in 10, and some argue it is already the world's costliest natural disaster (Wilhite, 2000). This is likely to worsen as the IPCC (2007) have identified drought risk as increasing during the twenty-first century particularly in tropical regions (Thornton et al., 2006) but with more extremes in precipitation and drier summers in the northern hemisphere the incidence of droughts is likely to increase across the globe (Lloyd-Hughes and Saunders, 2002). We have suggested above a key issue is whether the vulnerability of people and ecosystems to drought has changed not only due to climate change but also through changes in their socio-economic circumstances. This has implications for how we might plan to manage water resources under the impact of climate change.

Managing climate change

A growing number of indicators point to the impact of human activities upon the environment. The WWF's living planet report (2008) concludes that human demand has already reached the point by which it exceeds the Earth's 'biocapacity'. The human global footprint exceeds the earth's capacity for regeneration by 30 per cent and if current consumption trends continue over the next two decades it is expected we will exceed the capacity of the earth by two times. Superimposed upon anthropogenic changes, Wamsler (2007) notes that natural disasters appear to be more frequent, and in 2005 alone there were 360 natural disasters causing over 90,000 deaths and $160 billion in material losses affecting 160 million people. The IPCC (2007) present an equally worrying scenario of increased storms, sea-level rise, seasonal precipitation changes, increased drought, heat waves, and higher flood risk. The Stern report (2006) on climate change predicted impacts would cost 5 to 20 per cent of global GDP whereas mitigation costs would be less than 1 per cent GDP if acted upon now (assuming stabilisation of GHG 450 to 550 ppm requires emissions to be cut to 80 per cent or more, – stabilisation rates which, as we noted above, Anderson and Bows, 2008 contest).

The IWA (water) Vienna conference (Jowit, 2008) reported that World Bank investment to combat climate change would need to more than double the current annual rate of $80 billion just to cope with the current situation. Parry et al. (2009) have examined the costs of climate-change responses following the recent UNFCC report, and note a convergence of predictions of the expected costs of climate change adaptation for developing countries with figures falling between US$10 to 100 billion per annum but they conclude the UNFCCC forecasts may be a two-to-three fold underestimate.

Strategies to tackle climate change have emerged around mitigation (to curb emissions and reduce atmospheric greenhouse gas concentrations) and adaptation to the impacts. In practice these two strategies merge, with adaptation often being cited as part of mitigation measures, e.g. in energy consumption and waste reduction, while technical solutions are advocated under both headings. Mitigation measures typically include:

- Economic responses (e.g. carbon trading, carbon offsets)
- Consumption reduction including lifestyle and travel changes
- Waste reduction (biodegradeable)
- Technology investment.

There is then a combination of market-based mechanisms alongside technology transfer and human behaviour and lifestyle changes, but Ahmad and Opschoor (2009) report that achievements have been disappointing and aggressive action is urgently required. Funding change is a major problem and the discussion embraces not just the most appropriate strategies, but also responsibility and equity, given that the highest emissions arise from the industrialised north yet the most vulnerable are the poorest communities found in the south. One of the key issues over mitigation is then who should pay, while there is further controversy over the effectiveness of the economic tools available. Cosbey (2009) also notes the potential conflict between mitigation and economic growth given the latter is linked to energy consumption, but only if scale of production increase is not matched by technology efficiencies.

Mitigation and adaptation strategies both rely upon changes to human behaviour, whether through economic measures (see Chapter 6 on water-demand management), or through technical and community action. There are a range of measures that can be adopted, as listed by Bates et al. (2008) who note the adaptive measures deployed in the 'West' include flood defences, water demand management and water transfers, but caution that effectiveness is influenced by technical requirements, cost, political will, and management capacity. These are often designed to improve resilience in the management of water resources, i.e. to be able to withstand the climate shocks of both drought and flooding. The concept of resilience (Holling, 1973) is useful when applied to the possible impacts of climate change upon water resources because it introduces the notion that the same climate change may have different impacts. For example, carbon fertilisation can enhance vegetation growth in some crops (Box 3.4) but only if there are additional water resources, e.g. through irrigation, so while parts of North America can be expected to adapt effectively the same is not likely for large areas of Africa in the absence of substantial investment to improve water availability to agriculture.

Resilience has also been incorporated into the work by ILRI on mapping climate vulnerability and poverty in Africa (Thornton et al., 2006) Their analysis identified as areas particularly prone to climate change impacts: the Sahelian region; arid/semi-arid rangelands in eastern Africa; the Great Lakes region of eastern Africa; coastal regions of eastern Africa; and many of the drier zones of southern Africa. This is consistent with analyses by the IPCC (2007), and concerns over desiccation and desertification in the Sahel throughout the twentieth century. Against this consistent message about the vulnerability of dryland tropics, particularly in Africa, both Raynaut (1997) and ILRI caution that there is much variability in culture and demographic change at the sub-regional scale in Africa, suggesting that there are different degrees of resilience at more local scales, so that care is needed to avoid over-generalising the impacts of climate change. Impacts will be variable as human responses to drought are not linear and there are various forms of adaptation. TERI (2003) suggest three forms of drought vulnerability exist and that these are affected by climate change and globalisation: biophysical vulnerability; social vulnerability; and technological vulnerability. In their analysis on drought and poverty Thornton et al. (2006) argue that we need to move beyond considering only water supply when deciding upon drought-management strategies and to enhance resilience through developing adaptive strategies.

Interrelationships between climate change, impacts and poverty have informed the adoption of 'sustainable livelihoods' (SL) frameworks, for example by the International Fund for Agricultural Development (IFAD) in its strategy for achieving food security in the poorest parts of Africa. The FAO (with DfID) have similarly developed the Livelihood Support Programme (LSP) that has been evaluated against water-resources development by the ODI (Nicol, 2000), who note that a change to treating water as a commodity with an economic value means vulnerability assessments must include some idea of assets and capital at household level, and not just household ability to make environmental responses to climate change. This emphasis upon household adaptive capacity is also found in Wamsler's (2007) analysis of disaster risk reduction in El Salvador. She listed three types of coping strategies:

Reducing risks
   Technical measures (construction)
   Environmental control (run-off reduction)
   Economic (diversifying income)
   Organizational (social networks)

Insuring against risks
   Self insurance (education, ownership, family)
   Institutional insurance (community engagement)

Recovering from risks
   Loss financing
   Community action

However, this long-term disaster management approach is at odds with previous attempts to combat climate extremes through short-term disaster relief. Pereira (2007), who has written at length on drought mitigation, distinguishes between a proactive (preparedness) approach and a reactive (emergency) response through the cycle of disaster management. Figure 3.5 (adapted from Pereira, 2007) shows a 'classic' disaster response cycle. It is suggested that more should be done on planning, including risk management with less reliance on crisis responses.

Pereira lists ten possible measures to tackle drought including enhancing supplies through water storage, use of non-conventional sources; and managing demand through water pricing, allocation and raising

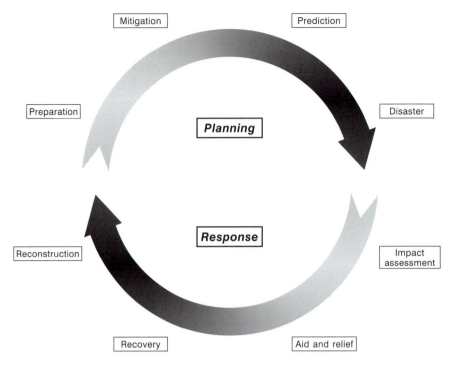

Figure 3.5 Drought planning and drought response

Source: adapted from Pereira, 2007

public awareness. These strategies will be discussed in Chapter 6 on managing water consumption. In reviewing Australia's National Drought Policy, Botterill (2005) also argues that it has been important to move from an attitude of disaster relief which arguably did not work, to one which requires land users to manage all risks, of which a variable climate is but one. She writes, 'At the heart of the policy (ANDP) is a recognition that Australia's climate is highly variable so although the implementation of risk management approach is fraught with difficulties, it is equally difficult to argue that a disaster response is appropriate' (p. 180).

In Europe, EU Directive 2007/60/EC on the assessment and management of flood risks notes the increased likelihood of flood events due to climate change and calls for a focus on prevention, protection and preparedness. The Directive stresses the need for integrated river-basin management and requires the preparation of flood-hazard maps and risk-management plans. In Chapter 2, we saw how the past four decades have seen a radical re-thinking of flood management on the Rhine. This is paralleled elsewhere in Europe, for example in the UK, where Johnson et al. (2007) identify a significant recent change in attitude to flood risk management, moving from flood defence measures to management of risks in much the same way as reported above for drought management. Rather than seeing nature as something to be controlled and dominated, the new attitude places more emphasis upon adaptation that ensures ecological integrity, i.e. 'living with floods'. De Graaf et al. (2007), through examination of flood risk in the Netherlands, note that there has been a tendency to focus upon technical and engineering solutions which they term threshold capacity, with much less attention being paid to community engagement and adaptive capacity (see Table 3.1).

De Graaf et al. report the IPCC (2001) list of determinants that affect a community's adaptive capacity includes,

- Range of technical solutions available
- Resources available and their distribution
- Institutional framework
- Capital (social and human)

Table 3.1 *Adaptive measures to combat flood risk*

Adaptive strategy	Typical measures
Threshold capacity	Reducing hazard and preventing damage through construction and planning, e.g. levees and flood-plain storage (reservoirs for drought) with clear responsibilities and an emphasis upon past occurrences. Requires high degree of social organisation and capital
Coping capacity	Reducing damage and disturbance, e.g. evacuation plans and warning systems (alternative water supplies for drought). It focuses upon immediate responses to a disaster but there can be complications over communication and responsibilities
Recovery capacity	The ability to rectify damage with an emphasis upon short-term action, i.e. returning to the previous state and not considering alternative longer-term solutions through insurance and disaster funds. Again access to capital determines ability to recover
Adaptive capacity	The ability of a community to cope and adjust in the future which may involve significant changes in land use

Source: after De Graaf et al., 2007

- Ability of those with decision-making responsibilities
- Attitudes of the community.

Conclusions on climate change and water resources

Climate change will alter the occurrence of floods and drought in different parts of the world. But much uncertainty remains. The work of the IPCC is identifying those who are most at risk but the measures to combat hydrological extremes have long been established. Figure 3.6 illustrates that for much of the last 100 years (and beyond) we have relied on technical and engineering solutions with a disaster relief response when these have failed. The water-management strategies that have emerged since the Dublin (1992) conference place greater emphasis upon environmental sustainability and community engagement, themes to be discussed further in Chapters 6 and 7 but this is not an attempt to recreate the past. Writing about responding to climate change in Africa, Washington et al. (2006: 1363) observe, 'Coping strategies for climate variability that existed perhaps 150 years ago, including migration, diversification, and conflict are no longer relevant in an era of international boundaries, globalization, urbanization and growing populations.'

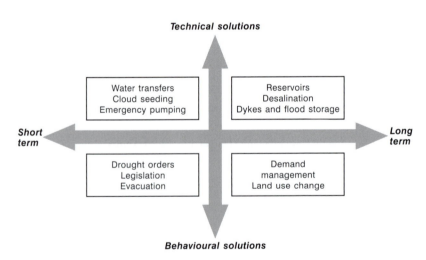

Figure 3.6 Illustrating the range of different options for tackling hydrological disasters over different time scales

Source: after Agnew 2002

In summary, there are four key points concerning drought and flood:

- The definition of the environmental problem must be understood (and stated) at the outset, for example distinguishing between pluvial and fluvial floods, or making the distinction between meteorological and agricultural drought.
- Floods and drought are normal in many parts of the world and should be expected from time to time.
- Climate changes are superimposed upon a hydrological environment that has undergone significant changes during the twentieth century as a result of water-resources developments.
- There is a need to move away from reacting to drought and floods (tactical strategies) to a more proactive approach (strategic measures) for management and mitigation.

Recommended reading

The IPCC remains the key source for information on climate change which is available online:

IPCC 2007 Fourth Assessment on Climate Change. IPCC, Geneva. Climate Change Synthesis Report. Intergovernmental Panel on Climate Change, Geneva. http://www.ipcc.ch/publications_and_data/publications_ipcc_fourth_assessment_report_synthesis_report.htm

There are many books summarising the science and possible impacts of climate change, such as:

Houghton, J.T. 2009 *Global Warming, the Complete Briefing*. 4th edn. Cambridge University Press, Cambridge.
Oldfield, F. 2005 *Environmental Change*: *Key Issues and Alternative Approaches*. Cambridge University Press, New York.

However, the field of climate change is changing so rapidly that websites remain one of the easiest ways to maintain updates, in particular:

Climatic Research Unit (University of East Anglia): http://www.cru.uea.uk/
Canadian Climate Centre: http://www.cccma.bc.ec.gc.ca/eng_index.shtml
IPCC: http://www.ipcc.ch/
Meteorological Office (UK): http://www.metoffice.gov.uk/index.html
Tyndall Centre for Climate Change: http://www.tyndall.ac.uk/
UK Climate Impacts Programme: http://www.ukcip.org.uk/

For those who wish to consider the debates over climate-change science uncertainties, see:

Dessler, A.E. and Parson, E.A. 2006 *The Science and Politics of Global Climate Change*: *a Guide to the Debate*. Cambridge University Press, Cambridge.

Oppenheimer, M. O'Neill, B.C., Webster, M. and Agrawala, S. 2007 Climate change: the limits of consensus. *Science* 317: 1505–6.

Schiermeier, Q. 2010 The real holes in climate science. *Nature* 463 (7279): 284–87.

The Royal Society 2008 Climate change controversies: http://royalsociety.org/ uploadedFiles/Royal_Society_Content/News_and_Issues/Science_Issues/ Climate_change/Climate_booklet_RS1420_reprint_Dec08.pdf

4 ▸ Water resources in colonial and post-independence agricultural development

Introduction

In Chapter 2, we considered the relationship between water use and economic growth, typically driven by industrial and urban expansion. A characteristic of this pattern of development is that it is uneven, the process of urbanisation generating contrasting, non-urban, areas from which population may migrate to urban centres of economic growth. As we observed in Chapter 2, water tends to follow economic activity, being diverted and stored to supply the urban population. As well as supplying water, however, rural areas remain important as sources of supply of food and agricultural raw materials for industry. Governments may therefore be concerned to increase agricultural activity as part of the industrialisation process, investing in irrigation to raise farm productivity. This was the case, for example, in nineteenth-century Japan (Hayami and Ruttan, 1985; Karshenas, 2004). In other instances, where good transport infrastructure connects commercial agriculture to large regional or international markets – the case of California observed in Chapter 2 – it may constitute a principal driver of economic development. In both instances agricultural intensification will demand improved control of water to enable higher crop productivity. In drier climates, and especially the drier tropics, water is the main limiting factor in agricultural production, and governments have seen irrigation as a key investment to increase productivity. In wetter climates, such as the UK, the key investment in water management to increase agricultural productivity took the form of drainage schemes.

However, in many developing countries industrialisation is insufficient to absorb rural labour, and large numbers of people remain in rural areas engaged in activities with relatively low productivity. In these contexts, investment in water management may be undertaken not only, or even principally, to increase agricultural output for urban and industrial consumption, but also to improve the well-being of a rural population at considerable risk of impoverishment, or even of being unable to generate their own subsistence from the land. For these reasons, in less industrialised economies irrigation is often the single largest use of water – typically on the order of 80 per cent or more (Anand, 2007: 36–37)

According to Postel (1992) the land area under irrigation increased five fold during the twentieth century to reach over 200 mha by 1990 and then 277 mha by the Millennium (World Bank, 2009). Postel notes however that by the 1980s the amount of irrigated area per head of world population had started to fall, matched by a decrease in the production of cereal grain with per capita grain production also falling (Hunt, 2004). The push to raise agricultural production through irrigation is evidenced by the doubling of irrigated areas in developing countries since the 1960s (Figure 4.1). There are important contrasts, however, with vast growth in Asia and comparatively little in sub-Saharan Africa. Around 18 per cent of the world's cropland is now under irrigation but for many countries, often those in arid and semi-arid regions, irrigation is far more significant (Figure 4.2).

In this chapter we are concerned with irrigation as a means of combating poverty. We consider the use of irrigation as a means of 'rural development' where opportunities for industrial employment are relatively low, focusing particularly on arid and semi-arid climates, where water is the key factor limiting agricultural output. This combination of conditions is commonly encountered in tropical areas that experienced colonial administration, and the chapter draws particularly upon cases of irrigation in India, in north-east Brazil and in sub-Saharan Africa. For technical advice on irrigation planning and practice, see DfID (2010) and Smith (1992) and Walker (1989).

The case studies in this chapter highlight a number of underlying themes. The first of these is the role of the state in funding and constructing irrigation infrastructure. Frequently, the state also meets the cost of operation and maintenance of irrigation, either directly or in the form of subsidies (such as cheap electricity tariffs for water pumping by farmers in India). This central role of state investment testifies to the strategic importance with which irrigation is regarded, but raises questions about

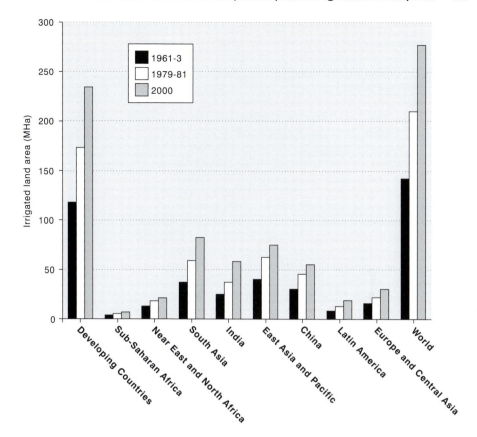

Figure 4.1 Growth in irrigated areas for selected countries and regions including total for developing countries

Source: after World Bank, 2009

its purpose and the criteria used to evaluate it. Poverty-reduction goals imply that all involved in irrigation should benefit. The cases considered here suggest on the one hand irrigation has mostly (though not always) achieved increases in agricultural output of food or strategic raw materials (e.g. cotton) at an aggregate level – as in 'national self-sufficiency'. On the other hand, however, overall increases in output may disguise great variation in outcomes at the level of individual households working in irrigation schemes, whether in terms of 'household food security' or household income above a notional minimum consumption expenditure 'poverty line'. In almost all cases, government irrigation designers sought to resolve this tension by attempting to standardise the conditions for each household working on irrigation schemes: plot sizes were allocated according to household size; water allocations were

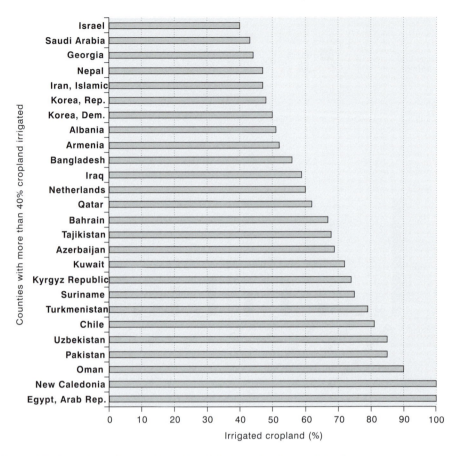

Figure 4.2 Countries with more than 40% of their cropland under irrigation

Source: after World Bank, 2009

controlled through the design of canal infrastructure; and, particularly in Africa, government agencies played a major role in the supply of inputs and control of output marketing, so as to regulate the productivity of the production process and the distribution of the proceeds.

A second theme apparent in these case studies is the unravelling of this engineering-dominated model of irrigation design, in parallel with a ceding of control by government planning staff to, on the one hand, the forces of markets, and, on the other hand, notions of greater democratic control through increasing involvement of irrigators in the planning and management of irrigation schemes. Where irrigation design was exclusively concerned with regulating water *supply*, and with efficiency measures concerning land (e.g. cropping intensity – the number of crops that can be grown on the same land in a year) or water (see Box 4.1),

Box 4.1

Water-use efficiency in irrigation

The simplest measure of irrigation efficiency is at the field-application level (ea) which compares the crop-water requirement (the evapotranspiration (Et) minus any precipitation (p)) with the irrigation applied (Va). Calculation of crop-water requirements needs good-quality climate data, which is often missing. Account also needs to be taken of more site-specific changes in humidity and energy balances after inundation (see Chapter 3 on evaporation, and Doorenbos and Pruitt, 1977 and Smith, 1992). Water may often be over-applied, giving rise to losses via drainage beyond the crop-rooting depth, additional evaporation from storage in depressions or via excessive surface run-off. Such losses can make field-application efficiencies quite low. However, there will have been several earlier opportunities for water to be lost, during storage and subsequent conveyance, even before the water is applied to a field.

Vn = crop-irrigation need (Et – p)
Vt = total water supplied to project
Vf = water delivered to farm
Va = water applied to cropped area

ep = project efficiency = Vn/Vt = (ea* ed)
ef = farm efficiency = Vn/Vf = (ea* eb)
ed = distribution efficiency = Va/Vt = (eb* ec)
ea = field-application efficiency = Vn/Va
eb = farm ditch efficiency = Va/Vf
ec = conveyance efficiency = Vf/Vt

The following table shows some examples of irrigation efficiencies (after Agnew and Anderson, 1992):

	ep	ef	ed	ea	eb	ec
Egypt	0.3	0.46	0.46	0.66	0.70	0.66
India	0.40	0.57	0.58	0.70	0.82	0.70
	0.14	0.20	0.34	0.40	0.50	0.67
	0.25	0.32	0.47	0.53	0.60	0.78
	0.16	0.24	0.34	0.47	0.51	0.67
USA	0.26		0.66	0.40	0.80	0.83
	0.33		0.70	0.58	0.80	0.88
	0.28	0.53	0.52	0.55	0.97	0.54
	0.33		0.50	0.59	0.80	0.63

Even though conveyance efficiencies can be improved through lined ditches and the removal of vegetation, and field efficiencies have improved through practices such as trickle irrigation, overall schemes may still run at low water-use efficiencies with values well below 50 per cent.

it has also become increasingly concerned with institutional and political aspects of regulating water *demand*, through water pricing and decentralised management structures, such as 'water users' associations.

These two themes of state investment in irrigation engineering and the subsequent erosion of central control interact with a third theme: that of social inequality and change. Issues of pre-existing (in)equality of land and water rights play an important role in the impacts of irrigation development on rural poverty. In the cases reviewed here, colonial authorities' early development of irrigation in sparsely populated areas usually required government planners to import population from elsewhere, resettling them as tenants of the state. More recently, the development of irrigation has greatly increased the value of land and set in train new processes of competition to control it, often pitting existing residents against incomers, whether immigrants or (increasingly) investors seeking to take advantage of the increase in potential productivity. A factor that influences the outcome of such growing competition is the possibility of technological change (e.g. borehole and pumping technology, high-yielding crop varieties, small-scale threshing and milling equipment) and the advantages it confers on those with access to capital with which to invest in the higher productivity that it offers.

A consequence of the processes highlighted by these three themes (government investment through technology, decentralised control and equity) is that, even where government investment in irrigation has had explicitly egalitarian goals, outcomes of irrigation 'success' understood in terms of aggregate output have tended to produce impoverished 'losers', as well as wealthier 'winners' among the rural population. A fourth theme, however (to which we will return at the end of the chapter), highlights the wider political and economic context in which even those who benefit from irrigation development may find themselves in competition for water with other economic activity, notably in relation to hydroelectric power generation, or for energy (for pumping), which seems set to change the terms of water use in irrigation in the coming decades.

Indian irrigation: protective vs productive?

In common with much of Asia, agriculture in India has to contend with periodic severe water shortage, and irrigation has historically featured as an element of farming strategies. On the dry plains of northern India the relatively shallow water table made feasible small-scale irrigation from

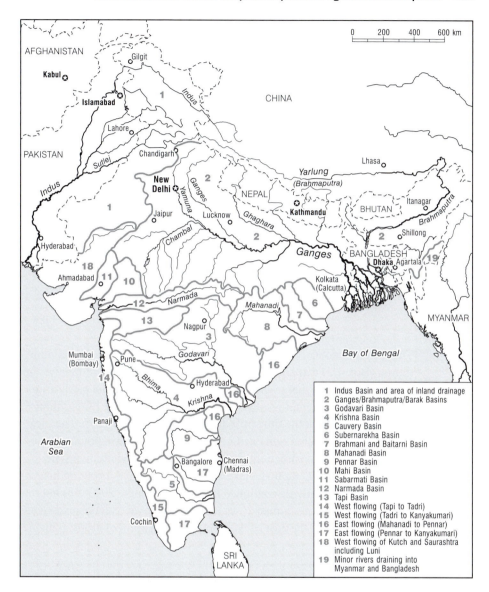

Figure 4.3 River basins of India

shallow earthen (*kachha*) or masonry (*pakka*) wells. In addition, larger water diversion schemes were constructed, notably the 'Grand Anicut' weir dating from the eleventh century on the Cauvery River in southern India which may have irrigated as much as 600,000 acres (Stone, 1984:14), and the Doab canal system on the Ganges and Yamuna floodplains constructed by Mughal rulers.

A major expansion of large-scale canal irrigation was one of the hallmarks of the British administration in the nineteenth century. Initially, this was a matter of British military engineers rehabilitating the existing Mughal works, funded out of the colonial administration's current budget. Following the 1857 uprising against colonial rule, a more expansionary policy of canal construction recognised the need to raise capital investment. While irrigation, along with transport, was a goal of large-scale canal-construction schemes undertaken by the British, this needs to be understood in a context where

> Colonial rule was always a juggling act: there were too few hands, not enough money, and too many considerations to attend to at once to permit policy-makers to realise – or, for that matter, even clearly define – their intentions in any specific area. … As far as irrigation alone was concerned, on a policy level it was simultaneously linked with famine prevention, revenue stabilisation, the settling of unruly tribes, expansion of cultivation, extended cultivation of cash crops, enhanced taxable capacity, improved cultivation practices, and political stability.
>
> (Stone, 1984: 8)

The idea of 'protective irrigation' developed under the colonial administration (Box 4.2) was substantially aimed at preventing rural famine and the political unrest that might follow. It was funded at least partly from famine-relief funds and resulted in the irrigation of some 13.6 million hectares annually at the time of Indian independence in 1947. The idea of protective irrigation remained a strong influence during the acceleration of irrigation construction by the post-independence Indian state (Box 4.3). The separate financial criteria for protective irrigation investment were eventually discontinued in 1964 (Mollinga, 2003: 58) but the design characteristics, aimed at spreading water over as large an area as possible, though at the cost of 'sub-optimal' irrigation of crops, remained

---

Box 4.2

'Protective irrigation' in colonial and post-independence India

The colonial government in India initially sought to expand irrigation by raising capital for canal building through a public–private financing arrangement, exemplified by the Madras Irrigation Company, formed in 1863. The government guaranteed an annual 5 per cent return on the company's

£1 million capital investment, but large cost overruns and the failure of the company to complete more than a small fraction of the project led to the government buying out shareholders for £1.18 million in 1882 and writing off a further £1.5 million of bad debts. Thereafter, the government used loan funding for canal construction by its own public-works department, and by 1921 had invested £69 million in infrastructure that irrigated an annual area of 18 million acres (7.2 million hectares), generating an annual profit to the exchequer of over £3 million (Stone, 1984: 31). By Indian independence, in 1947, some 13.6 million hectares were irrigated annually.

While the majority of this investment had been made subject to conventional criteria of financial rates of return, since the 1880s an element of irrigation investment had been funded from the government's £1.5 million Famine Relief Fund on the grounds that irrigation had wider social benefits, including saving expenditure on famine relief in times of drought. This was used to justify investment in 'protective irrigation' in which lower rates of return were permissible, averaging 1 per cent compared to an average of 9 per cent on 'productive irrigation' schemes (Stone, 1984: 31).

The 'protective' component of irrigation construction increased in the last two decades of colonial administration. Mollinga (2003) has observed that the concept came to acquire three distinct meanings:

- a political commitment to spread access to a scarce resource widely and maximise the social benefit in terms of 'crop per drop';

- an administrative–financial class of public works subject to distinctive criteria (such as lower financial rates of return) for approval of expenditure;

- a specific technical design of irrigation characterised by relatively long distributary canals delivering water to dispersed irrigation areas (both of which contradicted conventional design norms concerned to minimise conveyance losses), and relatively rigid water distribution structures determining the proportion of water moving from primary to secondary canals and from secondary canals to tertiary (i.e. farmers' fields) parts of the system.

Box 4.3

Irrigating India's Green Revolution

Hussain (2005) has identified four periods in post-independence irrigation development:

Independence to 1968: curtailment of the landlord (*zamindari*) system to provide a measure of tenure security for tenants, and a further acceleration of construction of large-scale irrigation by the state, delivering water to

22.6 million ha in 1950–1 and 32.7 million ha in 1966–7. Agricultural production remained extensive, however, with low investment (average fertiliser application 7 kg/ha), low yields and recurrent food shortages following droughts in 1957–8, and 1965–7, forcing the government to seek international aid, notably under the US PL480 scheme.

1968–1980 Green Revolution: introduction of short-straw (high-yielding or HYV) wheat varieties and credit to fund fertiliser and other inputs resulted in a rise in fertiliser use to an average of 45 kg/ha in the early 1980s. Aggregate agricultural production increased from 31 million tons in 1966 to 43 million tons in 1972 and 54 million tons in 1979, providing the Indian government the opportunity to amass grain reserves and declare the country self-sufficient in food. Irrigation was a key factor underpinning the productivity of HYV wheat, and irrigation expansion continued over this period, reaching 52 million ha in 1984. However, for the green revolution centred on the alluvial plains of northern India, the more significant development was the use of credit by individual farmers to develop irrigation from boreholes with diesel pumps. This made possible a shift from 'protective' to 'productive' (yield-maximising) irrigation for those farmers with the means to invest in boreholes and pumps. This action by individual irrigators was accompanied by a shift in emphasis in irrigation policy resulting in the creation of Command Area Development Authorities (CADA) in a number of Indian states to improve productivity on canal-irrigated schemes.

1980–1990 'pre-Reform': Government measures, such as drought and flood relief, and the Public Distribution System for food, attempted to address rural poverty and unemployment that remained undiminished by the aggregate increase of agricultural output. Government investment in new large-scale irrigation peaked between 1980 and 1990. Concerns with the failure of irrigation to significantly diminish rural poverty led to electricity subsidies for water pumping by farmers and to initiatives to improve the organisation of water allocation (and hence irrigation productivity) through 'water users associations' (WUA) at the tertiary level of irrigation schemes. However, efforts by government to increase GDP growth by increasing state expenditure and fiscal liberalisation precipitated a balance of payments crisis in 1991.

1991– 'post-Reform': liberalisation of the Indian economy following 1991 was followed by a period of overall economic growth driven by industrial and service sectors, with agriculture's share falling from the 55 per cent it had contributed in the mid-1950s to 26 per cent by 1999 and 19.5 per cent for 2003–5 (WDR, 2008). Reflecting wider international concerns to devolve resource management to resource users and diminish the involvement of government agencies, this most recent period has been characterised by initiatives to involve farmers in irrigation management through Participatory Irrigation Management (PIM) and Irrigation Management Transfer (IMT). Simultaneously, however, the expansion of groundwater pumping by individual farmers using subsidised electricity reached such proportions as to raise fears that it had become unsustainable.

important in Indian irrigation for the rest of the twentieth century. Nonetheless, as we explain below, the fourfold increase in irrigated area that transformed the country's food supply from dependence on external food aid to national self-sufficiency was accompanied by a growth in private irrigation. This most commonly used groundwater from privately owned boreholes and pumps, even within areas served by publicly funded canal systems.

By the first decade of the twenty-first century India had become a major force in the global economy, with per capita GDP growing at 7.7 per cent annually (2005–6). Rural poverty had dropped from 37 per cent in 1993–94 to 30 per cent in 1999–00 (WDR, 2008), and irrigated agriculture was estimated at around 55 million ha (Hussain, 2005: 19; WDR, 2008). Comparisons between states in India have been used to claim that increased irrigation is associated with reduced rural poverty, and related research has been used to claim a 20–30 per cent reduction in poverty levels within irrigated areas relative to adjacent non-irrigated areas (Hussain, 2005: 32; HDR, 2006: 174–75). Yet, by comparison with other Asian countries, irrigation in India is judged to be inefficient and unproductive. Hussain's (2005) synthesis of research undertaken by the International Water Management Institute (IWMI) found that, compared to canal irrigation in China, Vietnam and Indonesia, large-scale systems in India and Pakistan were less productive in terms of cropping intensity and in value of output, and in terms of the added value of irrigated production compared to neighbouring rain-fed production. Moreover, poverty levels (percentage headcount below poverty line) tended to be

Table 4.1 *Comparison of ranges of values for indicators of irrigation productivity in different countries in Asia*

Country	Farm size (ha)	Land distribution	Crop intensity[1]	Productivity (gross value) US$ha^{-1}y^1	Irrigation benefit[2]	Poverty headcount %
Bangladesh	0.92–0.93	Skewed	180–212	293–448	127–151	35–58
India	1.3–3.0	Skewed	68–127	231–637	35–194	16–73
Pakistan	2.5–6.5	Moderately– highly skewed	124–183	230–500	23–206	40–71
China	0.5–1.0	Fairly equal	152–198	1141–1444	n/a	6–9
Indonesia	0.25–0.77	Fairly equal – skewed	253–296	665–851	292–478	37–44

Source: adapted from Hussain, 2005

Notes:
[1] ratio of annual gross cultivated area to design command area
[2] the net value of farm production per unit area from irrigated fields minus the net value of farm production per unit area from adjoining rain-fed fields.

higher on Indian and Pakistani than on south-east Asian or Chinese systems.

The research identifies as the reason for this disparity the much higher levels of inequality of land holding and water allocation that characterise south Asian irrigation systems, and the much lower charges levied for water on the Indian (and Pakistani) systems than on those in China and south-east Asia. This latter factor is exacerbated by the low rate of collection of water charges, estimated at 21–55 per cent (Hussain, 2005: 118), on Indian canal irrigation. This leads to the conclusion, acknowledged in the National Water Policy (2002), that poor performance on Indian irrigation could be improved by increased cost-recovery to support better canal-system management:

> Disparities in availability of water between head reach and tail-end farms and between large and small farms should be obviated by the adoption of a rotational water distribution system and supply of water on a volumetric basis subject to certain ceilings and rational pricing.
>
> (Hussain, 2005: 103)

The problem of improving equity in the benefits derived from Indian canal irrigation, and its role in rural poverty reduction, has animated a considerable water-management literature for the past thirty years. The main themes arising from this are found in Indian irrigation policy: a focus on improving farmers' organisation at tertiary level in the form of water users' associations, and the participation of those organisations in the development of management strategies for the primary and secondary canals, in some cases even extending to a complete transfer of canal operation and maintenance to farmers' organisations. In part, these moves to increase farmer participation in irrigation management are informed by a perception that canal irrigation has been excessively controlled by officials and engineers with too little account taken of the needs of water users, the 'farmers'. There are reasons, however, to argue that this may be an inadequate analysis of the problem of inequitable distribution of water in Indian irrigation. More specifically, detailed studies of irrigation practice, such as that by Mollinga (2003), suggest that farmers are not disorganised, that inequitable water allocations are not the result of anarchy, and that irrigators are not without power to influence government irrigation officials (Box 4.4).

It is important to note that the principle of 'protective irrigation' that for so long informed Indian canal-system design incorporated within it a rigidity of control that minimized the amount of intervention – and hence personnel and costs – needed for the operation of the system.

Box 4.4

Equitable design, inequitable outcome; the Tungabhadra system

On the Tungabhadra system a centralised process known as 'localisation' designated which areas could be irrigated and which crops could be grown. This ensured that for each zone of irrigation the areas of crops needing large amounts of water (such as sugar cane) was restricted.

Initial inequality derived from the settlement of the scheme at its construction in the 1970s, principally as a result of land purchases by commercial rice farmers from densely farmed areas on the coast, for whom land prices in the new irrigation area were comparatively cheap (a tenth to an eighth of the price in their areas of origin). For existing residents in the area, in which irrigation had previously amounted to less than 1 per cent of the cultivated area, land sales to the new arrivals presented an opportunity to generate cash, to try and escape poverty or to invest on the remainder of their land. For the incomers, land purchases in favourable 'head-reach' locations immediately conferred advantages in securing water, but their comparative wealth and experience in commercial farming also made them well placed to establish relationships, as suppliers of credit and inputs or employers of farm labour, that provided them with considerable power over their poorer neighbours.

As a consequence, although a rotation of water is established among all fields within the same tertiary area, the amount taken by individuals is determined by the size of their land holding and the crops they grow. Where individuals opt to grow lucrative but highly water-demanding crops, or to irrigate them to 'optimal' levels (rather than the sub-optimal levels for which the system was designed) then they effectively reduce the water available to others within that tertiary area. However, further water could be made available by tampering with the outlet structures, thus appropriating water that would otherwise have gone to other outlets. Through these mechanisms, wealthier irrigators were able to use the system to produce commercial yields of sugar cane, rice and cotton, although at the cost of introducing severe inequity of water allocation through the system.

The relations of dependency that many poorer farmers had with their wealthier neighbours meant that they avoided confrontations over the latter's technically illegal crops (i.e. those contravening 'localisation' rules). Nor are the government managers of the scheme in a strong position to intervene. Violation of the prescribed cropping pattern is punished by fines, but farmers are able to file appeals in the courts against such fines, and such appeals are invariably won on procedural points, so that 'there is no example of the government having won a case' and, as a consequence, 'the legal and institutional mechanisms to control the cropping pattern and thereby the water distribution, are paralysed' (Mollinga, 2003: 75). More generally, it is wealthier farmers that tend to represent farmers' concerns to government authorities and local politicians, who, in turn, are obliged to put pressure on government agencies where necessary to deliver on electoral campaign commitments to their electorate.

The operation of commercial agriculture within this 'protective' irrigation scheme has seen a rapid transfer of land to commercial producers, primarily, though not exclusively, from outside the area. These may thus be seen as primary beneficiaries from the scheme. Within ten years from the opening of the canal a third of the land in one village had been sold to migrants while in another 41 per cent was owned and a further 4 per cent leased by incomers (Mollinga, 2003: 140–1).

The amounts of water distributed from primary to secondary canals was to a large degree fixed by the system hardware in order to deliver a fraction (generally between a third to a half) of the water optimally required by 'dry' crops (sorghum, millet, pulses). In addition, distribution of water across very large areas was achieved by a variety of methods. In the *warabandi* systems in north-west India and Pakistan, water was allocated by rotation to each tertiary area ('outlet') as a timed share of flow in the distribution canals. In the Tungabhadra system in the Krishna River Basin in southern India, described by Mollinga, the areas in which irrigation was allowed, and for which crops, was centrally designated by a process known as 'localisation' (Box 4.4). The overall objective was to eliminate decisions – and hence cost – from the operation of the canal system up to the point where water was delivered to the 'tertiary' level and needed to be allocated among the fields of different farmers. By design, therefore, the system was not 'demand-led' but delivered water according to available supply, but water costs were also low. Mollinga's study suggests that the inequality in access to water arises both between farmers within the same 'outlet' or tertiary area and between groups of farmers using different outlets. A critical conclusion from the findings of this study is that inequity of access to water in canal irrigation is shaped not so much through inequity of supply, but through inequity of demand arising from differentials of power among potential water users.

In the wider context of Indian irrigation, however, the widespread use of groundwater (not available in the Tungabhadra case) has provided an opportunity for individuals to acquire exclusive access to irrigation that is 'on-demand', and thus water their crops at optimal levels to achieve commercial yields. For those with the means to invest in borehole and pump, this presents an escape from government prescriptions to share water scarcity in the name of equity. There can be little doubt that this has been a major factor driving the growth of Indian agricultural production. Shah (2007) observed that in an IWMI survey of Indian agriculture undertaken in 2002, only 5 per cent of villages had no irrigation, and only 10 per cent relied on canal irrigation alone. Of the rest, he says nearly half had 'groundwater-dominated' irrigation (Shah, 2007: 18). There is also evidence that, in the areas with the most accessible and productive aquifers in the alluvial plains of the north and north-west, groundwater may be supplying a majority of the water for irrigation, even within canal irrigation systems. This is evident in a detailed study of crop water balances on farms in the Pakistan Punjab (Jehangir et al., 2007), which showed groundwater pumping from a depth of 4.3 to 5.6 m was supplying 75 per cent of the irrigation to a rice-wheat system. Moreover, this exceeded the water actually used by the crop by 60 per cent in the case of rice (hot – *kharif* season) and by 30 per cent for wheat (*rabi* season), much of this being lost to evaporation or returned as seepage to the groundwater. Shah (2007) suggests that this level of groundwater contribution to irrigation within canal systems is not uncommon, and, prompted by the subsidised electricity tariff for irrigation pumps, has been adopted as a model for irrigation even in areas of southern India where geology makes aquifers less productive. The low cost of electric pumping also means there is little cost constraint for continued pumping despite increasing water table depth, to 250m in the case of northern Gujarat (Shah, 2007: 29). In these circumstances, the role of canal irrigation can only be understood in terms of groundwater recharge, albeit a very inadequate one.

Water withdrawals in India amount to some 34 per cent of total renewable water resources, constituting a high level of pressure on the resource (FAO, 2008). Water use is so intense that some river basins, such as the Krishna, in southern India, are undergoing 'closure' whereby little or none of the river flow reaches the sea (Venot et al., 2007). Irrigation is responsible for 90 per cent of this withdrawal, leading some writers (Shah, 2007) to suggest that the current rate of groundwater pumping will prove unsustainable, leading to a collapse in agricultural productivity. North Gujarat is cited as a case of a declining, post-boom, agrarian economy where the better-off, who 'foresaw the impending doom, forged a generational response and made a planned transition to a non-farm, urban livelihood' (Shah, 2007: 23). In the face of this scenario, and the incapacity of government to confront the farming lobby and eliminate the electricity subsidy, a 'groundwater recharge movement' in India has begun to campaign for renewal of historical, dispersed, water-capture techniques such as small reservoirs ('tanks') on watercourses.

Large-scale schemes in sub-Saharan Africa: irrigating the Sahel

South Asia, with 22 per cent of the global population, has only 6.8 per cent of the world's renewable water resources, of which it is currently estimated to be using 25 per cent (Anand, 2007: 31) or 34 per cent in the case of India (FAO, 2008). By contrast, sub-Saharan Africa, with some 11 per cent of the world's population, has around 10 per cent of the water, of which 2–3 per cent is currently being used (Anand, 2007; FAO, 2008). In aggregate, despite having three times the total renewable water resources relative to population, per capita water use in sub-Saharan Africa is about a quarter of that in South Asia (Anand, 2007: 31). One reason for this low aggregate water use in Africa is that that water resources are very unevenly distributed, with 48 per cent of water located in a central African Region accounting for only 18 per cent of the area and 11 per cent of the population (FAO, 2008). The remainder of the continent is much drier, with rainfall generally between 200 and 1200 mm concentrated into a four- to six-month period each year and annual potential evapotranspiration of 2000 mm per year, making water availability a key factor limiting agricultural productivity.

In pre-colonial Africa there was little formal irrigation, in the sense of engineering infrastructure, although examples of furrow irrigation from stream diversion existed in parts of East Africa (Wittfogel, 1957; Watson et al., 1998) but water was managed through the organisation of cultivation and livestock herding. In particular, cultivation was spread over upper- and lower-lying sites to take advantage of varying soil and water regimes and seasonally waterlogged valley bottoms (variously termed *fadamas, dambos, machongos, mbugas, bas-fonds*, etc.) used to grow rice in the rainy season and a variety of vegetables on residual moisture in the dry season. Similarly, pastoralism evolved patterns of seasonal herd migration between rainy season (upland) and dry season (low-lying) pastures. In more arid areas, where water resources were more limited and localised, management of such wetland patches in predominantly dry-land landscapes was necessarily subject to stricter organisation and control by local political authorities.

Significant water resources, such as major river valleys, were the focus of a number of different, and potentially competing, activities: fishing, livestock grazing and watering, and cultivation. Some of the best-documented cases of regulatory systems governing such multiple claims on water resources are from the great river valleys of the Sahel, an area whose modern society and culture is strongly identified with successive phases of government, pre-dating European administration and stretching back two millennia. Until the seventeenth century, the area was governed by a series of powerful states whose wealth was based on control of trans-Saharan trade, particularly between the Mediterranean and the gold-producing areas of the Guinean zone: the Ghana empire (fifth–eleventh century) ruled by the Soninké, the Mali empire (thirteenth–fourteenth century) whose Islamic rulers are identified as 'Mande' or 'Marka', and the Songhai empire (fifteenth–sixteenth century) centred at Gao. From the end of the sixteenth century disruption of the trans-Saharan trade began a 300-year period of instability in the region that coincided with the growth of European-controlled maritime trade from the West African coast. During this period the area was dominated successively by waves of conquest by the Bambara kingdoms centred on Ségou (1670–1810), the Islamic Fulani state centred on Masina (1810–63), and the Islamic Toucouleur empire which stretched from Senegal to Timbuctou (1863–96). This history left its mark on the regulation of access to land and water, particularly where they were most productive: in the great river valleys that looped northwards into the southern Sahara. Two of the best-documented instances of pre-colonial

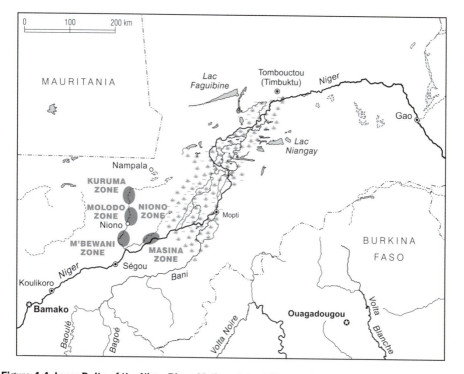

Figure 4.4 Inner Delta of the Niger River, Mali, and the Office du Niger irrigation area (shaded)

water management are those of the 'inner delta' of the Niger, and the Senegal River valley (Box 4.5 and Box 4.6). In both cases, patterns of land use and their control by ruling hierarchies were geared to the management of the annual flood which transformed the floodplain into a succession of different resources: fishery, arable land and pasture.

Irrigation developments by colonial administrations in the Sahel were originally motivated by strategic needs of the European economies for industrial raw materials (cotton, rice). With independence, the strategic goals became those of supplying the urban national market (notably with rice). In each case, the strategic imperative was reflected in a strong degree of state control over irrigated production. The Office du Niger was created in 1932 by the French colonial authority to meet a strategic goal of reducing French dependence on the United States for the supply of cotton. It was modelled on the rapidly expanding Gezira scheme

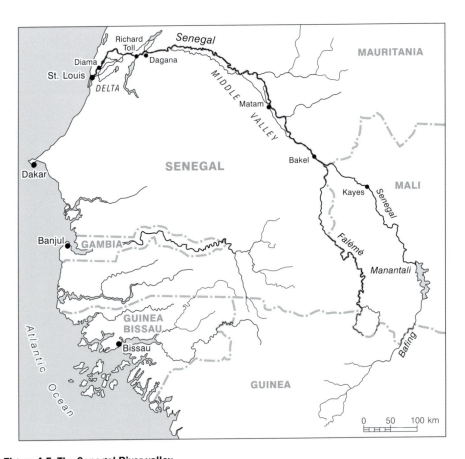

Figure 4.5 The Senegal River valley

Box 4.5

The Dina code in the inner delta of the Niger, Mali

On the river Niger, the 'inner delta' is one of the largest wetlands in Africa covering more than 40 000 km² in central Mali, from Segou to Timbuktu (Ramsar, 2004). The river's annual flood peak moves from one end of the Delta to the other during a five-month period from October to February, followed by a human migration to exploit fishing and grazing resources in its wake (Moorehead, 1989: 262; Davies, 1996:155). In 1820 the pastoralist Fulani established a theocratic state centred on Masna in the inner delta. Despite their military success, the Fulani constituted a minority: at present they are estimated to be between 20 and 38 per cent of the Delta population (Moorehead, 1989: 261; Crowley, 1991: 29). Moreover, as transhumant pastoralists they needed to secure access to pastures in the face of land tenure claimed by existing villages of cultivating (e.g. Bambara, Marka) and fishing (e.g. Somono, Bozo) people (Cissé, 1985: 142; Moorehead, 1989: 264). Finally, the Masina state needed to minimise conflict over pasture between rival Fulani clans.

The Dina was an administrative code that introduced a degree of sedentarisation of the Fulani clans in the Delta by partitioning them into administrative units (*ouro*) and allocating to each a 'home range' (*leyde*) containing a mix of wet-season (*harrima*) and dry-season (*burgu*) pastures as well as residence sites (Moorehead, 1989: 264; Crowley, 1991: 22; Ould Sidi Mohamed, 1992: 193). The Masina state was divided into 37 *leyde* (Crowley, 1991: 28), each administered by a centrally delegated official (*jowro*) whose principal duties were to regulate access by outside herds, including the levy of grazing fees (*tolo*), and to organise the transhumance of all livestock from the *ouro*, including the negotiation of access to other *leyde*. The *jowro* of all the Masina territories together constituted a central council to coordinate herd migrations, by defining transhumance routes and schedules, and negotiating reciprocal relations of access between specific *ouro* (Crowley, 1991: 23–7). The *ouro* embraced all villages within its boundaries, including those of captives (*Rimaibe*) and non-pastoralist people, as well as those of the Fulani themselves. Land and pasture was controlled by village and lineage authorities, who shared with the *jowro* the fees paid for grazing by outside herds.

Since its original formulation, the Dina has been substantially modified by subsequent governments. The Toucouleur conquest in the 1860s was followed by exile for the Fulani officials. Thirty years later the French administration reinstated the exiled Fulani and recognised the Dina code, but modified it to allow non-Fulani pastoralists access to the pastures during severe droughts in 1913–19 (Riddell, 1982: 42), and to permit the colonial government to expand the cultivation of rice. Post-independence governments have sought to varying degrees to replace 'customary' authorities such as the *jowro* with centrally nominated (*prefet*) or locally elected government. In practice, customary authorities not only often retain substantial control over resource use, but have increasingly moved to convert their customary control of 'community' resources such as the inner delta pastures to exclusive (private) property rights (Moorehead, 1989: 265–7, Crowley, 1991: 31, Cotula, 2006: 69–70).

Box 4.6

Historic patterns of flood management: Senegal River Valley

The Senegal River flows along the southern boundary of the Sahara desert, forming the frontier between Senegal, on the south (or 'left') bank, and Mauritania on the north bank. The river is derived from two main tributaries: the Bafing in Mali, and the Falemé, close to the Senegal-Mali frontier. As in the case of the Niger River, the flow is generated largely in the river's headwaters in Guinea, where annual rainfall is 2000 mm. Prior to its alteration by construction of dams in the 1980s, the river's flow at Bakel, just downstream of the Senegal-Mali frontier varied from 10 m³/sec, in May, to 3500 m³/sec in September (Gould, 1981). This huge seasonal variation in flow was manifest in an annual flood along a 600 km stretch of the river valley downstream of Bakel, covering on average 459,000 ha (de Georges and Reilly, 2006) and extending across a floodplain up to 20 km wide in the 'Middle Valley' between Dagana and Matam.

Whereas rain-fed cultivation and grazing are feasible in the 'upper valley' (upstream of Matam), this is not the case in the more arid conditions in the northwards loop of the middle valley into the Sahara where the annual flood from September to February dictates a sequence of different uses of land and water. During the height of the flood, the floodplain is used for fishing, with an estimated annual catch of 50–60 kg per flooded ha. Where the duration of the flood is 45 days or more, on average (1946–1971) on 312,000 ha of the lower-lying clay soils (*waalo* in the local Halpulaar), the soil retains sufficient moisture to support a cereal crop (sorghum) planted following the receding floodwaters. Some 100,000 ha are estimated to have been cultivated annually in the middle valley using this 'flood-recession' farming (de Georges and Reilly, 2006). Following the harvest of the cereal crop, livestock are moved from the dryland grazing areas (*jeeri*) beyond the floodplain to graze on the crop residues during the dry season. As in the case of the inner delta of the Niger, control of resource use the middle valley of the Senegal River was defined by historical hierarchies by which land ownership was reserved for members of particular local Halpulaar (Toucouleur) lineages and castes and excluded immigrants and those descended from slaves or captives taken in episodes of conquest in the nineteenth century. Following independence in 1960, a land-reform law nationalised some 97 per cent of land in Senegal and substituted customary land rights by land allocation through elected councils of newly formed 'rural communities'. In practice, members of traditional landholding families were often elected to the rural councils and 'the modern law practically reinforced the local standing of the traditional chief by conferring upon him a new state function' (Engelhard et al. 1986: 62). Diemer and van der Laan (1987: 153–4) noted that in the Middle Valley the 1964 nationalisation law remained effectively ignored, and, even in 1984, some cultivators of flood recession land were paying rent to traditional landholders.

begun by the British a decade earlier to produce cotton in Sudan.
A barrage was completed at Markala in 1947 to raise the level of the
Niger by 5 m in order to divert water into distribution canals created from
'fossil' watercourses to irrigate an ancient alluvial floodplain to the west
of the inner delta of the Niger. Average flow at Markala varied between
4000 and 50 m^{3}s^{-1}, considered ample to irrigate a projected million ha of
cotton. Unlike the inner delta to the East, the area to be irrigated was
sparsely populated and was initially cultivated by settlers identified by
the chiefs in their areas of origin and forced to migrate up to 500 km to
the scheme, which was irrigating some 22,000 ha by 1945 when forced
settlement was ended and 40 per cent of settlers promptly left (Aw and
Diemer, 2005: 11).

For much of the colonial period, the Senegal River was peripheral to the
colonial economy, and the valley's population was left to its historic
patterns of land use governed by the annual flood, supplemented by
increasing migration abroad in search of wage income. The wartime
experience of vulnerability arising from dependence on rice imports
(from French colonies in south-east Asia) prompted a number of colonial
efforts to develop partial control of the river flood to irrigate rice at
Richard Toll, where the 'middle valley' flood recession agriculture gives
way to the 'delta': a broad flat expanse affected by saline intrusion from
the sea during the season of low flow in the river. As in the case of the
Office du Niger, the area selected for irrigation was sparsely populated
and required settlement by migrants from elsewhere. Little formal
irrigation had been established, however, before the end of colonial
administration.

Within a decade of Senegal's independence two factors had combined to
reinforce efforts to create irrigated agriculture in the Senegal River
valley. The first was the succession of drought years in the 1970s that
decimated rain-fed agriculture and restricted the extent of flooding. The
second was the resulting food shortages, particularly of rice – the main
staple of the urban population. In the delta, large-scale schemes switched
from partially controlled flooding to canal irrigation fed by electric
pumping from the river. With the exception of 7500 ha operated as a
commercial sugar plantation at Richard Toll, land within these schemes
was divided into smallholdings, allocated to heads of households at the
rate of 0.5 ha per working-age adult. A state development corporation
(SAED – *Société d'Aménagement et d'Exploitation des terres du Delta
du fleuve Sénégal*), formed in 1965, undertook design, construction and
operation of the irrigation infrastructure, supply of inputs and services, to
be repaid by plotholders from the value of the rice harvest, also marketed

through SAED. With the drought increasingly threatening the food supply of villages in the middle and upper valley, small-scale irrigation (20–50 ha) was initiated from 1975, initially pumping water direct from the river using diesel pumps purchased by migrants from the valley working in Europe. These village irrigation schemes (PIV – *Périmètres Irrigués Villageois*) were constructed on the levees (*falo*) on land adjacent to the villages, divided into plots of between 0.1 and 0.5 ha per household. All cultivation was typically undertaken manually and water control was managed by the farmers (Adams, 1981; 2000; Engelhard et al., 1986). The initial success of the PIVs was quickly followed by support from the state, with SAED expanding its supply of credit and services throughout the valley, and funding for further irrigation construction from a variety of overseas aid agencies.

In parallel with this rapid expansion of irrigation, the drought had prompted the governments of Mali, Senegal and Mauritania to set up a river basin development organisation (OMVS – *Organisation pour la Mise en valeur de la Vallée du Sénégal*) which embarked on an internationally funded programme to build two dams: the Manantali, in Mali, on the Bafing, a tributary supplying 50–60 per cent of the flow of the River Senegal; and at Diama, 27 km upstream from the estuary. The Manantali dam and reservoir was to stabilise the flow of the Senegal River to allow year-round irrigation, navigation and 200 MW hydro-electric generation, while the Diama Dam was to prevent movement of saline water upstream from the estuary. Construction of the dams was completed in 1987 and they were officially inaugurated in 1992. Hydroelectric capacity was installed at Manantali in 2001 and power generation supplying the capitals of Senegal, Mali and Mauritania (though not the Senegal River valley itself) began operation in 2002. Under the OMVS plan, the Manantali Dam would permit the irrigation of over 300,000 ha in the Senegal River valley, of which 224,000 on the Senegalese bank. In practice, irrigation since the completion of the dams (Table 4.2) has failed to expand significantly beyond the level reached in the late 1980s, with only a third of the 94,000 ha nominally irrigable area actually being used annually (OMVS, 2003).

The main factor behind this failure is the withdrawal of state (SAED) support for the maintenance of irrigation infrastructure and equipment and subsidised credit and fertiliser as part of Senegal's new agricultural policy (NPA) introduced in 1984. This envisaged that commercial entities would take over the supply of inputs and services and that irrigation management (water and land allocation) would be undertaken by farmers' organisations. In the event, the rapid rise in production costs

Table 4.2 *Net irrigable area (ha) for different types of irrigation in the Senegal River valley*

Year	1976	1980	1984	1987	1988	1995	2003
Senegal (left bank)							
Large-scale	7083	9465	12577	12989	12940		
PIV – middle and upper valley	352	3577	7271	11991	12783		
PIV/PIP delta	691	1191	2191	4315	5853		
Total (left bank)	8126	14233	22039	29295	31576	71751*	94000**
Mauritania (right bank)	510	2205			13929		

Source: data for 1976–88 compiled from OMVS; Woodhouse and Ndiaye, 1990

Notes:
 * of which, 29,792 ha cultivated (Adams, 2000)
 ** of which 35,000–40,000 in production (OMVS, 2003).

could not be supported by the small-scale schemes (PIVs) in the middle and upper valley, producing mainly for household consumption and with very little marketed production with which to cover costs. Most were simply abandoned (Adams, 2000). On the larger schemes in the delta, unreliability of water supply through the canal infrastructure coupled with severe labour bottlenecks at rice harvesting meant that many farmers were unable to produce high enough yields to cover costs, let alone the two crops per year envisaged by government planners, resulting in indebtedness and eviction or selling-on of their plots to others.

In addition to eliminating government subsidy, the NPA was hoped to encourage entrepreneurial investment in private irrigation in the Senegal River valley. As observed above, this has yet to significantly increase the irrigated area beyond that under state administration twenty years earlier. On the contrary, it is arguable that the policy's failure to take account of the complexity of pre-existing land rights in the Senegal valley contributed to international conflict and environmental deterioration that displaced hundreds of thousands of people (Box 4.8). For these refugees, as for the existing population of the valley, economic conditions have been poor. In the middle and upper valley, the operation of the Manantali dam has largely suppressed the annual flood, despite plans to maintain an artificial flood to sustain about 30,000 ha of flood-recession agriculture over a ten-year period until it could be substituted by irrigation. In practice, during its 20 years of operation, the dam has either suppressed the flood or made it more unpredictable (Adams, 2000), with consequent loss of cultivation, grazing and fishing, and impoverishment of those who hope to make a living from them.

The operation of the Office du Niger (ON) irrigation scheme in the 1960s and 70s shared many of the characteristics of the SAED in the

Box 4.7

Making irrigation work? The Office du Niger

The Office du Niger (ON) was created in 1932 and barrage was completed at Markala in 1947 to raise the level of the Niger by 5 m in order to divert water into distribution canals. Average flow at Markala varied between 4000 m^3s^{-1} in the wet season and 50 m^3s^{-1} in the dry season, considered ample to irrigate a projected million ha of cotton. The Markala dam stored only 10 million m^3 but diverted water into two canals, the 'Sahel' and the 'Masina', each conveying water some 20 km to discharge into fossil watercourses, Molodo Fala and Boky Wéré Fala respectively, that were embanked to serve as canals and also to provide a further 100 million m^3 storage. Water was distributed from these two main 'canals' via primary canals and secondary canals to tertiary land units of 12 ha.

During the colonial and early post-independence years, control over marketing of cotton and rice from the scheme ensured a flow of revenue from the scheme's rice mills. However, low productivity, arising from difficulties of water control on the large (12 ha) plots, required continuing government subsidy. Cotton production was abandoned in favour of rice in 1968. Annual production of rice more than doubled, to 101,000 tons, between 1969 and 1978, making Mali self-sufficient, with the ON supplying 80 per cent of the domestic market.

This prompted the government to seek international investment to expand the scheme and to rehabilitate the existing infrastructure to increase productivity, which remained low (1 ton paddy ha^{-1}). However, potential funders (coordinated by the CILSS – *Comité Permanent Inter-états de Lutte contre la Sécheresse dans le Sahel)*, while recognising the potential of the ON, remained cautious because they perceived the tight control of the production process by the ON management as leaving few incentives for the settlers to increase productivity. ON went ahead with the construction of a third main canal without external investment, but by switching expenditure from operation and maintenance of the existing infrastructure. As a consequence rice output from the ON declined from 101,000 tons in 1978 to only 56,000 tons in 1983, amid an escalation in coercion by the scheme management to prevent settlers holding back rice from sale.

This collapse in rice production maintained pressure on the government to engage with external funders to find ways of increasing productivity. External funding (mainly from the Netherlands and France) for infrastructure rehabilitation projects was used to leverage changes in the production process. Over a ten-year period from 1982 these included:

- improved water control (large fields (12 ha) subdivided into 0.5 ha compartments, within which further levelling and construction of bunds for water control was undertaken by the farmers themselves);

  - introduction of transplanting of rice to minimise weed infestation;
  - credit for fertiliser;
  - equipment purchased to replace the ageing ON machinery for threshing and milling the rice was supplied as small-scale, mobile units which could be managed by farmers themselves – thus weakening the ON management's control over the marketing of rice.

These technical changes were accompanied by steps to improve farmers' organisation through the creation of *Associations Villageoises*, which were assigned responsibility for the operation and maintenance of, and management of the income from, the new threshing and milling equipment. Land-tenure security was improved for plotholders and a land registry was set up for farmers on rehabilitated sections of the scheme. Operation and maintenance was funded from increased water fees paid by plotholders, on the basis that rice yields had increased sufficiently that even the higher water charges constituted a reduced fraction of the value of the rice harvest than previously. Oversight of the operation and management expenditure was assigned to joint ON–farmer committees. A coup in 1991 brought a new government with broad popular support, less politically dependent on a constituency of civil servants and other government employees. External funders seized this opportunity to lobby the new government to reform the ON itself, reducing its staff from 1200 to 400 by 1994. A year later, a doubling of water fees paid by irrigators provided the basis for covering the cost of running the scheme. The expanded area following completion of the third main canal is divided into five zones, each of 12,000 ha and governed by a joint ON–farmer committee. Water fees are paid in to ON offices, and maintenance is largely undertaken by ON, but in each case the performance of such functions is governed by three-year contracts, monitored and audited by the joint committees.

Source: Aw and Diemer, 2005

Box 4.8

Destabilisation and stagnation of production in the Senegal River valley

An expectation of the NPA was that the prospect of year-round irrigation following the regularisation of the flow of Senegal River by the Manantali Dam would encourage entrepreneurial investment in irrigated agriculture. Many authors report land allocations in the Senegal River Valley to investors from urban areas and government agencies – not least retrenched SAED staff (Adams, 2000; Sylla, 2006).

However, much of this investment focussed on privately owned versions of the PIV model: 25–50 ha irrigated by a diesel pump set pumping from a canal or drainage channel. Moreover, such 'PIP' systems were concentrated where communication infrastructure was better developed, in the delta. The outcome of rudimentary and poorly managed irrigation applied to the delta soils overlying saline groundwater was a rapid rise in soil salinisation and consequent poor productivity.

Worse, the encouragement of a political climate of entrepreneurial investment in irrigation undermined the security of existing customary land rights and fed a perception that investment in irrigation infrastructure provided a means to secure rights over land. As a consequence, holders of customary rights in villages in the middle valley competed to claim land for small-scale irrigation development, contesting the rights of others, for example to seasonal grazing. Disputes of this nature had, by 1988, led to violent confrontations between villages on both the Senegalese and Mauritanian sides of the valley. In April 1989 this conflict erupted after two Senegalese farmers were killed by the Mauritanian military during a dispute over grazing rights in the middle valley. In the weeks that followed, riots in Nouakchott resulted in the deaths of several hundred Senegalese and corresponding violence in Senegalese towns killed 60 Mauritanians.

The crisis rapidly assumed not only international, but racial, dimensions. The 'repatriation' of 200,000 Mauritanians – many small traders who had never lived in Mauritania – and the deportation of 90,000 Senegalese from Mauritania was accompanied by the expulsion of some 60,000 black farmers – most with Mauritanian nationality – from the northern bank of the river.

Senegal River valley: a state corporation controlled not only the hydraulic operation of the system but also the supply of agricultural inputs and machinery services and monopolised the marketing of output. Despite continued departures of plotholders from the scheme, the availability of water presented the ON as a refuge during the prolonged drought of the 1970s, and population rose from 35,000 to 53,000 (Aw and Diemer, 2005). A coup in 1968 brought a new government keen to increase supplies of rice for its key urban political constituency, and cotton production in the ON was abandoned in favour of rice, which was more easily irrigated given the abundant supply of water. Within ten years Mali was self-sufficient in rice, with the ON supplying 80 per cent of the domestic market. Under changing international and domestic political pressure, the state management of the ON was subsequently reformed, with a stronger role for organisations accountable to those cultivating plots on the scheme (Box 4.7). This has been translated into a steady increase and diversification of output by the ON in the past two decades (Table 4.3).

Table 4.3 Evolution of irrigation on the Office du Niger

Year	Population	No. of holdings	Net area (ha)	% Rice transpl.	Paddy yield T/ha	Vegetables (ha)	Total cropped area (ha)	Cropping intensity %	Total paddy (mt)
1982–3	62,895	5,484	47,080	0	1.6		35,310	0.75	56,524
1986–7	109,604	9,245	47,080	2	2.2		40,018	0.85	88,011
1991–2	117,120	10,465	47,080	48	4.1		44,255	0.94	180,909
1996–7	151,839	13,970	47,991	94	5.1	2,519	54,988	1.09	245,365
2001–2	221,548	21,827	56,506	99	6.1	4,592	67,080	1.19	332,078

Source: compiled from Aw and Diemer, 2005: 65–67

With abundant water and a main infrastructure designed to irrigate an area many times larger than that cultivated, the achievement of self-sufficiency in rice and the financial viability of the irrigation scheme offer the prospect of continual expansion for both new settlers and the children of existing plotholders. Substantial increases in incomes have also been achieved, as an increasing amount of land is occupied by higher value output, such as vegetables and fruit orchards. By 2002 these had added a further 46 per cent to the value of the ON output, relative to the main wet-season rice crop. Aw and Diemer (2005) note that an addition of over 3866 ha had been co-financed between the government and farmers at a cost of US$2518/ha – a quarter of the price conventionally quoted for irrigation in Africa.

However, not all settlers in the ON have benefited equally from this success story. A majority have increased their agricultural incomes through investing in higher-value production, or by diversifying from agriculture to non-farm activities such as crop processing and transport. An estimated one-third of settler households struggle to achieve more than subsistence from irrigated production on Office du Niger (Aw and Diemer, 2005: 68). For such households, rising costs of water fees pose a threat of eviction from the scheme – one that Cotula (2006: 16) suggests materialised after 2003, and one that Aw and Diemer (2005) acknowledge is a factor driving an informal (and technically illegal) market in irrigated land, as those with a prospect of eviction sell their tenancy rights before they lose them. This threat is certainly heightened by the perception that, as in the Senegal River valley, the expansion of irrigated farming in Mali requires capital investment that only established commercial agribusiness can supply.

In both the Senegal and Niger river valleys, post-independence governments have sought to develop irrigation as a key to stabilising food supply, especially to the politically important constituency of urban consumers. As a consequence, irrigation has focused on the preferred urban staple of the colonial era, rice, rather than the more drought-tolerant millet or sorghum that historically formed the bulk of Sahelian cereal consumption. Patterns of water management for irrigation were thus shaped fundamentally by political choices that to an extent reflected aspirations moulded by colonial experience. From a strict water-efficiency standpoint, the decision to grow a crop with very high water demands in a very arid climate appears questionable. On the other hand, transplanted rice tolerates relatively poor water control, as long as fields receive abundant water. This means that irrigation can be undertaken over large areas without sophisticated (and expensive) land levelling to avoid damage to crops more sensitive to waterlogging (as in the case of cotton on the Office du Niger). It is also important that rice is the staple of urban consumers in the Sahel, so its production supplies a relatively strong (i.e urban) market. This is even more true of vegetable and fruit production, and explains the diversification pattern on the Office du Niger. The cases of the Senegal River valley and the Office du Niger suggest both possibilities but also pitfalls in the use of irrigation to reduce rural poverty. In each case the role of state policy has had a profound influence on outcomes, and in particular on which social groups have gained or lost opportunities from irrigation investments. We shall return to reconsider this point at the end of the chapter.

Brazil's Nordeste: water and citizenship

The semi-arid north-east of Brazil was the centre of initial Portuguese colonisation based on sugar plantations in the sixteenth century, and the location of the colony's first capital, Salvador. This development was largely limited to the coast, however. The interior remained sparsely populated, notionally subdivided into a number of large land concessions (*sesmarias*) awarded by the Portuguese crown to military commanders and explorers. Despite the existence of substantial water resources, of which the largest is the São Francisco River, extending 2800 km northwards from the uplands of São Paulo and Minas Gerais, the north-east interior remained peripheral to the Brazilian economy for two hundred years. For most of this period the region known as the *sertão* remained a zone of feudal landlords, extensive livestock production, and a refuge for slaves who had escaped the coastal sugar plantations. Small towns and settlements of ex-slaves (*quilombos*) 'lost in leagues of unoccupied land' (James, 1948: 658) that were still described at the end of the twentieth century as 'demographic voids'(SAP, 2004: 27), in which small-scale farmers practising flood-recession agriculture along river margins (*vazantes*) remain subject to labour rent and sharecropping contracts with the local landowner. Periodic multi-year droughts were associated with starvation and emigration to the growing urban and industrial zones of São Paulo and Minas Gerais, to the south.

The construction of the new Federal capital at Brasilia in the 1960s was followed by the development of road links through the *sertão* connecting Brasilia to the major north-eastern state capitals of Salvador, Fortaleza and Recife, spurring the development of hydro-electric power generation with the construction of dams on the São Francisco river. This, in turn, prompted state action to dynamise agriculture, offering subsidies to private investors and investing public funds in irrigation in order to

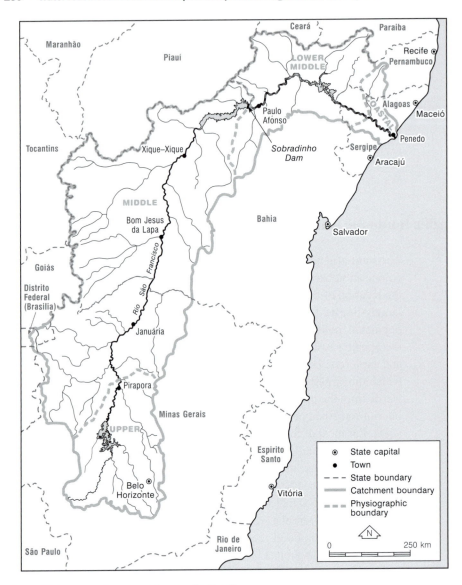

Figure 4.6 The São Francisco River basin, Brazil

resettle people displaced from the site of the largest reservoir, the Sobradinho Dam, in the mid-1970s. Despite initial investment in 258,000 ha of irrigation for some 7000 displaced people settled in 23 'agrovilas' at Serra do Ramalho 10 km south of Bom Jesus da Lapa, government technical support was found inadequate and private investment proved more oriented to land speculation than to agricultural production. The scheme largely failed and many displaced people preferred to return 700 km to live at the margins of the reservoir that had displaced them (Germani et al., 2007).

The end of more than a decade of military dictatorship, in 1985, marked a political trajectory of increasing democratic pressure on the Brazilian state and a growing expectation of rising living standards. The poverty of the north-eastern *sertão* became increasingly visible in official statistics, with human development indices (HDI) typically of 0.5–0.6 (as low as 0.34 in some administrative areas or *municipios*), compared to 0.77 for Brazil as a whole (SAP, 2004: 27). Further investments in irrigation explicitly aimed to raise the living standards of the poor, but in publicly funded irrigation built since 1980 in São Francisco valley 73 per cent of the area was exclusively or predominantly occupied by companies, against 27 per cent occupied by small-scale farmers, or *colonos*. Germani et al. (2007: 99) trace the dynamics of inequality that generate such out-comes through a shifting frame of selection that favours commercial over social criteria in allocation of land in new irrigation. Thus, even in the Formoso irrigation scheme surrounding Bom Jesus da Lapa built in the middle valley of the São Francisco in the 1990s and allocated to irrigators from 1989 to as late as 2004, half of the area was being farmed by companies, and as many as 40 per cent of plots had changed hands unofficially, often via middlemen. However, whereas the areas farmed by *colonos* are completely cultivated, some 69 per cent of plots and 54 per cent of the area occupied by companies were not in production in 2004 (Germani et al., 2007: 104). While this testifies to the continuing incapacity of government departments to resist acquisition of public resources by speculative local capital, the past 20 years have also witnessed two sets of political dynamics that suggest a greater capacity on the part of the poor themselves to overturn their historic disenfran-chisement from the management of natural resources.

The first of these is the growth in the agrarian reform movement in Brazil. Growing popular campaigning and agitation by movements of landless and rural workers,[2] drawing on support from the Church (*Comissão Pastoral da Terra*), political parties and civil society organisations, have during the past two decades resulted in a strengthening commitment by successive governments to the principle of re-allocation of under-used land to those without land but with the will and means to cultivate it. The popular mobilisation of thousands of people involving occupations of farms and government offices has resulted in land transfers of some 334,000 ha involving 8000 families in the middle São Francisco valley, dwarfing in scale the transfer of land by government-assisted land purchase schemes (*Cedula de Terra*, and *Crédito Fundiário*) totalling

14,000 ha and 520 families in the same area (Germani et al., 2007: 125). However, this relative success of the agrarian reform movement has exposed the insufficiency of land transfer as a basis for improved living standards. The considerable political mobilising capacity of the agrarian reform movements has increasingly focused pressure on government to deliver infrastructure, social services and technical support to the newly established land reform settlements. It has also campaigned to regularise the land tenure of existing settlements whose legal status has been historically precarious, notably the case of the *quilombos* established centuries before by ex-slaves. The outcome of such pressure, in terms of significant reduction of rural poverty through successful small-scale irrigated agriculture, is not a foregone conclusion, however. This is particularly because irrigable land is increasingly the object of acquisition by Brazil's booming agribusiness sector. Nonetheless, the increasing assertion of political power 'from below' in relation to land rights raises questions about the future direction and control of government efforts to decentralise and democratise the governance of the other key natural resource in the north-east: water.

The acute effects of drought in the north-east reached the point of threatening water supplies of major urban centres, prompting state government reforms of the control of water allocation from water reservoirs. The evolution of these reforms from the late 1980s has been documented particularly for the state of Ceará (Lemos and Oliveira, 2004), and became the model for the 1997 legislation and organisation that underpins Brazil's national water reforms. The centrepiece of this legislation is the River Basin Committee – a 'water parliament' for a defined river basin made up of representatives of water users and of government departments, tasked with making decisions on water allocation each year. The literature on the operation of River Basin committees in Brazil has suggested that current distribution of representation on the Committees allows less scope for water users' priorities than before. In particular, it has criticised officials' control of decision-making procedures that may be restricted to choosing between scenarios generated by hydrological models simulating predicted rainfall and reservoir discharge and recharge characteristics (Lemos and Oliveira, 2004; Broad et al., 2007). We shall return to examine these issues in Chapter 7. However, it also seems that this democratisation of water governance 'from above' has yet to feel the force of popular political mobilisation by the agrarian reform movements. If current trends succeed in consolidating agrarian reform resettlement projects into broader social and economic programmes of 'territorial development', then they will also need representation as water users in Water Basin Committees, raising questions as to the political parameters which will govern water allocation on such bodies.

Irrigation and poverty reduction

In this chapter we have focused on the role of irrigation as a means of reducing poverty by raising productivity and reducing risks for people primarily dependent on agriculture in less industrialised economies. Here, irrigation development is perceived in terms of meeting social, rather than narrowly commercial, goals. There are resonances here with the original mandate of the US Government's Bureau of Land Reclamation (see Chapter 2) at the start of the twentieth century. However, in the US this social remit was quickly overtaken by a rapidly expanding industrial economy and commercial agriculture. In the semi-arid contexts we have considered in this chapter, industrialisation has been absent or distant and water is the key constraint not just to raising income, but to achieving bare subsistence.

In these contexts, and the cases we have considered from India, the African Sahel and the Brazilian nordeste, it is possible to identify two key areas of tension for policy. One relates to the most effective use of water. Specifically, should irrigation be designed to meet 'crop water requirements' at an 'optimal' level (e.g. in Box 3.3 and Box 4.1) in order to maximise crop productivity, or should water be spread over as wide an area as possible on the understanding that most irrigation will be 'sub-optimal' and most crop yields well below maximum? the latter approach – 'some for all' – was the basis of 'protective irrigation' designs used for large-scale canal systems in India. Lankford (2004) has argued that this principle should underlie the design of irrigation in river catchments where growing demand for irrigation is being generated by large numbers of small farmers dispersed in the catchment. Drawing on studies in the Usangu Plains in southern Tanzania, he observed that, in the absence of large water-storage reservoirs to smooth out seasonal and inter-annual river flow fluctuations, diversion structures (e.g. weirs and intakes to canals) designed to optimise water delivery to crops within a specified irrigated 'command area' would often exacerbate water scarcity for downstream users at times of low river flow. In contrast, he argued, intake structures that delivered irrigation water proportionately to the total flow in the river would be more consistent with farmers' own agricultural management strategies, that varied the area cultivated according to water availability. Such strategies accept that at least part of the area will be sub-optimally irrigated. However, this may not be of overriding concern, for example if such marginally productive areas enable the growing season to be extended and food supplies to be strengthened at critical times of the year.

The key point here, as with the protective irrigation designs in India, is that all farmers gain proportionately the same when water supply increases (and lose proportionately the same when water supply decreases). Lankford (2004: 41) recognises that for such a policy to work at the level of a river basin a high level of coordination among different irrigators would be needed: ' … it is for local participative decision-making to plan land use and adjust flows between intakes accordingly'. This, however, raises a second area of tension for policy. As Lankford (2004: 43) also recognises 'In water scarce situations … farmers are less likely to share risks with others'. The experience of the cases reviewed in this chapter is that the role of the state is critical in managing the distribution of water scarcity among water users. In particular, the commitment and capacity of the state to defend the interests of more vulnerable water users has a profound influence on the outcomes of irrigation in terms of poverty reduction. In this regard, a major question is the extent to which state irrigation agencies seek to regulate the ability of wealthier or otherwise more influential groups to gain a larger share of water through private control for commercial production. It seems clear that for the most part government policy has either overtly or tacitly fostered private capital investment as a basis for improved access to water in order to maximise aggregate output. This was the basis for borehole irrigation in India's green revolution. It was also evident in Senegal's New Agricultural Policy, and in the more recent irrigation expansion on the Office du Niger. It has also been the policy of land allocation on irrigation schemes in the São Francisco valley. However, this latter case also offers a view of an alternative political pressure – that of organisations of the rural poor – that may conceivably push water allocations in a different direction. It needs to be underlined, however, that the global context is increasingly one in which concerns about the security of fossil fuel supplies have multiplied the demand for agricultural land to produce biofuel. It is therefore likely that commercial demand for irrigated land will intensify in the foreseeable future.

This discussion has followed irrigation's transition from dominance by technical considerations of agronomic and hydrological potential to social and economic questions about relations of production (who benefits from irrigation) and exchange (agricultural markets). A further transition (linked with the first) is one of substituting centralised and hierarchical state control by decentralised management, either structured by the organisation of water control (as in the Office du Niger or in Brazil's water-basin committees), or structured by the operation of individualised control of water (as with groundwater exploitation in India, or the entrepreneurial development of small-scale irrigation in Senegal). This raises further questions about how tensions between individualised water use and collective interests are managed. In particular, it raises questions about the role of the state in managing such tensions and establishing social priorities within which local 'democratic' management structures will operate. We will return to consider these broader questions of water governance in Chapter 7.

Further reading

The FAO provides technical advice on irrigation-water requirements and scheduling in the following publications:

Doorenbos, J. and Pruitt, W.O. 1977. Guidelines for predicting crop water requirements. FAO Irrigation and Drainage Paper 24, Food and Agriculture Organization of the United Nations. Rome.

Smith, M. 1992. CROPWAT: a computer program for irrigation planning and management. FAO Irrigation and Drainage Paper 46, Food and Agriculture Organization of the United Nations. Rome.

Walker, W.R. 1989. Guidelines for designing and evaluating surface irrigation systems. FAO Irrigation and Drainage Paper 45, Food and Agriculture Organization of the United Nations. Rome, online at http://www.fao.org/docrep/T0231E/t0231e00.htm#Contents

A perspective on the change from technical considerations to the social and economic dimensions of irrigation development can be found in:

Anand, P. (2007). *Scarcity, Entitlements and the Economics of Water in Developing Countries*. Edward Elgar, Cheltenham.

Hayami, J. and Ruttan, V. (1985). *Agricultural Development: An International Perspective*. Johns Hopkins University Press, Baltimore.

Hussain, I. (2005). *Poverty in Irrigated Agriculture. Realities, Issues and Options with Guidelines. Pro-Poor Intervention Strategies in Irrigated Agriculture in Asia*. Colombo: International Water Management Institute.

Mollinga, P. (2003). *On the Waterfront*. Wageningen University Water Resources Series. Hyderabad: Orient Longman.

⬤5 Water supply

Introduction

In Chapter 1 we demonstrated that, at global and regional levels, water
withdrawals do not exceed renewable resources, with often less than
10 per cent extracted. However, at a more local level, there is much more
variability, such that in places consumption exceeds local supply. Such
imbalances between supply and demand can be met through one of two
possible responses: augment the (sources of) supply, or seek to reduce
demand. In this chapter, we will focus on the first of these alternatives
(Chapter 6 will consider strategies to reduce demand). Figure 5.1
(similar to Figure 2.1 but here we display the relative proportions of
water withdrawn rather than absolute amounts) shows that, globally,
agricultural demand dominates water withdrawals. This is particularly
pronounced in water poor regions, whereas in industrialised areas of
Europe and North America municipal and industrial demand for water
is much higher. As we showed in Chapter 1 (second section Water
Scarcity: a Water Crisis, p.7), manufacturing industry is generally a more
efficient user of water than is agriculture. The different composition of
demand means that, although industrialised countries have high levels
of water consumption, heavy use of irrigation water means that in
'less-developed' Asia and the Middle East overall consumption figures
are comparably high. Furthermore, in industrialised regions rates of
withdrawal from renewable water resources are reduced through
recycling and re-use of domestic and industrial effluents and use of
technology to augment natural water supplies. Where recycling is

concerned, we need to note that wastewater becomes effectively a new source of local 'supply', while also providing a means to reduce rates of abstraction from the wider water resources. To some extent, therefore, recycling may make it possible to sustain relatively high rates of individual water consumption while reducing environmental impacts.

In industrialised countries, then, demand for increased water supplies tends to be characterised in terms of increased levels of recycling, innovative technologies, large-scale integrated projects and high capital investment. These characteristics are not unique to industrialised economies but this chapter focuses on a technology-driven capital-intensive approach which emphasises increasing water supply, rather than reducing individuals' water consumption, and the environmental problems associated with this type of strategy. As we saw in Chapter 2, such strategies are often characterised by an escalation in scale of authority (e.g. from neighbourhood to municipality to national government) and of geographical unit (e.g. from river or well to catchment or aquifer), leading to integrated management and a hierarchical 'top–down' approach. They can also be identified with an emphasis upon environmental control through 'hard' engineering. These projects are, then, frequently undertaken with government backing and multi-criteria objectives of which

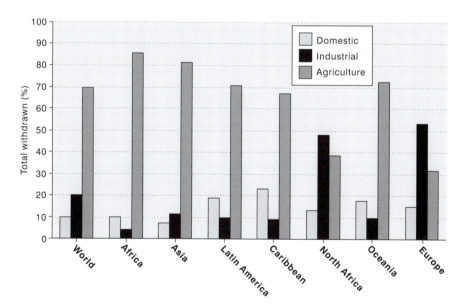

Figure 5.1 Freshwater withdrawals (2001) for agricultural, industrial and domestic purposes
Source: after FAO-AQUASTAT, 2010

water-supply enhancement is but one. The approach was exemplified in Chapter 2 in links between water and economic development via cheaper energy (hydroelectric power generation (HEP)) and flood control. These were the basis of the Tennessee Valley Authority scheme in the 1930s, and also of the more recent Three Gorges scheme in China (Sutton, 2004), while the Grand Anatolia Project combines irrigation and HEP (Agnew and Anderson, 1992).

Figure 5.2 presents a framework for the comparison of different water-management strategies. It contrasts the engineering and technology-rich approaches to those which are based on modifying human behaviour, e.g. through economic and legislative means. Figure 5.2 is directly comparable to Figure 1.14 used to illustrate the organisation of this book. We use the term (water) engineering to include those strategies which emphasise water management through construction of a physical artefact, while technology refers to the means by which it is constructed and operated (i.e. the water infrastructure). The term 'engineering' embraces a wider sense, such as 'social engineering' – the social organisation of people necessary to make the infrastructure work (see Chapter 8). This chapter, then, deals with the upper two quadrants of Figure 5.2, while Chapter 6 will focus on the lower two quadrants. This framework is not a comprehensive description, but serves to illustrate a range of strategies to enhance water supply. It should be clear that forcing all

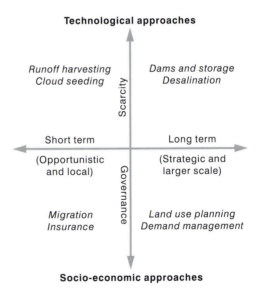

Figure 5.2 Water management strategies grouped by temporal scale and degree of technology

possible strategies onto only two axes risks some oversimplifications. For example, not all demand management is strategic (long term) but may also be based upon short-term measures such as drought orders to combat immediate and local water shortages. Figure 5.2 should also not be seen simply as a list of alternative techniques, as it is likely that a range of solutions will be utilised by any nation facing water shortages. Rather, our purpose here is to explore four key questions that confront the application of a technological approach to responding to water scarcity by increasing water supply:

- Will the benefits be equitable?
- How much will it cost?
- Will it work?
- Is it sustainable?

These questions can be applied to four case studies used in this chapter:

- Can mega projects be made to work for the benefit of all (especially those who are most vulnerable)? The case of large dams
- Do the benefits outweigh the costs ? The case of desalination
- What is the scientific proof? The case of cloud seeding
- Does it stand the test of time without degradation? Rainfall harvesting

Finally we give special consideration to the specific problem of supplying water to rural inhabitants and the urban poor, who, as Chapter 1 demonstrated, often face the worst conditions in terms of safe supply and sanitation.

Long-term and large-scale supply enhancement

Abstracting water directly from water courses and raising it to artificial canals for distribution has long been practised through mechanisms as the shaduf, water wheel and even the Archimedes screw. These can use human, animal or even wind power capable of raising substantial amounts of water when used in series, as demonstrated by the draining of many European wetlands even before the introduction of steam pumps. Abstraction, storage and distribution of water from surface lakes and channels has been practised since King Menes of Egypt was perhaps the first to build a significant barrier across the Nile some 5000 years ago (Overman, 1976). The development of water and agriculture in the Euphrates–Tigris basins also goes back as far as this, and possibly even earlier (Altinbilek, 2004). Walski (2006) traces the development of water distribution infrastructure to at least two and possibly four thousand years ago (see urban supplies below), yet there is growing awareness of the pressing need to improve water supply systems in developing countries (Page, 2005). Parry et al. (2009) calculate that two-thirds of the costs for adaptation to climate change in DCs fall under the heading of 'infrastructure improvements'. The water engineer can now select from a vast array of different types of pumps, dams and distribution systems (Twort et al., 2002). We have seen in earlier chapters that technological development, notably the development of centrifugal pumps, powered initially by diesel engines and later by electricity, had a major impact on the use of groundwater for irrigation in the twentieth century, first in California (Chapter 2) and later in northern India (Chapter 4). This was of major significance not only because it enabled individual appropriation of water (because large-scale infrastructure was not required to store and distribute water), but also because it opened up much larger sources of water for economic use.

Indeed, it can be argued that there is too much emphasis upon surface water management through dams and inter-basin transfers, when far greater amounts can be found below the Earth's surface. However, Cook (1998: 110) notes: 'Modern groundwater development can seldom be separated from consideration of local surface flows.' That is, the interaction between groundwater levels, groundwater recharge, surface infiltration and surface flows needs to be viewed as a hydrological cycle and not as separate surface and subsurface systems. Estimation of renewable groundwater resources is problematic, however. It can take months, often years, for surface precipitation to seep through overlying soil and rocks to recharge groundwater. Some subsurface storage is deemed 'fossil', as it lies in aquifers that have not been recharged for hundreds, if not thousands, of years, and is effectively a non-renewable resource. Such aquifers are difficult and expensive to exploit, but development of technology for deeper boreholes, and the rising cost of water, have made it profitable to develop fossil groundwater resources. Many parts of the Middle East rely on groundwater for their water resources, often exploiting non-renewable sources. Figure 5.3 shows the ten countries with highest amounts of extracted groundwater (WRI, 2007), most using less than 10 per cent of annual renewable supplies with the notable exception of Saudi Arabia which relies heavily upon water stored thousands of years ago. One of the controversial examples of 'mining groundwater' because of cost and arguments over the renewable nature of the groundwaters is the Libyan transfer from interior boreholes to coastal irrigation sites in the Great Man Made River project (Allan, 2001).

Figure 5.4 presents groundwater withdrawals as a function of per capita extraction and per capita renewable resources (WRI, 2007).

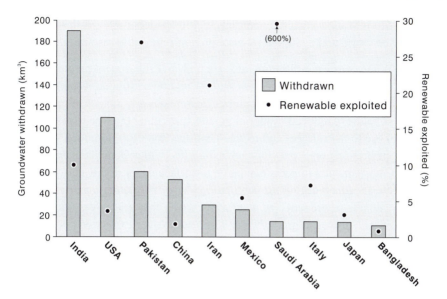

Figure 5.3 Ten countries with highest annual groundwater withdrawal rates, which are compared as percentage of total of renewable freshwater resources

Source: after WRI, 2007

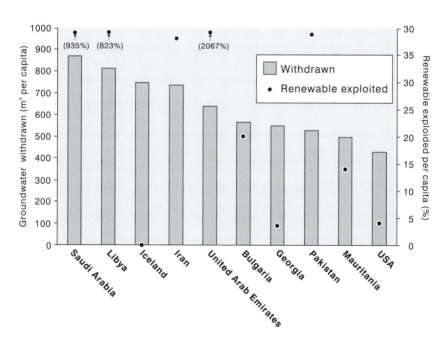

Figure 5.4 Ten countries with highest annual per capita groundwater withdrawal rates, which are compared as percentage of renewable freshwater supplies per capita

Source: after WRI, 2007

More Middle Eastern countries are now evident amongst the top ten in the world and again the reliance on non-renewable resources is evident.

Where surface infiltration rates are high and subsequent percolation rates are rapid, often the case with shallow aquifers, artificial recharge can augment groundwater resources. The obvious advantages of groundwater storage are reduced evaporation rates (providing the water table is not too close to the surface) and reduced construction costs, although wells and infiltration ponds may have to be installed. Other benefits include: reduced environmental impacts compared to surface dams, and water-quality enhancement during the percolation phase. This latter is particularly valuable where recycled water is used to recharge aquifers. More commonly, however, groundwater is subject to over-extraction (as in northern Indian irrigation discussed in Chapter 4). In the worst cases this can lead to land subsidence, deterioration of water quality and, in coastal areas, intrusion of saline sea water. It is not surprising, then, that one of the oldest approaches to water-supply enhancement, the impounding of surface flows has regained significance since the turn of the century.

Dams and mega dams: issues of scale

In September 2006 *New Scientist* (Pearce, 2006) reported that 'Mega-dams were back on the agenda' with the World Bank identified as supporting this strategy after a period when environmental damage had led many to question whether the benefits out weighed the costs. Since the eighteenth century demand for water has grown 35-fold whereas global population has increased eight-fold (Altinbilek, 2002). Dams are constructed to enhance industrial and domestic water supplies but they are also constructed for a variety of additional purposes including the regulation and augmentation of surface flows, to improve river navigation, to prevent floods, to provide water for irrigation, to conserve water resources, to preserve ecological systems, aesthetic reasons possibly promoting tourism, for power production and as a political statement of progress. Takahasi (2004), for example, writes that Japan's 2700 large dams (greater than 15m) were constructed from the 1950s in the hope of increasing food production, HEP and flood control. We have seen in Chapters 2 and 4 that the construction of new dams, and in particular huge 'mega-dams', were seen as keystones for industrial development, intensification of agriculture and to support rapidly expanding cities. In the case of the USA (see Chapter 2) the dam-building boom from the 1930s to the 1970s was a political response by federal government to agricultural and urban lobbies for irrigation water, flood control and hydroelectric power generation (Reisner, 1993; Billington and Jackson, 2006). Bandyopadhyay (2002) presents the contemporary case for using HEP and additional water resources as the mechanism to promote economic development in the Ganges-Brahmaputra-Meghna basin. Altinbilek (2002) notes that 12 per cent of large dams are designed to enhance domestic water supplies, whereas 50 per cent were constructed for irrigation producing 12 to 16 per cent of world food production. In addition, 19 per cent of the world's electricity is generated through hydro power. Figure 5.5 illustrates that dams in Asia and Africa were mainly built for irrigation, that in Austral-Asia water supply is the primary motivation while in Europe and North America dams tend to be multi-purpose.

The construction of dams is not without a price, including environmental and social costs, exemplified in Chapter 4 by the Manantali Dam on the Senegal River. Nevertheless, enthusiasm for damming rivers has continued throughout the twentieth century such that Cantor (1985) predicted that between the 1970s and the start of the twenty-first century river discharges regulated by reservoirs world-wide will increase from one-tenth, to two-thirds. In Chapter 2, we noted the increasing HEP-generating

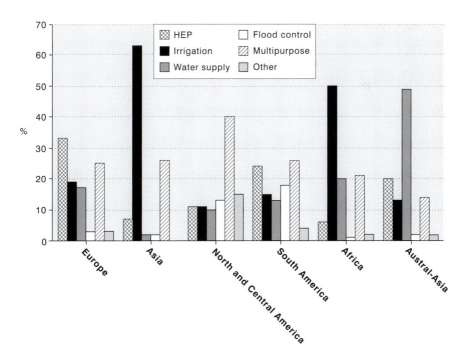

Figure 5.5 The various purposes for dam construction as a percentage of total number
Source: after WCD, 2000

capacity of dams constructed during the twentieth century. A similar comparison applies to scale of construction. The first Aswan dam was built over a hundred years ago at 27 m high. It was regarded as a 'wonder', but is dwarfed by many present-day dams. The WCD (2000) reported that the average height of large dams was 31 m with an average reservoir area of 23 km^2. However, by the start of the 1980s, two-thirds of India's water storage came from just 26 major dams including the Bhakra Dam (226 m high) and the Tehri Dam (260 m high). Goldsmith and Hildyard (1984) note that the Volta dam in Ghana is large enough to impound a reservoir the size of Lebanon. The Three Gorges dam in China, rising 185 m above sea level, created a reservoir which may extend over 1000 km^2, whilst the Ataturk dam in Turkey, among many examples of similar size in the Euphrates–Tigris basin, rises to a height of 166 m impounding an area of 817 km^2 (Altinbilek, 2004). Figure 5.6 shows that the average heights of dams are similar across the world but the volumes impounded are much greater in the Americas and Africa while HEP is generally lowest in Africa and Austral-Asia.

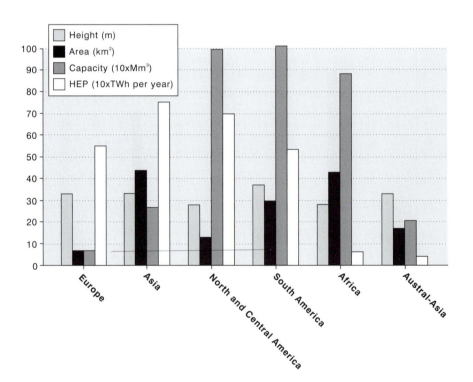

Figure 5.6 Large dam characteristics by region

Source: after WCD, 2000

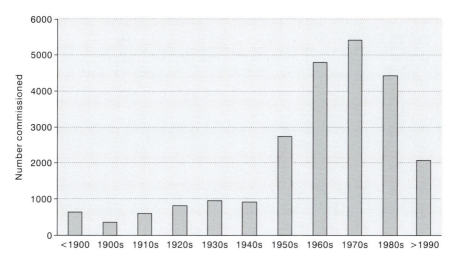

Figure 5.7 Numbers of dams commissioned each decade during twentieth century
Source: WCD, 2000

Figure 5.7 shows the growing commitment for building large dams for much of the latter half of the twentieth century up until the last two decades. Altinbilek (2002) reports that industrialized countries dominated dam building up to the middle of the twentieth century. By the year 2000 three countries dominated large dam construction, China, USA and India (Figure 5.8), but there were 45,000 large dams in over 150 countries with 625 in Turkey alone.

Dam construction in the Middle East began in earnest in the 1950s in the wetter highlands of Algeria, Tunisia, Turkey, Iran and Iraq, peaking in the early 1970s, (Beaumont, 1989). The Grand Anatolian Project, in south-eastern Turkey, is a major irrigation scheme to provide 50 per cent of the irrigated area and 50 per cent of the power generation of the country (Agnew and Anderson, 1992, Altinbilek, 2004) with 22 dams and 19 hydroelectric plants. A major element is the Ataturk Dam, the lake behind which was filled in January 1990 and reduced the flow of the Euphrates to a trickle for one month, raising fears that Syria's share of the Euphrates would be cut by up to 40 per cent and Iraq's by 80 per cent. Although many ideal sites have already been utilised, there is still much potential for large dam construction. Altinbilek (2002) states that Africa has exploited around one percent of its potential for HEP, so that the entire African continent produces only one sixth of the HEP generated by the USA.

Large dams bring multiple benefits and there remains much enthusiasm for this strategy of enhancing water supplies and contributing to economic

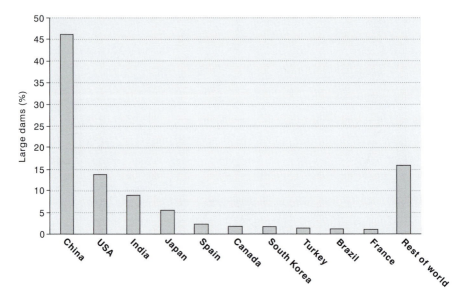

Figure 5.8 Locations of large dams by country

Source: WCD, 2000

development (Altinbilek, 2002; Schultz, 2002). The question of scale is important as large dams serve as icons for development and progress. As we saw in Chapter 2, the large 'multi-purpose' dam allows significant impact on a regional scale through the enhanced supply of energy and water, with the sale of the former allowing the latter to be supplied at subsidised rates.

Other large-scale engineering approaches to increase water supply include inter-basin transfers. Some areas benefit from 'natural' inter-regional water transfers (IWT), for example the Nile for Africa, the Rio Grande for North America and Euphrates for Asia (see Chapter 7 on the geopolitical issues). There have been many proposals to emulate nature by the artificial construction of canals and the rerouting rivers. Large scale dams create water storage that enables its transport to areas of high demand. Water transfers by canal can be traced back to at least 4000 years ago in Egypt (Lake Qarun) while Romans built aqueducts in 200–300 BC. up to 90 km in length delivering more than 1000 lcd to Rome (Overman, 1976; Walski, 2006). In Africa, plans have been drawn to direct the Congo river into lake Chad (Balek, 1983), we have already referred to the transfer of water from the Libyan desert to coastal communities while in Australia the Snowy Mountain scheme transfers water from the Snowy River across and through the Great Dividing Range to augment flows of the Murray River in the drier interior.

More controversially, the Lesotho Highlands Water Project was implemented in the last years of the apartheid government to store and divert water from the Lesotho watersheds northwards to the Johannesburg conurbation. The chronic shortage of water in California (Chapter 2) has also generated ever larger-scale proposals for inter-basin transfers, such as the North American Water and Power Alliance (NAWAPA) plan to divert water southwards from Canada (Reisner, 1993). In scale, the proposal is only slightly less fantastic than a pilot scheme to bring 100 Mm^3 of iceberg to the Californian coast in the 1980s, estimated to cost US\$20 million, or US \$0.2 m^{-3} (Hult, 1982), which compares favourably with other water-supply options if the ice can be melted and the water harvested. In Africa, however, the unfinished Jonglei Canal, designed to divert upper Nile flows, underlines the vulnerability of such large schemes to political instability (Pemberton, 1987; see also Chapter 7). Opposition to the ecological impacts of major water transfers has become more vociferous in the last two decades and the World Resources Institute (1989) reported the abandonment of proposals to combat the falling levels of the Aral sea. Biswas (1979: 87) quotes Howe and Easter (1971) saying, 'Large scale transfers of water are likely to cost more than they are worth to a nation except in certain resource operation cases.'

Apart from environmental consequences, involving seepage losses, possible climatic changes and alterations of water quality, there are immense political and legal obstacles and the World Resources Institute (1989: 133) stated, 'In the United States the era of building huge federal dams and long aqueducts and canals appears to have ended.' Both *Nature* (2006) and *New Scientist* (2008) have published editorials pointing out the hazards associated with dams and their potential failure. The issues, reviewed by Altinbilek (2002), are summarised in Box 5.1 (environmental costs) and Box 5.2 (social costs).

Criticism of dams was in many ways brought into focus by the World Commission on Dams (WCD, 2000) that identified many dam projects as failing to reach their goals of HEP production and irrigation water supply. Two decades previously the Committee on Safety Criteria for Dams (1985) stated that dams generally do not protect property and people downstream, and flood control is more challenging than claimed. Earlier still, FAO (1975) published its concerns under the title 'Water: available but mismanaged' and, in the same publication, Vohra (1975) voiced opposition to large-scale water projects with the headline, 'No more Gigantism'. Carruthers and Clark (1983: 2) were equally emphatic: 'The era of the big dam is, with one or two notable exceptions now over.'

Box 5.1

The environmental costs of dam construction

Several studies have reviewed the ecological and hydrological impacts of dams (Church, 1995; Kondolf, 1997; Ligon et al. 1995). While most concern attaches to impacts downstream of dams, changes in the upstream catchment can cause greater inputs of sediment into the dam reservoir. Coupled with lower releases of suspended sediment downstream, this leads to siltation which greatly reduces lifetime operation, and expected benefits, of the dam. Pre-construction predictions of the amounts of material transported by the fluvial system often underestimate those experienced post-construction, and more reliable prediction of siltation rates are needed for accurate cost–benefit analysis. For example, the Tahri dam in India is expected to silt up in 40 years, rather than the 100 years predicted. Elsewhere, observed rates of sedimentation are four times those predicted pre-construction (Dogra, 1986).

Changes in downstream flow regime cause both agricultural and ecological impacts. Braatne et al. (2008), in their study of the Snake River, USA, noted adverse impacts as arising from altered timings of water flow, changes to water quality (especially suspended sediments) and fragmentation of the river system. A striking example of downstream changes is provided by the dams of the Grand Anatolian Project (GAP). Turkey pledged the release of 500 m^3s^{-1} Euphrates water throughout the year at the Syrian border, rather than the seasonal variations of the past. Altinbilek (2004) states this minimum guaranteed flow has been maintained, and often exceeded. Peak flows have been greatly reduced, however. After construction of the GAP Ramadi dam, peak flows downstream in Iraq decreased from 2594 m^3s^{-1} to 831 m^3s^{-1} but minimum flows increased from 272 to 575 m^3s^{-1}. 'Normalising' flow in this way may seem a major management advantage. However, as in the case of the Senegal river (see Chapter 4), it will undoubtedly cause problems downstream where flood peaks have been used to irrigate through gravitational means rather than using pumps. Conversely, higher minimum flows provide water for irrigation at other times of the year. By 2004 there were 32 dams on the Euphrates and Tigris Rivers, with the possibility of 8 to 21 more even though the current reservoir capacity is five times larger than the river's annual flow (Altinbilek, 2004).

Dams create a number of other changes. *Water quality* can deteriorate due to thermal stratification in deep reservoirs, through pollution by artificial fertilisers as irrigation expands, and through industrial and urban development promoted by additional availability of water and HEP. Bodies of still water are habitats for vectors of *harmful diseases*, and water-resource projects are often accompanied by increases in schistosomiasis, bilharzia and malaria (Agnew and Anderson, 1992). Furthermore, *water losses* may increase through greater evaporation rates as advection is higher around lakes and reservoirs. For example, of the average annual flow of 84 Bm³ in the Nile at Aswan, 17 Bm³ reaches the sea and 13 Bm³ is lost through evaporation from the major reservoirs in Egypt and the Sudan, (Chesworth, 1990). Dam reservoirs also cause *loss of land* through impoundment of waters, including fertile valley floors and wooded hillslopes, and loss of 'land-forming' sediments downstream. Sediment loss downstream of the Aswan dam, for example, has caused greater use of artificial fertilisers and the retreat of the delta at the mouth of the Nile. In terms of *ecosystem changes*, WWDR (2006: 176) states: 'Dams play a major role in fragmenting and modifying aquatic habitats, transforming lotic (flowing) ecosystems into lentic (still) and semi-lentic ecosystems.' While larger wildlife will relocate from reservoir sites, dams will affect interactions and dependency within the immediate ecosystem, and in adjacent areas. Inland fisheries, on the other hand, can benefit from dam construction through the release of nutrients from rotting of submerged vegetation. Fish stocks in both Lake Volta and Lake Kariba soared after impoundment but then reduced as this supply of nutrients inevitably fell.

Failure of a dam is also an environmental concern, and large dams pose a serious threat of damage and loss of life. May and Williams (1986) note that between 1970 and 1980 in the USA dam failures caused 500 deaths and $2 billion of damage. The Committee on Safety for Dams (1985) noted that there are 10 significant dam failures each decade. Reasons include: faulty design (including inappropriate sites – see Reisner's (1993) account of the Teton dam), inadequate maintenance or mismanagement, and unanticipated floods or earthquakes, as the weight of impounded water can result in earthquakes. Goldsmith and Hildyard (1984) note increased seismic activity has been associated with the Hoover dam (USA), Kariba dam (Zimbabwe) and Konya dam (Israel).

Box 5.2

The social costs of large dams

Resettlement of peoples displaced by large dams and the ensuing social disruption has long been observed (Scudder, 1973), although Nakayama and Fujikura (2006) suggest that more detailed research is needed. In Africa, for example, during the 1970s, some 350,000 people were displaced, while the Three Gorges Dam in China has displaced a staggering 1.4 million people. Social disruption arising from relocation of population can threaten the success of water development projects (Adams, 1988). Relocation concerned not only sites of reservoirs, but also forced migration of people to resettle on irrigation schemes (see Office du Niger in Chapter 4; also DeWilde, 1967).

The disruption of village life and breakdown of social networks resulted in large numbers of settlers abandoning the scheme. It is apparent that when such schemes were designed in the past, the views of local populations were largely disregarded, with little local participation and inadequate compensation. Takahasi (2004) observes the rise in opposition to dams in Japan, from the late 1950s, led to three primary measures of state response: (1) compensation, (2) development of infrastructure, (3) resettlement. It was noted that these did not fully address the social impacts because financial compensation did not deal with impacts of resettlement extending beyond the physical, including a sense of victimisation among those resettled. Controversy has surrounded the Brazilian scheme to build dams, in particular the Altamira (formerly Babaquara) dam on the Xingu River for HEP and flood control. Fearnside (2006) identifies lack of communication and weak engagement of stakeholders as causes of increasing tensions. On the pro-dam side are engineers, the Ministry of Mines and Energy, and the industries that will benefit from HEP such as aluminium producers. On the anti-dam side are indigenous peoples who will be displaced, represented by myriad NGOs and international movements. It is worth noting that dam construction generated social disruption for much of the twentieth century, as exemplified by Ortolano and Cushing's (2002) study of one of the first mega dams, the Grand Coulee, in the USA, the adverse impact of which upon the indigenous population continues some 70 years later.

Goldsmith and Hildyard (1984) concluded their study on the social impacts of large scale dams with a call for organisations to cut off funds from all large-scale water-development schemes regardless of how advanced those schemes might be. This view was not unchallenged, however. The International Rivers Network (IRN, 2003) paper 'Dammed rivers, dammed lies', for the Third Water Forum in Kyoto met with a mixed reception (Biswas, 2004). The International Commission on Irrigation and Drainage (Schultz, 2002: 160) stated: 'dams have played and will continue to play an important role in the development of water resources, especially in developing countries. A balance needs to be found between the requirements based on the needs of society, acceptable side effects and a sustainable environment'. They go on to argue that WCD (2000) failed to identify appropriate and sustainable solutions to achieve this balance. Others have criticised variously the WCD's technical competence, the evidence and range of studies it used, and the commission's membership and mandate, leading to claims that the report was biased against dams (Bandyopadhyay, 2002; Bird, 2002; Biswas, 2004; Nakayama and Fujikura, 2006).

Despite such polarised views, there is recognition that, in practice, investment moved away from large-scale water projects (Figure 5.7).

In 1989, the World Bank shelved a US$500 million grant constructing dams in the Amazon basin, and instead initiated talks on grants for environmental projects (Cummings, 1990). Support from development banks for dam construction fell from $4.4 billion a year in the 1980s to $2.6 billion a year by the late 1990s (Bird, 2002), and in Africa public investment in irrigation came to a halt for a decade from the mid-1980s, so that loans for irrigation and drainage were lower in 2002–5 than they had been in 1978–81 (CAWMA, 2007: 73). In the twenty-first century, however, it seems likely that dam construction will gain new impetus. The World Resources Institute (1990) argued that the potential for water storage and HEP, particularly in Africa, will attract investment for dam construction, while the recent construction of the Three Gorges Dam in China illustrates that many remain convinced that the benefits outweigh the costs, and Pearce (2006) noted a recent increase in World Bank support for large dam projects.

Desalination: the engineers promise but at a price

We have already seen that, through dam construction and deep bore-holes, engineering solutions have enhanced water supplies over the last century, despite environmental concerns. Postel (1992) wrote about the pitfalls of relying solely upon – often costly – engineering solutions, but noted that desalination was a feasible technology although, at the time she was writing, it contributed a small fraction of the world's water supply. As 97 per cent of the world's water is saline, the ability to create fresh (potable) water from non-useable saline or brackish water is an attractive option, particularly where water resources are in high demand or where water quality is a key consideration (e.g. industrial processes such as brewing or pharmaceuticals). The problem with desalination is that it can be expensive compared to other water-enhancement strategies.

Desalination is the separation of pure (potable) water from a saline solution. It should not be confused with desalinisation, the recovery of saline soils. Small-scale operations have been practised since historic times (Khan, 1986; Popkin, 1969), but the first land-based commercial plants of 5000 m^3d^{-1} (approx. 1 million gallons per day) were not constructed until the 1960s. This is therefore still a relatively novel technology. By 1985 the world's desalting capacity had reached at least 6.8 Mm^3d^{-1} (World Resources Institute, 1987), and arguably as high as 10 Mm^3d^{-1} (Hornburg (1987). By 2005, however (see Table 5.1 and Figure 5.9), this had increased to 24.5 Mm^3d^{-1} of which 58 per cent

Table 5.1 *The development of world desalination capacity (capacities in Mm³d⁻¹)*

	1968[1]	1977[2]	1985[3]	2005[4]
Middle East	0.33 (28%)	1.82 (49%)	3.795 (69%)	18.87 (77%)
Americas	0.24 (21%)	0.65 (17%)	1.13 (12%)	1.72 (7%)
Rest of World	0.56 (51%)	1.24 (33%)	1.89 (19%)	3.91 (16%)
Total	1.11	3.71	6.81	24.50

Sources: 1 from der Leeden (1975: 479), only listing plants producing more than 100 m³day; 2 from Heathcote (1983: 71); 3 from World Resources 1987 where 'Middle East' also includes North African countries; 4 from Lattemann and Hopner, 2008

Note: FAO-AQUASTAT database (2009) estimates the total Middle East consumption of desalination water to be 8.82 mm³d⁻¹ (i.e. 3,225 km³year⁻¹) led by Saudi Arabia, the United Arab Emirates and Kuwait.

is from sea water, 22 per cent from brackish water and 5 per cent wastewater (Lattemann and Hopner, 2008).

Desalination has increased rapidly since the 1960s and two-thirds is now located in the Middle East, where desalination makes an important and growing contribution to national water supplies, as shown for Israel by Table 5.2. Following periods of drought and rising demands Dreizin et al. (2008) report that Israel had exhausted all the lower-cost alternatives. It embarked upon a strategy of desalination growth such that by 2020 30 per cent of potable water is expected to be provided by this means.

Desalinated water is produced mainly by processes of distillation (thermal), or the use of membranes (reverse osmosis), although other methods (e.g. freezing and solar humidification) have also been developed

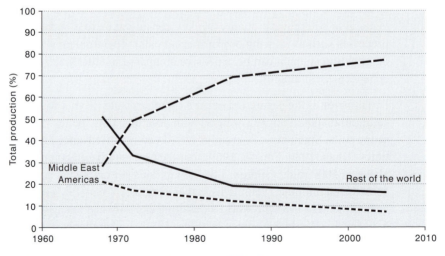

Figure 5.9 World desalination capacity based on Table 5.1

Table 5.2 *Israel water supply, consumption and desalination (Mm³y⁻¹)*

	2005	2010	2015	2020
Agriculture	990	1170	1270	1370
Industry	125	135	148	155
Domestic	20	840	960	1080
Total demand	2060	2405	2658	2805
Total supply	2090	2425	2710	2910
Renewable	1470	1470	1470	1470
Desalination brackish	30	50	80	80
Desalination sea water	50	315	500	650
Brackish use	160	140	140	140
Wastewater	300	450	520	600

Source: after Dreizen et al., 2008: 139

(Tleimat, 1980). Delyannis and Delyannis (1980) provide a comprehensive listing of the technical developments and patents for desalination, while Agnew and Anderson (1992) present the different technologies which are summarised in Box 5.3.

Table 5.1 shows that during the 1970s while the number of desalination plants doubled from 700 to over 1500, capacity trebled from 1 Mm³d⁻¹ per day to 3 Mm³d⁻¹. In fact, initial rapid growth of desalination – 23 per cent between 1961 and 1968 (Office of Saline Water, 1970) – had projected a much higher worldwide capacity of 9 Mm³d⁻¹ by 1977. The reason for the much lower growth trend, with global capacity still only 6.8 Mm³d⁻¹ by 1985 (World Resources Institute, 1987), was the large rise of oil prices in the early 1970s. As a consequence, by 1985, 80 per cent of capacity was to be found in the Middle East, North Africa and USA regions characterised by availability of cheap fossil fuels and/or a sufficiently high demand (Table 5.3). Due to the comparatively high cost of

Box 5.3

Desalination processes

Thermal (distillation) processes evaporate water from saline solutions and then condense the water vapour. The first land-based plant was constructed 100 years ago (Khan, 1986), but most growth has taken place since the 1950s (Hornburg, 1987). Distillation remains one of the main types of desalination as it can operate with water of high salt content and can be built to a large scale. Through lowering pressure, heated brine will boil at a lower than normal temperature. This 'flash distillation' process enables higher energy efficiencies, especially when several low-pressure chambers are linked to produce 'multi-stage-flash' (MSF) distillation (El-Sayed and Silver 1980). First patented in 1957, MSF distillation went into commercial production in 1960 in a plant with 4555 m^3d^{-1} capacity (1mgd) (Silver, 1978). By 1970 there were 59 desalting plants of at least this size, 57 employing distillation techniques, (Leeden, 1975). Today, MSF supplies all desalinated water in Kuwait (Darwish et al., 2008), but has higher capital and operating costs than reverse osmosis.

The use of membranes to separate water from solutions was reported by ancient Chinese writers and mentioned in Hebrew scrolls (Dickey, 1961), but commercial development based upon a technique known as reverse osmosis required high flow rates at high pressures and temperatures which were developed in the USA through the 1950s and 1960s (Channabasappa, 1982; Dresner and Johnson; 1980 Hoornaert, 1984). Saline solutions exert an osmotic pressure of around 2.5 million Pa (sea water). Reverse osmosis separates pure water from the solutes (salts) by passing the saline solution, under pressure, through a semi-permeable membrane, which retains low molecular weight salts but allows water to pass through. Alternatively an electrical current is applied to ion-selective permeable membranes, such that dilute and concentrated solutions are produced on opposite sides (electrodialysis). In a recent review of the latter (capacitive deionization) by Oren (2008) it is noted that, while the approach has comparable energy consumption to reverse osmosis, it is not suitable for seawater and suffers from scaling and fouling drawbacks. Lower-pressure requirements of modern techniques of reverse osmosis result in lower-energy inputs such that comparative energy costs are much lower than for flash distillation or vapour compression. CADT (2008) caution, however, that it is very difficult to obtain directly comparable estimates of costs (and benefits) for desalination, citing three examples where costs range from US $0.43 to 0.83 m^{-3} of water using inputs with salinities of less than 15,000 ppm and greater than 25,000 ppm.

In 1979 global reverse osmosis capacity was 1.5 Mm^3d^{-1}, with individual plant capacities of up to 200,000 m^3d^{-1} (Dresner and Johnson, 1980; Wade, 1985). Reverse osmosis requires less energy than distillation and it is possible to employ alternative (solar and wind) energy sources (World Resources Institute, 1989). It is scale flexible and capital costs are lower. Kamal (2008) argues that despite the circulation of 'myths' about reverse osmosis, notably in relation to lack of flexibility and energy costs, and persistent enthusiasm for thermal desalination plants in the Middle East, reverse osmosis has become the more competitive technology because reductions in membrane costs have cut the cost of desalinated water to half that from thermal plants. The government of Israel has approved new reverse osmosis plants to produce 305 Mm^3y^{-1} by 2010, or 17 per cent of potable supply. The largest plant at Ashkelon was already producing 100 Mm^3y^{-1} in 2005 (Safrai and Zask, 2008), while a second plant with 30 Mm^3y^{-1} commenced operation in 2007. By 2020 production is projected to increase to 650 Mm^3y^{-1}, supplying 30 per cent of potable water (Dreizin et al., 2008).

Table 5.3 *Water resources for selected ME nations*

	Water (km³yr⁻¹)	Water (km³yr⁻¹ cap⁻¹)	Groundwater (%)	Total use (%)
Kuwait	0.02	8	0	2227
Saudi Arabia	2.4	96	92	722
Qatar	0.1	86	94	554
UAE	0.2	49	80	1538

Source: after Darwish et al., 2008: 343 (groundwater % is contribution to total available water and % total use is based on renewable water resources)

desalinated water, desalination plants are intended for domestic (municipal) and industrial demands, although there are examples of desalinated water supplying high-value horticultural production. Even in Kuwait, which uses desalination for most of its water supply, 'the sustainability of energy use and desalted water in Kuwait is questionable' (Darwish et al., 2008: 342).

The high energy cost of desalination (Box 5.3) makes it an expensive option for increasing water supply, and Postel states (1992: 46), 'desalination remains a solution of last resort'. However, the cost has fallen. A cubic metre of pure water from desalination cost between US$0.88 and US$0.28 in the early 1960s, and still averaged US$0.22 in the 1970s, with a lower limit of US$0.15 for a plant at Tijuana, Mexico (Pereira, 1973: 16). More recently, development of reverse osmosis technology (Box 5.3) has dramatically reduced the energy costs of desalination. In membrane plants energy is typically 19 per cent of the total costs with 45 per cent on capital and 19 per cent on staff, compared to figures of 59 per cent, 32 per cent and 6 per cent for MSF – distillation (Elhassadi, 2008).

In addition to energy costs, desalination creates an environmental problem in the form of pollution created by the production of salt waste. Amounts of 300 to 500 tonnes of solid waste are not uncommon from distillation plants (Channabasappa, 1982). Safrai and Zask (2008) list possible additional environmental impacts as the release of anti-scalants, ferric contamination and higher levels of nitrogen leading to eutrophication. Darwish et al. (2008) also calculate significant thermal and atmospheric pollution from plant discharges. Desalination is also strategically a vulnerable supply for water. The Gulf War (1991) demonstrated the vulnerability of coastal desalination plants to water pollution, especially oil spillages from military actions or collisions, (Elshorbagy and Elhakeem, 2008). Therefore, security measures may have to be included in costs.

Despite these constraints, desalination remains an important supply of water in some parts of the arid world and recent reports are optimistic over future growth, especially using reverse osmosis (Burbano et al., 2007). Elhassadi (2008) quotes costs for membrane desalination in Lybia of 0.7$ m^{-3}, against the 0.5$ m^{-3} cost of fresh water transferred to the coast by Libya's Great Man-Made River (GMMR) project (Allan, 2001). Qdais (2008) reviews an ambitious project to link the Gulf of Aqaba to the Dead Sea, employing the difference in elevation to generate HEP to run desalination with an expected water yield of 850 Mm^{3}y^{-1}. The study concluded the benefits would outweigh the likely adverse environmental impacts if appropriate mitigation measures are adopted.

Desalination may prove a cost-effective way of purifying wastewater, and municipal authorities worldwide are increasingly considering this technology to solve their water supply problems while also enhancing wastewater treatment. This may meet cultural objections, however. Wilson and Pfaff (2008: 3) noted that in 1978 the Council of Leading Islamic Scholars in Saudi Arabia concluded that treated wastewater can be used for drinking providing there is no health risk. However, in Durban, South Africa, while there were no religious objections to the recycling of wastewater for potable supplies, they found there was an 'emotional' negative response. GWI (2005) list the 100 largest desalination plants planned, in construction, or in operation. These account for 40 per cent of global capacity, and, as expected, the list is dominated by the Middle East, North Africa and USA who between them have 89 per cent of plants. However, other countries listed include Spain, Singapore and Australia. Spain, for example, opened in 2009 a 200,000 m^{3}d^{-1} desalination plant to supply 24 per cent of Barcelona's water (BBC News 21 July 2009). A similar-sized plant is under construction for Sydney, and a plant is also under construction in the UK for London. This suggests desalination is becoming a viable option for enhancing water supply in urban areas beyond those of its historic application in hyper-arid oil-rich nations.

Short-term (opportunistic) responses to water scarcity

Thus far we have considered approaches to increasing water supply that require major capital investment and are therefore necessarily long term. This section considers two examples of smaller-scale investments to increase water supply. The sub-section below on rainwater harvesting illustrates that not all water-supply enhancement projects are undertaken at the 'grand' scale. National water policies increasingly recognise that local community engagement is effective in managing water, perhaps best demonstrated by *qanats* that have sustained dryland communities of the Middle East and North Africa over hundreds, if not thousands, of years. The role of local management in cloud seeding is less clear, as clouds are by their very nature ephemeral and dynamic. Most cloud-seeding programmes have been conducted at some time by governments at the regional level, even though commercial enterprises (in the 'west') have become more significant in the latter half of the twentieth century. The rationale for dealing with cloud seeding in this section is that it is often a response to short-term need because of rainfall unpredictability.

Precipitation enhancement (cloud seeding): scientific and legal dispute

While the technical expertise associated with cloud seeding might be considered to lie beyond all but the most developed countries, we argue that cloud seeding is a well-established technology (see box 5.4). Practical techniques include the use of aircraft, also ground-based emitters that do not require significant investment or infrastructure. The more significant problem is the identification of appropriate cloud conditions and then proving that the approach actually works. South Africa has a long-established engagement with cloud seeding and this has led to recent developments in the use of airborne flares, while China is also reportedly investing in this technology (Moseman, 2009).

Precipitation includes rainfall but also other forms, from frozen (snow, hail, rhyme) through to aerosol occult precipitation (dewfall, mist and fog).

It can be enhanced either by stimulating more than would have otherwise have reached the ground (e.g. by cloud seeding), or by more effective use of that actually received (e.g. run-off harvesting, or collecting dewfall and mist – 'occult precipitation'). Here we deal with the first of these strategies.

Box 5.5 summarises key stages in the development of scientific under-standing of precipitation enhancement. Experimentation and government support can be traced back to the nineteenth century but enhancement through cloud seeding began to emerge in the in the late 1940s (Agnew and Anderson, 1992) and the latter half of the twentieth century witnessed a rapid growth in scientific and commercial interest in the possibility of augmenting water resources through artificial weather modification.

CAS (1980) reported that by the early 1950s possibly as much as 10 per cent of the USA's land area was under commercial cloud-seeding operations at an annual cost of several million dollars. The President's Advisory Committee on Weather Control announced in 1957 that in the

Box 5.4

Cloud-seeding practices

Glaciogenic (cold cloud) seeding

In glaciogenic clouds, temperatures are below zero, typically –10 to –25°C. Precipitation is mainly generated by ice crystals acting as hygroscopic nuclei plus (Bergeron–Findeisen) preferential condensation over ice surfaces. 'Seeding' is performed through the introduction of an ice crystal agent, normally silver iodide or dry ice. This is known as static seeding. Recently, Wakimizu et al. (2002) suggested these traditional agents cause ice crystals to become too small. They suggest using liquid carbon dioxide at slightly below 0°C to overcome the problem. The effectiveness of this method may be limited by (lack of) detailed scientific information on atmospheric conditions required for selection of the seeding agent and appropriate timing.

Static glaciogenic seeding in Texas yielded 50 per cent more precipitation (Woodley and Rosenfeld, 2004) than comparable cloud seeding programmes in Thailand. Conversely Bruintjes (1999) found that static seeding in the US High Plains (HIPLEX), South Africa, Australia and Canada, produced inconclusive results: 'considerable scepticism exists as to whether this method provides a cost-effective means for increasing precipitation for water resources' (p. 812): A more detailed review of static seeding by Silverman (2001), drawing upon work from Israel between 1961 to 1991 and USA from the 1960s, found inconsistencies between statistical and physical evidence.

As an alternative to static seeding, dynamic seeding involves the introduction of a seeding agent into clouds around minus 10°C where super-cooled water is present, resulting in conversion of super-cooled water droplets into ice, thus releasing latent heat which adds to the buoyancy of the air. Despite observed changes in cloud heights, Bruintjes (1999) cautions that evidence for areal rainfall increases is lacking and describes this as unproven technology. Similarly Silverman (2001) drawing from experiments from Florida, Texas, Cuba and Thailand concluded for glaciogenic seeding experiments over the last four decades: 'it has been found that they have not yet provided either the statistical or physical evidence required to establish their scientific validity' (p. 919).

Hygroscopic (warm) cloud seeding

In hygroscopic (warm cloud) seeding, water-droplet growth is dominated by collision and coalescence. Hence air buoyancy and updrafts are important to sustain the development of precipitation that reaches the ground. Condensation (the release of latent heat) is enhanced through water sprays, but the extra cost of transporting water led to alternative hygroscopic nuclei (such as sodium chloride) being used. As with cold-cloud seeding, there is concern that the benefits have been overstated, and Cotton, (1986) suggests that a reported 42 per cent increase in rainfall in India achieved with this method is questionable. Bruintjes (1999) concluded that the historic evidence for effective hygroscopic seeding was inconclusive but recent work with airborne flares in South Africa dispersing salt particles suggest more promising results, (Terblanche et al., 2005). Silverman's (2003) review of hygroscopic seeding noted experiments in South Africa, Thailand and India produced statistically significant results but retained concerns over the data, and concluded (2003: 1226) that hygroscopic seeding experiments 'have not yet provided either the statistical or the physical evidence required to establish that hygroscopic seeding of convective clouds to increase precipitation is scientifically proven'. However, he continued to advocate cloud seeding 'in situations where it is scientifically and operationally appropriate'.

mountainous areas of the Western USA a 10–15 per cent increase in rainfall was due to cloud seeding. Cotton (2007) suggests a 10 per cent increase in ground-level precipitation is a conservative estimate, with many studies producing much higher figures. Similarly, the South African Rainfall Programme of 1997–2001 reported an average doubling (108 per cent increase) in radar-estimated rainfall, which, over an area of 10,000 km^2, translates to a 10 per cent increase in area seasonal rainfall, at a cost of R0.04 m^{-3} from a budget of R12.5 m over the period (Terblanche et al., 2005). Numerous similar reports of successful seeding programmes are found across the globe. Since the 1940s, rising interest and three decades of experimentation up to the 1970s have been followed by a period of decline in scientific interest, although commercial cloud seeding has continued worldwide (Cotton and Pielke, 1995), particularly as a response to drought.

There appear to be two elements to the scientific problems of cloud seeding. The first relates to greater scrutiny of the scientific evidence. The second relates to changing public attitudes to human-induced changes to weather patterns. With respect to the quality of the scientific evidence that cloud seeding is effective, Silverman (2001; 2003) noted that results for glaciogenic seeding were still not conclusive as to whether they had been successful or not, but there was some optimism concerning hygroscopic seeding experiments. At the heart of the debate over whether or not seeding is successful are questions of experimental design and statistical analysis. CAS (1980), reviewing projects from the 1950s and 1960s, concluded they showed generally positive effects for cold winter orographic clouds, mixed results for convective clouds, and no clear effects for non-orographic and non-convective storms. A study of Kings River (California) showed an increase in run-off of 6 per cent, while the Skagit River Project (Washington State) resulted in a significant increase in stream flow. Criticisms arose that these studies lacked randomisation in the selection of seeding opportunities, and that the experimental evidence was ambiguous. Breuer (1980) noted that only 23 cloud-seeding studies were statistically rigorous, and, of these, only six showed increased precipitation, seven cases were ambiguous and ten showed reduced precipitation. These conclusions reflect the difficulty of proving that seeding actually causes enhancement of precipitation, due to highly variable atmospheric conditions and poor observational networks. Braham (1986) summarises the problems as follows:

1) the physical mechanisms of rainfall are complex,
2) results from seeding clouds can be both positive and negative,
3) the variability of clouds has not been appreciated,
4) rain-gauge networks and monitoring are rarely adequate, and
5) experiments must run for a long period of time and are therefore expensive.

The background variability of weather events makes it hard to attribute causality with confidence from statistical analysis, even for large effects. Decker (1978) cites an example where a 30–40 per cent increase in rainfall was not 'statistically significant'. Breuer (1980: 9) quotes Joanne Simpson's remark that, 'The most striking lesson these writers have learned from 30 years of cloud study is that a cumulus cloud can do virtually anything all by itself without any interference by man.' Finally, it should be evident that if humidity is low then there is little to seed! This is not, therefore, a method to combat aridity. It is a tool for the enhancement of precipitation, not the creation of precipitation.

The question of public attitudes to cloud seeding presents an interesting comparison with evolving perceptions of human-induced climate change. According to Cotton (2008: 294):

> there is a human tendency to accept the results of assessments of inadvertent modification of clouds and precipitation even though those studies do not meet the standards of 'proof', while at the same time requiring that cloud seeding evaluations meet such 'proof' criteria.

However, cloud seeding has been beset with controversy from the outset. Byers (1974) recounts Langmuir's early attempts to seed a hurricane in 1947, which changed direction and consequently struck South Carolina and Georgia. This raised questions over who was responsible for the damage, although the US Weather Bureau at the time found little evidence of rainfall enhancement, and public concern over inadvertent weather modification was focused on the possible side-effects of atomic bomb testing. Examples of public disquiet over cloud seeding include it being outlawed as the 'crime of the century' in Pennsylvania (Dennis, 1980), and the bombing of a trailer used by a firm of weather consultants by farmers in Colorado opposed to cloud seeding (Breuer, 1980). As public opinion moved from 'engineering the environment' towards 'living with the environment', interest shifted from establishing whether cloud seeding 'could be done' to whether it 'should it be done' (Sewell, 1973). Concerns that pollution by seeding agents increased flooding may have been unfounded (Bartlett, 1979), but Haas (1974) cites examples of cloud seeding raising controversial questions of ownership and legal challenge. For example, cloud seeding to end drought in Florida was believed to be beneficial for citrus growers but not for growers of other crops damaged by heavy rainfall. Barley growers in Colorado wished to reduce hail storms but this would then reduce precipitation for ranchers. Opponents of cloud seeding in South Dakota believed raindrop size had been reduced, leading to more evaporation and less effective rainfall (Dennis, 1980). Davis (1974) provides three interpretations of 'cloud ownership': (a) when New York city's cloud seeding was challenged by a Catskills holiday resort, it was ruled the city had the right to seed clouds and there were no individual property rights in clouds; (b) when Texas cereal farmers attempting to reduce hail damage were opposed by cattle ranchers, it was judged that clouds belong to landowners; (c) when Pennsylvania cloud seeding was challenged, it was ruled clouds are common property, and, if seeding reduces rights of access to rain, then it must be illegal.

In summary, therefore, cloud seeding continues to be employed world-wide (Woodley and Rosenfeld, 2004), with those involved deciding potential benefits outweigh the costs – even though formal scientific evidence for this is lacking (Silverman, 2001). Bruintjes' (1999) review of cloud seeding noted that, while the process was well understood scientifically, whether it was deemed worthwhile depended upon the degree of proof required. Silverman (2001: 919), neatly summarised current cloud-seeding research: 'it has been found that they have not yet provided either the statistical or physical evidence required to establish their scientific validity' but then restated an earlier conclusion that 'cloud seeding was promising, unproven, and worth pursuing'. Much of the controversy surrounding cloud seeding, then, results from the difficulty in proving it changes rainfall when rainfall itself is so variable and hard to predict. As we have seen in the previous paragraph, this has not prevented legal challenges over the practice, such as who owns clouds (Gerik, 1973). However, while academic journals continue to debate the scientific evidence for cloud seeding, academic interest anticipated in the legal implications of the practice (Davis, 1975) has tailed off. In the twenty-first century, the idea of climate change brought about by human intervention is widely accepted, so it seems curious that we should need proof of human modification of the weather through cloud seeding. Over thirty years ago the UN deemed climate modification unacceptable as a military strategy through the Environmental Modification Convention. Cotton and Pielke (1995) link the decline of research interest in cloud seeding, in part, to environmental concern that sees anthropogenic climate change as something to be combated, rather than encouraged through government funding. Yet, current acceptance of scientific evidence of weather modification via global warming contrasts strongly with the half century of discussion over whether cloud seeding actually works. Furthermore, legal questions of ownership of clouds and precipitation suggest that the history of cloud seeding may enable us to draw further interesting contrasts with the paradigm of global warming.

Box 5.5

Development of cloud seeding

The development of cloud seeding has gone through four distinct phases (based on, Sewell 1973; Byers 1974, Agnew and Anderson 1992; Cotton and Pielke, 1995 and 2007),

- prior to nineteenth century, period of observation and superstition nineteenth and early twentieth century, period of meteorological investigation. This led to the establishment of official weather forecasting, such as the US Weather Bureau in 1870 (Bosart, 1985). By the end of the nineteenth century mechanisms leading to rainfall were sufficiently well described to encourage a number of attempts to enhance precipitation artificially. By the 1930s cloud seeding attempts were being organized in the USA and Russia.

- 1940s–1970s, cloud seeding experimentation and optimism. Irving Langmuir and his colleague Vincent Schaefer are often credited as starting the scientific development of cloud seeding based on dry ice, and subsequently silver iodide, as a seeding agent, in the late 1940s. In 1963, the US Federal Government allocated $2.7 million for weather modification, rising to $10 million in 1967 (Sewell, 1973). At the end of this period the US National Academy of Sciences Panel on Weather Modification reported that in some circumstances cloud seeding was effective and could produce increases of 20 per cent. Other nations pursuing this technology included Russia, which had run experiments since the 1930s, Australia and Israel, with projects running from the late 1940s.
- 1980s to date, the demise of scientific interest and national funding.

Cotton and Pielke (1995) chart US research funding, peaking at $19 m/year in the mid-1970s and subsequently declining rapidly to $5 m/year in the 1990s and to $0.5 m at the millennium (Bruintjes, 1999). Despite the attraction of influencing precipitation in water-scarce regions, even Israel, a long-time advocate of cloud seeding, reduced operations compared to the 1970s (Cotton 1997). During this time, the reliability of scientific evidence was increasingly questioned, and changes in public opinion following the IPCC reports fuelled growing political opposition to human activities perceived to cause climate change. Nevertheless, contemporary web pages describe the potential benefits of this technology including hail suppression (http://www.just-clouds.com/index.asp; http://www.weathermod.com/seeding_equipment.php).

Rainwater harvesting: the test of time

Rainwater harvesting is enhanced collection of precipitation through use of roofs, run-off plots and even dewfall (occult precipitation) collectors. Fleskens et al. (2005) report widespread government support for this strategy across semi-arid Mediterranean countries, where it has long been practised. In wetter regions, run-off plots may reduce the need for more expensive pumped water. Rainwater harvesting can supply either domestic (household) needs or agriculture (livestock and crops) (for reviews see Barrow 1987; 1999; Ben-Asher and Berliner, 1994; Gould and Nissen-Pewtersen, 1999). Patrick (2000) distinguishes between macro water harvesting, using large natural catchments and transporting water a significant distance, and micro water harvesting, normally on run-off catchments less than 1 hectare to collect water for local use. Cowden et al. (2008) make another distinction between formal, permanent, rainwater harvesting and informal, opportunistic systems. Kahinda et al. (2008: 788) define rainwater harvesting as 'the collection,

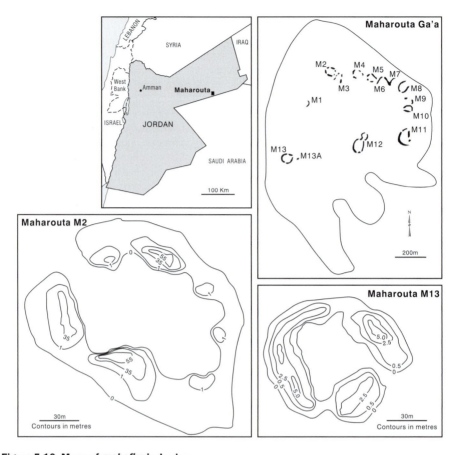

Figure 5.10 Maps of *mahafirs* in Jordan

Source: from Agnew et al., 1995; permission granted by Springer (*Geo-Journal*)

storage and use of rainwater for small scale productive purposes', and Rockstrom (2003) has argued that rainwater harvesting is the key to building more resilient agriculture in semi-arid areas. This requires land cultivation in anticipation of future rainfall, and reservation of areas from which surface run-off will be collected or diverted. Schiettecatte et al. (2005), analysing 'catchment to cropped area ratio' (CCR) in Tunisia, found values of at least 7 to be effective, although much higher ratios have been reported elsewhere (Fleskens et al., 2005). Rainwater-harvesting catchments of whatever scale demand high levels of local organisation to manage them, and we therefore regard this approach as quite distinct from more centrally managed developments such as large dams.

Indigenous water-harvesting practices in semi-arid Kenya are well documented (Patrick (2000) and a number of international meetings have explored exploiting and improving these techniques further (FAO, 1994; UNESCO, 1983, 2000). This literature contains relatively little scientific analysis (Patrick, 2002) except on the role of soil surface crusts and, although there is relevant research within the vast literature on dryland hydrology and arid zone geomorphology, there appears a dearth of understanding of ancient water-harvesting systems and how their operation was underpinned by social customs (Lancaster and Lancaster, 1999). The issues are exemplified by the case of water harvesting in Jordan, where, in 1996, annual water consumption was less than 200 m³/per capita, compared to 370 m³/per capita for Israel, 1200 m³/per capita for Egypt and 3500 m³/per capita for Turkey (Ahmed 2005). On current population projections, current water supplies in Jordan will in future provide only 91 m³/per capita, but there is evidence of water resources declining further through degradation and falling groundwater levels. Agnew et al. (1995) and Ahmed (2005) investigated the *mahafir*, an ancient run-off harvesting system in the Badia desert region of Jordan, consisting of depressions and barrages built onto the playa surface. The system, also described by Barrow (1987) is currently filled with sediment, but is one of several water-collection structures found in Jordan, including those dating back to Bronze Age settlement at Jawa (Helmes, 1981) and Byzantine dams at Burqu' (Betts et al., 1990). Agnew et al. (1995) calculated that a fully operational rain-fed mahafir at Maharouta (Jordan) would store over 100,000 m³ which, after taking account of seepage and evaporation losses could support a herd of 2500 sheep for six months (dry season). A multi-variate model predicting surface run-off onto the playa (Ahmed, 2005) provides fresh insights into the design and location of the mahafir and thus has linked scientific analysis with indigenous knowledge.

Another ancient form of water harvesting is the *qanat*, where water diverted from a spring or 'mother well' is transported via gravity channels, or even tunnels, to fields or houses. *Qanats* have been observed with flow rates sufficient to support a village of some 1000 inhabitants and require high levels of social organisation for their construction and maintenance. *Qanats*, described in detail by Wilkinson (1977), are found throughout the Middle East. Although their origins and terminology are disputed (Ron, 1986; Rowley, 1986), they appear to have been introduced to the Middle East from central Asia, and examples can also be found across North Africa, in Mexico and other semi-arid parts of Latin America and even China (Allen, 1987; Nir, 1974; Ron 1985). These ancient water-harvesting systems appear to have been in decline for much of the latter half of the twentieth century. In Iran, for example 40,000 *qanats* were reported in the 1960s (Bemont, 1961), but only half remained two decades later (Goldsmith and Hildyard, 1984). Similarly, in the 1970s, Nir (1974) noted it was unlikely that any new *qanats* were being constructed in the ME and those in Africa were clearly deteriorating. The prime causes appear to be social and economic change associated with declining agricultural markets (notably for dates) and alternative employment opportunities (Hills 1966; Allen 1987; Harrison, 1987). It underlines the fact that water supply systems have developed (and declined) in response to local economic opportunities.

Plate 5.1 shows the remnants of a water-harvesting system in Uganda that was reportedly operational in the 1950s, but by the 1990s was in a state of decay and disrepair. Household roofs may be a convenient means of collecting rainwater at the point of household consumption, but are only useful during the rainy season if water can be stored. This form of water harvesting also risks contamination, during both collection and storage. If the harvested water is used for drinking then disinfection is important, but if it is primarily for irrigation and cleaning, as found by Abdulla and Al-Shareef (2006, in Hlavinek et al.) in Jordan, then

Plate 5.1 Abandoned rainwater harvesting system for an old school house in the Mpigi district, Uganda

this is less problematic. Analysis of water quality for a Ugandan village (Agnew and Anderson, 1994a, 1994b), found levels of faecal coliform in rainwater barrels ranged from 1 to 50 FC/100 ml, demonstrating that, while some were relatively free from contamination, others were not (see Box 5.6 for case study). The study also found that contamination was very high (many in excess of 50 FC/100ml and some > 1000 FC/100 ml) in water stored by households using traditional pots. Kenya's Water Sanitation and Education for Health (WASEH) programme, using rainwater harvesting alongside education to reduce diarrhoea infections, advocated water disinfection (chlorination) before consumption, combined with modified household 'clay pots', and digging latrines in the appropriate location and at depth all decreased diarrhoea risk (Garrett et al., 2008; see also Chapter 2).

The potential contribution of rainfall harvesting to water supply is widely recognised, and numerous case studies of projects operating in sub-Saharan Africa have been documented (Cowden et al., 2008). However, the frequent lack of household storage means the design of such systems needs accurate assessment of daily precipitation characteristics. The same conclusion was reached by Ahmed (2005) for rural Jordan. In particular, the lack of data for the most important variable (rainfall intensity) makes it hard to assess risk and predict crop yield, although stochastic modelling can be used to address this problem (Tsubo et al., 2005). Arguably the first comprehensive model to link rainfall, run-off and yield for semi-arid areas was that of Parched-Thirst (Wyseure et al., 2002; Young et al., 2002). Srivastava (2001) notes that there is also scope for water harvesting in high rainfall areas but these have received much less attention. Other studies to determine the potential for increased rainwater harvesting include agro-climatological analysis of semi-arid China (Wei et al., 2005), use of sand ditches in Jordan (Abu-Zreig et al., 2000), assessments for olive growing in Tunisia (Fleskens et al., 2005; Schiettecatte et al., 2005), maize cultivation in Kenya (Barron and Okwach, 2005), maize cultivation in South Africa (Walker et al., 2005) and rice production in Kenya (Panigrahi et al., 2001). Such studies make use of agro-climatological models for assessing crop water requirements (see Box 3.2 and Box 3.3), but the relationship to slope run-off generation is problematic (Giakoumakis and Tsakiris, 2001; Bevan, 2003), so that predictive models may be weak.

Kahinda et al. (2008) employed environmental variables alongside socio-economic such as levels of poverty or unemployment to assess rainwater harvesting potential in South Africa. There is growing appreciation that agro climatic models only capture limited environmental factors, and that social factors underpinning community activity are as important. Fleskens et al. (2005: 614) note, 'hydrological research alone does not allow for a complete impact assessment of water harvesting techniques because it does not study the human dimension of water management'. The latter is complex, ranging from issues of land ownership and tenure through to regulation. For example it is believed that one of the obstacles to revitalising the Mahafir in Jordan is: who will own the water. Kahinda et al. (2008) attempted to map rainwater harvesting potential, employing a 'vulnerability map' to capture socio-economic criteria. They concluded that, despite a degree of subjectivity in the selection of criteria, the approach was useful, and 25 to 30 per cent of the area was found suitable for some form of water harvesting. Cochran and Ray's (2008) analysis of equity in rainwater harvesting in Rajasthan, India, identifies multiple benefits, such as improved health, income, and changing labour demand (especially for women), but reiterates a central role for the community in water harvesting projects. This theme of community engagement and social considerations regarding water consumption is explored further in Chapter 6.

Rural water supply

In Chapter 1 (see Figures 1.11 and 1.12) we observed that rural communities have lagged behind urban areas in improving water supply and sanitation. Rural water supply suffers from distinctive difficulties arising from the often dispersed location of households, making infrastructure more expensive per head of population to install than in urban areas. Relative poverty also results in a lack of capital both for new installations and for future payments. A World Bank study (Churchill, 1987) noted that to finance piped water required a population of at least 800. Thus, for many rural areas with low population densities, conventional large-scale water-development technologies will not be feasible without significant subsidies. It is also worth noting that a focus on 'urban' and 'rural' water development as two extremes, risks neglecting the challenge facing small towns for which water supply is neither community nor centrally based.

Box 5.6 summarises an investigation of water supply problems in a rural area of Uganda that exemplifies the problems of rural water supply found in many areas (Agnew and Anderson, 1994a and 1994b), including the

Box 5.6

Kanyike case study of rural water supply

The Kanyike refers to a collection of village settlements in rural Uganda, in the Mpigi district some 30 miles south west of Kampala. The project started as a small-scale cooperative farm in 1986, and then developed educational facilities, including a carpentry skills centre (1994) and a clinic (1992). There is, then, a high level of social organisation among some 250 households with an estimated population of 1500. Water-resources development is not new to the area (Plate 5.2 illustrates water-treatment facilities constructed in the 1950s) but past investments have failed through lack of maintenance and technical capacity, as did spring-capping, introduced at around the same time to protect water supplies from contamination. McNeill (1985) notes that rural water-supply projects frequently fail after only a few months of operation, and that (lack of) money to pay recurrent costs is one of the major obstacles for efficient operation of any water supply scheme. Bajard et al. (1981) identify additional problems of lack of motivation, and poor dissemination of training and information.

Plate 5.2 Water treatment facilities were constructed in Kanyike (Uganda), reportedly in the 1950s, but, as this picture illustrates, by the 1990s they had been abandoned and allowed to decay

In Kanyike gneiss bedrock is weathered into deep and porous sediments that store sufficient water for perennial discharges from the springs. Figure 5.6 shows the dispersed nature of settlements and eight major springs used for water supply: two at the head of a tributary valley and six along the lower

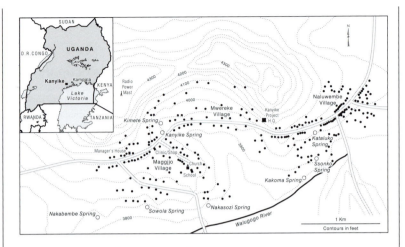

Kanyike household distribution and location of main water sources (springs)

slopes of the main valley (River Walugogo). Each morning and evening children walk up to 1km to collect household water (in many parts of the world much greater distances are required). Employing local school children and project staff, the project measured spring discharges, household collection and household consumption during August 1992. This was augmented by a questionnaire survey of 83 households. Questionnaire estimates of household consumption averaged 36.5 ld⁻¹, whereas those from direct observation averaged 42.6 ld⁻¹.

The table below shows that, in general, water demands are not over-taxing the amounts of water supply, although some springs are preferentially used irrespective of flow rates. It became clear during discussions that decisions about water collection had little to do with supply. This was explained by access problems, such as steep slopes around the spring (avoided by children tasked to collect water), or whether a threat from animals was perceived in the area. Daily per capita domestic consumption is low: 10 to 15 litres compared to recommended levels of 50 lcd (drinking water 5, sanitation 20, bathing 15, cooking 10) (Moriarty and Butterworth, 2003). In Kanyike 83 per cent of water collected from springs was used for domestic purposes with the remainder used for watering gardens and livestock. Personal hygiene and washing clothes accounted for 65 per cent of household consumption, and drinking amounted to 10 per cent (around 8 PL household⁻¹d⁻¹). As we observed in Chapter 2, substantial increases in water consumption are anticipated to follow improved access. For example, Cairncross (1987) found villagers in Mozambique with access to a standpipe used four times (12.3 lcd) as much water as those without access (3.2 lcd).

For households in Kanyike the main problem is contamination of these water supplies by poor sanitation. During 1994 all supplies: (springs, rainfall harvesting and household storage) were examined for faecal contamination using the membrane filtration OXFAM–Del Agua Water Testing Kit (Roebens Institute, University of Surrey, Bartram, 1990, Lloyd and Helmer, 1991, WHO, 1995). Over 200 samples were taken of all springs, household containers and rainwater harvesting vessels. Only sources with zero faecal colliform (F.C./100 ml) can be considered safe drinking water (Guillemin et al., 1991, Shaw, 1994 and WHO, 1984). Most samples were found to be contaminated, with values rising to 1000s of F.C./100 ml (see Table). It was expected that the ponds were contaminated and Kanyike households understand that that they should not collect water from ponds, although this still often happens. More worrying was that the springs themselves were also mostly contaminated (see Table). Spring-capping, that protects the source from waste and pollution after it issues from the ground, can eliminate the highest levels of contamination, but these results showed it to be ineffective. Moreover, comparison of the table results with the figure above reveals that rates of water collection from springs take no account of water-contamination levels, with the most heavily contaminated (and most heavily used) springs located in the most densely settled areas. This appears due to contamination from latrines built adjacent to the springs, which spring-capping cannot alleviate. The wider point for all water supply development is that safe sanitation and waste disposal must run alongside water supply enhancement.

The following table shows the faecal contamination (F.C./100ml) and consumption analysis of Kanyike water sources (see Figure above for locations):

Spring	Water Source (F.C./ 100 ml)	Water Pond (F.C./ 100 ml)	Spring flow (m³d¹)	Spring collection (m³d¹)	Withdrawal (%)
Sowola	2	228	1.642	0.495	30.1
Ssonko	2	277	–	–	
Katluko	2	4500	–	–	
Nakabembe	5	19	19.267	0.265	1.4
Nakasozi	9	112	15.034	0.490	3.3
Kakoma	27	7000	54.691	2.326	4.3
Kimete	117	87	4.234	0.595	14.1
Kanyike	139	2000	3.888	0.625	16.2

major problem of water contamination and the difficulty of financing improvements. Gine and Perez-Foguet (2008) chart a similar story of numerous efforts during the 1970s in Tanzania that proved unsustainable, so that, by 2002, only 30 per cent of schemes were functioning. Therkildsen (1988), in a key publication appropriately entitled 'watering white elephants', also noted that from the 1960s to 1980s $150 m was spent in Tanzania to little long term effect. By the 1980s, seeking to improve the outcomes of rural water-supply investment, the World Bank had identified (Churchill, 1987) as requirements for rural water development:

- ownership of resources
- local financing and cost recovery
- capacity building
- institutional strengthening
- important roles of women and children.

In line with this thinking, in Tanzania, the policy in recent decades has been to move away from thinking of water as a 'free' (i.e. subsidised) service, and to prioritise cost recovery. This has involved establishing property rights in water schemes, moving to a slower-paced, more sustainable introduction of new schemes, and emphasising the role of communities through decentralisation. This government strategy was described by Gine and Perez-Foguet (2008) as 'hands off, eyes on'. Thus, the National Rural Water Supply and Sanitation Program (NRWSSP), introduced in 2002, focuses upon community-based schemes, 'stakeholder ownership' and local contributions towards costs amounting to 5 per cent of capital investment. The NRWSSP specifies that a water point should serve no more than 250 people within a radius no greater than 400 m and a rate of 25 lcd. The programme is fairly representative of current rural water-supply efforts that provide external assistance but demand greater community engagement through local investment of labour and cash.

Unfortunately the literature continues to report failed rural water supply projects. Rietveld et al. (2008) examined the reliability (system failure) and durability (lifespan) of rural water-supply projects in South Africa based on community standpipes fed by groundwater. They found poor installation and maintenance resulted from disputes over payments, and recommended projects seek greater partnership with communities. This is not straightforward, however, and demands an understanding of what 'communities' want from water supplies. For Inupoaq villages in Alaska, Marino et al. (2009) noted differences in acceptance of new state-funded water supply, with some communities deciding to continue traditional methods of water collection from precipitation and ice. Their analysis revealed a complex array of factors, including differences in taste, problems of maintaining equipment in an extremely cold, arid climate, and financial constraints, but concluded that local perceptions are paramount, and local participation a prerequisite for success.

These conclusions are reflected in key requirements listed by Gine and Perez-Foguet, (2008) that show little change from the World Bank list, given above, some twenty years earlier:

- Management at the lowest appropriate level
- Communities own and manage schemes
- Maintenance and technical skills
- Full cost recovery for maintenance and operations
- Protection of water resources (from contamination and degradation)
- Balancing technology, provision and conditions
- Recognising key roles of women and inclusion of the poor.

A similar emphasis on local engagement is evident in DfID's (2001) priorities for water projects:

- Recognising water as an economic good
- Respond to demand
- Put people at the centre.

These policy prescriptions reflect a more general emphasis on using an analysis of household capabilities and assets as a first step in identifying appropriate water development interventions (Carney, 1998; Moriarty and Butterworth, 2003). Typically, such 'sustainable livelihoods approaches' would involve an inventory of:

- *Assets*
  Human capital
  Natural capital
  Financial capital
  Social capital
  Physical capital

- *Vulnerability*
  Shocks
  Trends
  Seasonal changes

- *Processes*
  Policies
  Institutions
  Regulation and governance

- *Livelihoods*
  Household activities
  Strategies
  Outcomes.

This type of analysis has often drawn on methods developed in earlier participatory rural assessment (PRA) approaches, such as stakeholder identification, community mapping and institutional mapping. Woolley et al. (2009) noted a need for more work on livelihood improvements in parallel with institutional changes within the Challenge Programme on Water and Food (Sri Lanka), that seeks to promote (p. 10) 'accelerating behavioural and institutional change through social learning', and to develop 'impact pathways'.

A focus on 'stakeholder engagement' and capacity building has high-lighted and reinforced the long-standing international recognition (e.g. ICWE and Agenda 21 – see Chapter 1) of women's role in the organisation of water collection, storage and hygiene. We will return to the role of women in household water provision in Chapter 6. Here we will simply note that any notion of local capacity-building for water management must be informed by an understanding of the differentiated, frequently gendered, existing water-management roles in rural communities. This is all the more important because such roles are not immutable, but will adapt to changing water supply circumstances. An example we consider in Chapter 7 is Cleaver and Toner's (2006) study in Tanzania that found local community-based Water Use Associations had improved water supply, but at the cost of raising inequality.

Finally, it is worth noting that the theme of 'stakeholder engagement' as key to rural water supply is not restricted to developing countries. A study of community participation in rural water management in Northern Europe (Hophmayer–Tokich and Krozer, 2008), following the EU Water Framework Directive's (2000) requirement for public participation, concluded that the approach had resulted in more successful adoption of plans through a raised sense of ownership.

Urban water supply

In Chapter 2 we noted the increasingly urbanised nature of population growth, and the likelihood that water consumption is likely to parallel this, particularly in cities of Africa and Asia. Cities are also significant because they drive the world economy, with 30 per cent of the population but contributing around 60 per cent of total gross domestic product (Lundqvist et al., 2005). This greater economic activity means urban inhabitants are, in general, far better provided with clean water and adequate sanitation (World Bank, 2006; see also Chapter 1) due to greater capital investment in infrastructure. However, aggregate figures mask huge problems of inadequate access to water due to poverty and contamination. Much of urban growth is constituted by the urban poor, who face the same problems of securing water supplies as their rural counterparts. The water-supply problems facing eight mega-cities (Jakarta, Dhaka, Johannesburg, São Paulo, Mexico City, Riyadh, Istanbul and Singapore), reviewed by Varis et al. (2006), present a mixed picture of some successes alongside an alarming picture of inadequate and, in places, worsening provision.

We have earlier noted differences between urban and rural water supplies, such as the increased costs of infrastructure in rural areas, and a higher intensity of demand and greater volumes of wastewater in urban areas. Technological solutions will be much the same, however, and Pumphrey et al. (2008) caution against making too much of the urban–rural distinction as there are problems of governance and financing water investments in both, as also in settlements that are neither villages nor cities.

Thomas Walski (2006) charts the evolution of water supply and distribution in cities over two thousand years ago in Asia and the Mediterranean, culminating in the extensive system of aqueducts and even lead pipes used in Rome around this time. The use of lead pipes in urban water supply continued through the Middle Ages in Europe, but much of the earlier knowledge of water distribution had to be re-learned in the eighteenth and nineteenth centuries in response to the rapid expansion of industrialised towns. Walski reports that by the middle of the eighteenth century, London had more than 50 km of water mains made of wood, lead and cast iron. Pipe manufacture was dominated successively by cast iron (nineteenth century), cement linings (early twentiethth century), and PVC (late twentieth century). Parallel developments in the manufacture of joints and pumps meant gravity-fed systems were increasingly supplemented by pumping using wind or animals, subsequently steam-powered pumps, superseded by electric centrifugal pumps during the twentieth century. By the mid-twentieth-century technology development was sufficient to ensure a reliable and high-quality supply to many major conurbations. The problem for urban water supply today is, therefore, frequently not about technology, but the financial resources needed to install and run it. Lundqvist et al. (2005) present this as a crisis facing many urban areas because increasing consumption, inadequate supplies and sanitation, and environmental degradation require increased investment, but are confronted with diminished ability of municipal governments to raise adequate funding from the urban poor. Varis et al. (2006) have suggested that rates of growth in new urban centres in the South coupled with relatively slow economic growth make it inappropriate for the new mega-cities (of the South) to try and follow the path of nineteenth-century industrialised cities such as London or New York. Gandy's study of Mumbai (see Chapter 2) suggests that local political factors need to be addressed before such a conclusion may be accepted.

It is nonetheless increasingly acknowledged that urban environments need to be better understood if urban water supply in the future is not to merely reproduce with ever-increasing intensity the problems of the past. A specific literature on hydrological processes in urban areas, reviewed by Marsalek et al. (2006), advocates an integrated approach embracing supply, demand, contamination and governance (GDRC, 2010; Hlavinek et al. 2006). Arnfield (2003) and Roth (2007) have noted that urban areas have a distinct climate compared to their rural surroundings. This is observed in the 'heat island' effect, higher incidence of rainstorms, lower average wind speeds and changes to the energy budget. It is important not to overstate these effects. For example, heat island effects are most noticeable in minimum temperatures following nocturnal cooling. Nevertheless, the urban fabric does influence climate and hydrological effects so as to create more rapid drainage and lower evaporation rates that, together, raise flood risks. As we noted in Chapter 2, risks of contamination of water are also high. Urban contexts, therefore, have a number of contradictory characteristics: a need to remove excess, often contaminated, water to reduce flooding; a lack of space to store water; and intense demands, frequently greater than local water supply. Urban water management will evidently need to take account of basic differences of hydrology between cities developed alongside major river systems, such as London (Thames), Cairo (Nile) and Rotterdam (Rhine), and those such as Beijing, Dhaka, Jakarta, Mexico City, that have relied on increasing extraction of groundwater resources (Foster et al., 1999). In the former, river management has been critical to satisfy the demands of urban expansion, as Gandy (2002) demonstrates in his study of water-supply development and the growth of New York. In the latter cases, over-abstraction causing lowering of the water table is a key constraint.

Swyngedouw (2004) has argued that cities' success has depended upon their ability to exert dominion over their environment, in particular through ensuring water supplies, or the 'conquest' of water. In Chapter 2 we explored the widening reach of water demand of expanding cities in the nineteenth and twentieth centuries. This trend continues, with Johannesburg, Mexico City and Istanbul among the many mega-cities now reliant upon long-distance water transfers, with Beijing being an acute case (Lundqvist et al., 2005). This raises sustainability questions, and in Chapter 2 we also identified an emerging counter argument, that the future of urban development will need to adapt to hydrological conditions – for example providing 'room for rivers' on floodplains – rather than seeking total control (see also Johnson et al., 2007; White, 2008).

Urban areas have been thirsty for much of the twentieth century. Their growing demand has often outstripped supply and fuelled technological development but at considerable financial and environmental cost. We noted above that desalination is now widely considered as a solution to the need to supply water to cities. It may be seen as consistent with past emphasis on engineering approaches to withdraw and convey water. Lundqvist (2000: 259) observes '... In some parts of the world there is simply no more fresh water to develop, in other areas there is no investment capital'. This suggests more attention should be paid to conservation and demand management, topics we will address in Chapter 6. From another perspective, water demand by mega-cities now frequently extends beyond the immediate river basin, requiring national and even international institutions to manage water resources, to which we turn in Chapter 7.

Water-supply summary

In this chapter, we have examined different aspects of the massive development of engineering to enhance the supply of water during the twentieth century. Some of these, such as dams and water-harvesting, are based upon pre-existing technologies. Others, such as cloud seeding and desalination were based on new science. In all cases, however, the science is subject to contestation, for example for cloud seeding, but also for dam siltation rates, evaporative losses and so on. However, as we have seen in this chapter, contestation of the effects of large dams is perhaps one of the sharpest areas of controversy in water management. It appears to have led to a pause in building large dams during the 1990s, but has been followed by renewed interest.

Two distinct policy directions characterise this renewed investment. One emphasises safeguards to ensure fewer negative impacts. Ortolano and Cushing's (2002) analysis of the impacts of the Grand Coulee Dam (Columbia River basin, USA) some 70 years after its construction exemplifies this, showing that, despite the evident benefits of electric power, irrigation, metropolitan growth and flood control, environmental and cultural costs continue to impose vast differences between the project's beneficiaries and cost bearers. Their recommendations for future projects include:

- Improved participation and reduced marginalisation of cost bearers
- Periodic reassessments of water allocations and prices in the light of changing technologies and social values
- Sharing a conceptual framework and terminology to enhance communication
- Irreversible decisions (e.g. major water diversions) require detailed and full study.

There is considerable overlap with WCD's (2000) framework for design of large dams, based upon:

- Gaining public support (participation of all stakeholders is generally advocated for water resource projects)
- Comprehensive options assessment (full review of alternatives)
- Review of existing dams (poor past performance needs to be explained and lessons applied in future)
- Sustaining river and livelihoods (impacts upon ecosystems, indigenous peoples and downstream activities need to be established and mitigated).
- Recognising entitlements and sharing benefits (past experience demonstrates the benefits are rarely shared equally)
- Ensuring compliance (rehabilitation and resettlement are too often not delivered as promised in the planning stage).
- Sharing rivers for peace, development and security (see Chapter 7).

The touchstone of this line of thinking is that more representative and democratic governance of water resources will improve the equity of outcomes from large infrastructure investments. It is consistent with the move to greater participation advocated by the World Summit (UNCED, 1992) and the International Conference on Water and the Environment (Dublin) and is embraced by many of the reforms to national water governance enacted since (see Chapter 7).

A similar goal is envisaged by the second policy direction, which advocates greater dispersal of benefits through smaller-scale infrastructure. Frost, (1987: 75) argued two decades ago that when developing water resources, 'Systems must be cheap, small scale and decentralized. … capable of being locally erected, controlled and maintained by the people themselves.' Similarly, Matlock argued (1988: 939), 'Small scale water management systems represent a modern technology; they need not be construed as being second class.' The World Resources Institute (1989) also noted a shift towards small dam construction as large-scale projects become unpopular. As with other 'community-based' natural resource management, there are those (Golladay, 1983) who advise caution, arguing that such approaches require time, resources and individual freedom, and that community participation is not spontaneous but requires organisation and incentives (see Chapter 6 for further discussion of community participation and social learning). These notes of caution also emerge from Brown's (2007) investigation of the institutional principal and stakeholder engagement in South Africa's recent reforms of water management. Moreover, it should not be assumed that smaller-scale necessarily means more efficient, as Moris and Thom (1985: 62) summarised somewhat bleakly in a report on African irrigation to USAID: 'If in Africa small-scale projects are not necessarily cheaper to build, they are nevertheless easier to withdraw from.'

To some extent there is a debate here over whether large dams are inherently more disposed to produce more negative impacts than smaller-scale dams or water harvesting technology. The discussion above suggests that answers can only be given for specific economic, political and ecological contexts. It is worth recalling, however, that in Chapter 3 we noted that increases in flood run-off are expected, due to global warming (IPCC 2007). More frequent extreme flood events will increase the risk of storage capacity of dams being overwhelmed, possibly leading to a dam breaching unless the water is allowed to run off downstream. This risk is increased because the need to maintain reasonably high levels of water behind a dam, for water supplies and water-quality regulation, limits the capacity of reservoirs to retain floodwaters. On large river systems, this risk is more complex because there may be a sequence of dams, and the failure of one would almost certainly risk failure of those downstream. The problem is further intensified where dams in a sequence may be in different national territories, as for example with the Kariba and Cabora Bassa dams, respectively in Zambia/Zimbabwe and Mozambique. Both dams are on the notoriously flood-prone Zambezi river, and the coordination of water releases to minimise (rather than accentuating) downstream flooding poses a major challenge to the achievement of positive outcomes from these investments. It is a challenge that global warming may intensify.

It is unsurprising that there is no consensus on costs vs. benefits of dams, cloud seeding, or choice of desalination technology. All are constructed across a wide range of environments and to meet a variety of objectives.

Decisions on what is a 'sustainable' water supply depend critically on economic resources available to fund it, as we saw in the case of desalination. The growing focus upon participation and local interests has reinforced rising enthusiasm for smaller-scale engineering solutions, or even rejecting engineering approaches altogether. This reflects both growing awareness of the need to avoid the adverse environmental impacts associated with large dams and other technology-driven water supply projects, but also a sense that better ecological adaptation may be possible by using smaller-scale dams and water storage systems. Biswas (2004) argues for a less dogmatic discussion of dam construction, allowing consideration of both mega-dams and small impoundments, or even rainwater harvesting, as alternatives or in combination. Water management might then be an opportunity to combine local practice and experience with modern technology.

Further reading

The American Association of Water Works published a number of review papers in 2006 that address key developments in water supply and distribution:

McGuire, M.J. 2006. Eight revolutions in the history of US drinking water disinfection. *Journal of American Water Works Association* 98(3): 123–49.

Symons, G.E. 2006. Water treatment through the ages. *Journal of American Water Works Association* 98(3): 87–98.

Walski, T.A. 2006. A history of water distribution. *Journal of American Water Works Association* 98(3):110–21.

Twort et al. (2002) provide a comprehensive account of water-supply options. More specifically the impact of dams was nicely summarised some time ago by Goldsmith, E. and Hildyard (1984 and 1986), with useful updates from Altinbilek (2002), Biswas (2004) and the WCD 2000. A review of cloud seeding can be found in Cotton and Pielke (2007), and urban-supply issues by Varis et al. (2006). Online advice and resources can be found at UNESCO (2000) for water harvesting and the World Bank (2010) for rural water and sanitation.

Altinbilek, D. 2002. The role of dams in development. *Water Resources Development* 18(1): 9–24.

Biswas, A.K. 2004. Dams: cornucopia or disaster? *Water Resources Development* 20(1): 3–14.

Cotton, W.R. and Pielke, P. 2007. *Human Impacts on Weather and Climate*, 2nd edn. Cambridge University Press, Cambridge.

Goldsmith, E. and Hildyard, N. (eds) 1984. *The Social and Environmental Effects of Large Dams*, Vol. 1 Case Studies, Wadebridge Ecological Centre, Cornwall.

Goldsmith, E. and Hildyard, N. (eds) 1986. *The Social and Environmental Effects of Large Dams*, Vol. 2 Case Studies, Wadebridge Ecological Centre, Cornwall.

Twort, A.C. Ratnayaka, D.D. and Brandt, M.J. 2002. *Water Supply*, 5th edn. Butterworth-Heinemann, Oxford.

UNESCO 2000. Rain water harvesting and artificial recharge to ground water http://www.unesco.org/water/ihp/publications/water_harvesting.pdf

Varis, O., Biswas, A., Tortajada, C. and Lundqvist, J. 2006. Megacities and water management. *Water Resources Development* 22(2): 377–94.

WCD 2000. World Commission on Dams Report: Dams and development, Earthscan, London. http://www.dams.org/report/earthscan.htm

World Bank 2010. Rural water supply and sanitation. Online resource: http://web.worldbank.org/WBSITE/EXTERNAL/TOPICS/EXTWAT/0, contentMDK:21630630~menuPK:4618329~pagePK:210058~piPK: 210062~theSitePK:4602123,00.html

6 Water demand

Introduction: water-demand management (WDM)

Demand for water has been rising for much of the twentieth century, at different rates in different parts. We saw in Chapter 1 that there is a quite variable relationship between rates of withdrawal of water from available supplies and rates of water consumption (e.g. by individual households, industries or cities). This suggests scope for managing water in ways other than always seeking to increase supply. The earth contains vast quantities of water but high intensity of demand, for example in large cities, will, as we saw in Chapter 2, inevitably outstrip local water sources that are readily available. Thereafter, further supply must be achieved from more distant sources or by using more sophisticated technology. In Chapter 5 we reviewed different ways of engineering larger supplies of water and concluded that more water can invariably be obtained, but at a rising cost.

This presents an obvious problem in poor rural areas: Figures 6.1 and 6.2 illustrate the strong association between water poverty and economic poverty. Of the ten countries with the poorest rural water supply, nine lie in sub-Saharan Africa. The problem of rural water poverty is exacerbated by the high water demand of irrigation, accounting for 70 per cent or even over 80 per cent in some places. In industrialised economies, water use for industry and energy is more significant (48–53 per cent see Table 1.1) with municipal supplies accounting for only 13–15 per cent. Moreover, much of this produces wastewater available for recycling. Thus, in terms of global withdrawal, industrial water use accounts for only 20 per cent and domestic water for about 10 per cent.

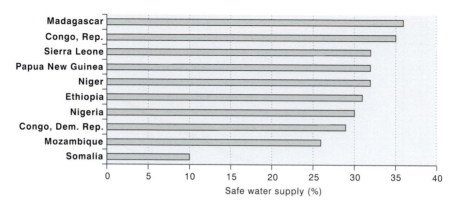

Figure 6.1 The ten countries with the poorest availability of safe water supply as a percentage of population served

Source: after World Bank 2006

In Chapter 2 we saw that, despite forming only a relatively small proportion of total water use, it is the intensity of demand for potable supplies in urban areas that has stimulated much of the investment in development and management of water supplies. Since 2008 more people live in urban than in rural areas, and the urban population is expected to rise from 3.3 billion currently to 5 billion by 2030, by which time 81 per cent of the urban population will be in developing countries (Figure 6.3). Of the 20 mega-cities (population of over 10 million) in the world, none are in Western Europe and only two are in North America (Figure 6.4). Furthermore, in many parts of the world urban inhabitants

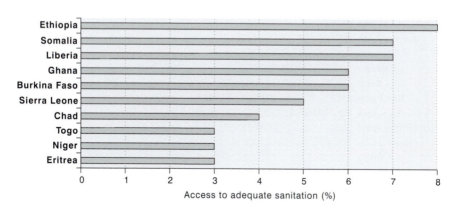

Figure 6.2 The ten countries with the poorest sanitation availability expressed as a percentage of population with access to adequate sanitation

Source: after World Bank 2006

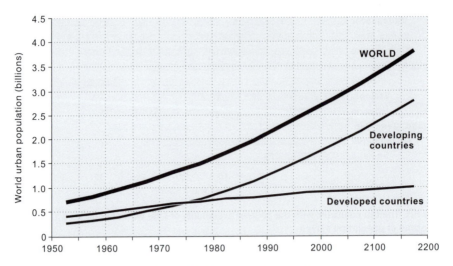

Figure 6.3 Observed and predicted world urban population growth

Source: after Earth Trends, WRI, 2009

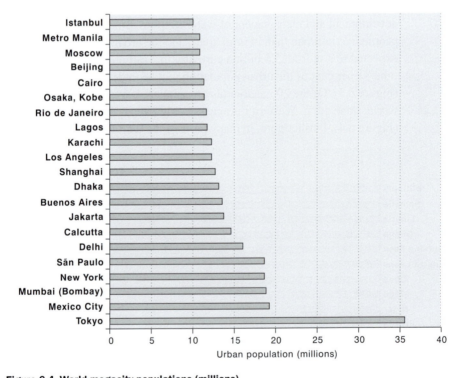

Figure 6.4 World megacity populations (millions)

Source: from CityMayors, 2006

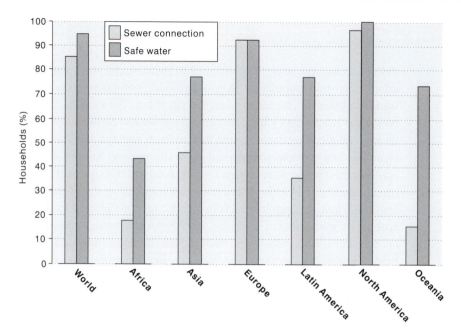

Figure 6.5 Urban households with safe water and sanitation (sewer) connections

Source: after WWAP, 2009

continue to lack safe water supplies and adequate sanitation, just like many rural dwellers (Figure 6.5).

Countries in the Middle East and North Africa typify the challenges of satisfying water demand, as they lie in an area with significant exogenous water resources but have limited local renewable supplies. In most areas water demand is at, or close to, the limit of renewable resources and a range of water-management strategies, some developed thousands of years ago, are used to combat water shortages for at least part of the year. Russell et al. (2007) list the following strategies for the region:

- *Increase supply*
  General measures e.g. dams and boreholes
  Desalination
  Loss reduction
  Run-off harvesting

- *Manage demand*
  General measures e.g. distribution infrastructure
  Economic instruments

- *Develop databases*
- *Strengthen institutions*

In the past two decades, the rising cost of increasing water supply for urban expansion, population growth and food production has prompted a significant shift in attitude to water management. The emphasis of water policy has moved away from a simple focus on increasing supply through public investment, to one that seeks to manage demand through economic instruments, and particularly water pricing. Renzetti (2002) believes this change in emphasis for water-supply management was led by economists and can be traced back to the middle of the twentieth century, when research in the 1960s first pointed to the need to take account of water costs. These ideas only influenced policy some three decades later, when neo-liberal thinking had largely supplanted the model of state-led development. This may be seen as the last of the four stages outlined by Swyngedouw (2006) for the longer historical evolution of urban water management:

- Small-scale supply enhancement through private investors for profit
- Large-scale municipal sanitation and potable supply development
- National-scale public provision of basic needs and economic development
- Privatisation and water an economic good.

The dates for these stages approximate for Western (OECD) countries to:

1 Up to 1850s – small private companies with an emphasis upon supply and profits
2 Up to 1920s – municipal bodies using taxation revenue to invest in water and sanitation measures due to health and environmental concerns rather than profit with opportunities for an integrated system
3 Up to 1970s – national control of water investment and regulation to improve public services, to fuel economic development and sometimes to support social welfare goals involving state subsidies for water and large-scale plans
4 Post-1970s – period of privatisation following economic recession and decline of state-led economic development.

In an important sense, therefore, the growing policy emphasis on Water Demand Management (WDM) needs to be seen as consistent with a broader change in perceptions of how water supply should be financed.

Specifically, there was a shift in expectation, from public funding out of taxation toward greater reliance on payments by water users. There can be seen to be two elements to this, however. The first is that of 'cost-recovery': that is, a greater proportion of the cost of supplying water (and sanitation) is borne directly by water consumers. The second element is that, by making water costs proportional to consumption (e.g. through a price per unit volume used), an incentive is created to reduce water use. Water pricing is relevant to both elements, but only in the latter can it be said to be 'demand management'.

This chapter explores the adoption of economic and social measures to manage the consumption of water through influencing human behaviour. Water pricing is one of such strategies, but not the only one. Demand management has been adopted by a wide variety of water-management agencies. Predictably, it is in water-scarce regions of high demand and low supply, such as the Middle East and drier (south west) parts of the USA that attention first focused on WDM. However, Sharp (2006) notes that, even in the relatively wet climate of England and Wales, parts of Southern England already face chronic water deficits and WDM is institutionally part of a water-management strategy supported by the Department for Environment, Food and Rural Affairs (defra), the Environment Agency and OFWAT (Office of the economic regulator for the water and sewage industry).

Emphasis on demand management has prompted a growing interest in recent decades over the nature of water consumption, and Jha and Murthy (2006) examine the role of consumption within the goal of sustainable development. They distinguish between three types of consumption:

1 Background consumption: the water required for normal biological functioning.
2 Over-consumption: the aggregation of impacts, which may individually be rational, that threatens water resources and hence the dependent human population.
3 Misconsumption: irrational use which leads to degradation and loss of the water resource.

This suggests that there are opportunities for changes in consumption that are beneficial for the efforts of demand management. Much of the WDM literature tends to be dominated by urban areas where consumption can be influenced by price, technology or legislation, particularly in North America, as the infrastructure of urban areas lends itself to strategies of recycling and conservation. However, WDM also appears in the very different context of the UN Millennium Development Goals

(UN–MDG, 2009, see also Chapter 1, section on Water Markets vs Water Rights, p. 35) and the International Decade for Action, Water for Life 2005–15 to meet the MDG (WHO–UNICEF, 2006). The six themes identified as underpinning current approaches to meeting the water elements of the MDG (Moriarty and Butterworth, 2003):

- demand management
- cost recovery
- decentralisation
- community management
- gender and equity
- integrated water-resources management.

Brooks (2006) argues for a different listing of strategies, with greater emphasis on water-conservation measures and sustainable water use:

- reducing the amount and quality of water required
- changing the task
- reducing losses
- shifting time period of use
- managing droughts more effectively.

In this chapter we shall adopt as elements of WDM three key strategies identified by Kolokytha and Mylopoulos (2004):

1 economic (water pricing)
2 technical (infrastructure, e.g. metering, recycling and retrofitting)
3 social (e.g. education, legislation, governance and regulation).

We shall consider each in turn with particular attention to their capacity to bring about a more efficient use of water.

Water pricing

Economic principles of water pricing

The OECD (2003) note that water can be regarded either as a 'public good' with everyone having a right to water in a non-competitive manner, as a 'private good' that is owned and to which there are access restrictions and a 'common good' to which there is non-competitive access only until the resource becomes scarce. Further economic distinctions are possible, but it should be clear that in the development of access to water different cultures have interpreted the nature of this resource in different ways and often the legacy can present twenty-first century challenges.

For example, in Canada (Schmidt, 2007) licences granted in the past to extract water are now regarded as having included too few safeguards for conserving the catchment resource. As a consequence, when the licences acquired commercial value, following privatisation of the water industry, over-abstraction became a serious threat. In a similar fashion Cook (1998) describes how in England licences to abstract water, granted in perpetuity through the Water Resources Act of 1963, later presented difficulties for water-management institutions (NRA and then the Environment Agency).

There is, however, widespread institutional support for water pricing with the stated policy objective of recovering costs: both direct costs of water provision and indirect costs, such as environmental conservation measures. In addition water pricing is seen as having the potential of increasing efficiency, which, according to Montginoul (2007) can enhance social welfare. The logic of this position is that under previous, predominantly state-provided schemes, water services were argued to have suffered mismanagement and inadequate investment, leading to under-provision and under-pricing of water. Cost-recovery strategies and sound financial management, linked to privatisation of water services, were widely promoted in anticipation that they would reduce 'political interference' in water pricing, make the management of water services more transparent, and put the sector on a firmer economic footing, ensuring financially viable companies and sufficient income for network expansion (World Bank, 1994; Dinar, 2000). The poor were expected to benefit as more effective formal providers would deliver safer and cheaper water than that bought from existing informal water sellers (Renzetti 2000; Chisari et al., 2001). This, then, establishes a link between 'privatisation' and commoditisation of water. However, as Bakker (2008) has pointed out, this is not a fixed linkage. In particular, she draws attention to a number of quite different reforms that may be undertaken within a neo-liberal policy framework:

- 'privatisation': introduction of private property rights for specific water resources, transfer of ownership of water supply infrastructure (e.g. UK).
- de-regulation: cessation of state oversight (e.g. in relation to water quality) (e.g. Canada)
- 'private sector partnerships': outsourcing of management of public water supply to private companies (e.g. France)
- 'marketisation': introduction of a water market (e.g. Chile)
- 'commercialisation': introduction of commercial principles, such as cost recovery (e.g. South Africa)

- 'devolution': devolving water management (e.g. water-quality monitoring) to more local organisations, such as irrigation boards or water users' associations (e.g. Indonesia).

It is possible to see, for example, that commercialisation (cost recovery) may be introduced by public water-management agencies (i.e. without privatisation), while privatisation may not necessarily involve the full 'marketisation' or commoditisation of water. It should also be clear that privatisation significantly increases the pressures for 'full cost-recovery' in provision of water, as was evident in the World Water Forum in 2000 (Chapter 1, section on Water Markets vs. Water Rights, p. 35).

The notion of cost recovery in pricing strategies for water is primarily a mechanism for ensuring financial means for continued investment in water supply, including (especially in developing countries) expanding supply to those whose existing access to water is inadequate. However, pricing strategies are also justified as a means of reducing demand, particularly where it is felt that water use is excessive. This applies principally in industrialised economies, but also – throughout the world – to water use in irrigation (see below). This is the basis of the principles for water pricing summarised in Box 6.1 which starts from the premise that if price increases by 10 per cent there will be an equivalent or even greater reduction in consequent demand i.e. an 'elastic' response. It is generally reported however that water use is relatively inelastic, hence a price rise of 10 per cent may cause only a 6 per cent decrease in demand (i.e. an 'elasticity' of demand of –0.6), which was the overall figure Kenny et al. (2008: 201) found for Colorado. This figure aggregated quite different behaviour among different types of water users, however. Those consuming more water tended to be more than twice as price-responsive (elasticity of –0.75) in their water use as those using less water (elasticity of –0.34), with an intermediate price response (elasticity –0.57) for mid-level users. We consider further in the next section the question of price elasticity of water and changes in water consumption behaviour.

Many of the issues, economic and social, concerning the pricing for water have recently been comprehensively reviewed by the OECD (2003) in light of their 2001 commitment to reducing the intensity of fresh water use while improving water quality. They concluded that water pricing can contribute positively to environmental, economic and social goals, but that it may face resistance due to negative economic and social impacts on some stakeholders. Although within the OECD

Box 6.1

Economic principles of water-demand management (WDM)

The economic principles of water demand and pricing set out by Renzetti (2002) are summarised here. Setting aside controversies over the commoditisation of water (see Chapter 1, section Water Markets vs Water Rights), we can start the discussion with the premise that, within the sphere of WDM, water is regarded as a commodity. Therefore, it is expected that the price charged may affect demand (domestic, agricultural or industrial use) as follows:

$$\text{The Price Elasticity of Demand} = -\frac{\%\ \text{change in demand}}{\%\ \text{change in price.}}$$

The calculation is not without problems. In particular, comparable data on price and consumption are needed, taking account of seasonal and annual fluctuations and different tariff structures. A range of approaches is reviewed by Martinez-Espineira (2002). Kenny et al. (2008) use average values for an historic analysis, but note a range of confounding variables, including other policy measures taken (in addition to pricing) over the five-year study period. Renzetti (2002) notes that marginal prices are preferable to average prices, but problems with obtaining marginal prices mean price-elasticity for water use is often calculated on the basis of average changes in demand and prices over time.

Examples of price elasticities cited by Renzetti are:

Residential

Household income	0.110 (high)	−0.530 (low)
Bill time period	−0.358 (short period)	−0.497 (long period)
Outdoors season	−0.305 (winter)	−1.38 (summer)

Industrial

Chemicals	−0.958
Food	−3.288 to −6.713
Beverages	−1.3 to −4.1
Non-metal minerals	−2.5

Commercial

Dept stores	−1.33
Grocery stores	−0.76
Hotels	−0.12 to −0.24
Restaurants	−0.174

universal access to safe water and sanitation has largely been achieved, especially in urban areas, not all households are connected to main piped supplies. Moreover, many countries have witnessed real increases in water prices, and, in approximately half of OECD countries, problems of affordability either are now, or are expected to become, significant, as low-income households have to pay a greater proportion of their income on water. OECD (2003) argue that affordability can be achieved through income support, and increasing block tariffs that differentiate between basic and discretionary water use.

This conclusion suggests that, even in wealthier countries, an element of subsidy is contemplated to make sure that everyone has some basic access to water. This points to a key issue in the discussion of water pricing, one that goes to the heart of the 'water markets vs. water rights' debate that we reviewed in Chapter 1. It is that water is essential for life but some uses are discretionary. Hence, we may expect that water use will be only partly responsive to price, as well as other factors such as climate, social organisation and public water-supply policies (Babel et al., 2007). In practice, some consumers, especially large consumers of water, may be subsidised (as frequently in the case of irrigation) or may be able to pass on any price increases. We need, therefore, to explore the link between water pricing and changing behaviour in a more detail.

Influencing behaviour through water pricing

Research on the price elasticity for water demonstrates the complex nature of responses across sectors and different consumer and environmental conditions. In order to manage demand effectively some understanding must be gained on the likely responses of consumers to changes in price. In addition there needs to be an infrastructure for monitoring and charging for water use (see section below on water meters) to which we will return later. Box 6.2 presents a review of the different pricing structures used around the world.

Renzetti (2002) reviews literature over four decades and notes that for domestic purposes water price elasticities range from –0.214 to –0.231, compared to –0.438 to –1.57 for outside (garden) uses. Kenny et al. (2008) and Jeffrey and Geary (2005) list a number of additional factors that influence demand, from the weather through to household characteristics (income, age, size, type of dwelling and numbers of children). Renzetti (2002) finds that for domestic users a one percent rise in

Box 6.2

Water-pricing mechanisms

It is difficult to generalise and to make comparisons over pricing structures as there are a multitude of different ways by which water is delivered and costed. Ayoo and Horbulyk (2008: 92) note: 'In some places, these water prices are the principal form of water allocation. In other places, these prices are combined with various forms of licences, permits, quotas, restrictions and other practices and customs that dictate how much water is used … '

Pricing structures are based upon the flat rate or a volume rate where the latter provides many opportunities to vary charges according to time (season) or amount (increasing or decreasing). Montginoul (2007: 863) presents water price β

$$\beta = \alpha X + b, \, (\$/m^3)$$

where
X is consumption (m^3) i.e. volume of water used

Flat rate structure	$\alpha = 0;\ b > 0$
Volumetric structure	$\alpha > 0;\ b = 0$
Two part structure	$\alpha > 0;\ b > 0$

Magiera et al. (2006), for the Middle East and North Africa, report rising block tariffs in all cases, with the following generalisations:

Low consumers	0.03 (Egypt) to 0.62 (Palestine) €/m^3
High consumers	0.04 (Egypt) to 2.4 (Cyprus) €/m^3
Average for cities	0.56 €/m^3
Irrigation water	0.01 (Algeria and Jordan) to 0.05 (Tunisia) €/m^3 (free in Egypt)

Swyngedouw (2006) provides examples of huge variations for 1997 (prices in US\$/m3): 0.84 for Dhaka, rising to 2.15 in Manila and then 28.94 in Bangkok.

A flat rate based upon household value is widespread in the UK, where cost recovery, rather than conservation (reduced consumption) is the priority. Where water is metered at point of entry, rather than of consumption, as in some apartments, flat-rate charges are found rather than volumetric. In France Montginoul (2007) reports 93 per cent of the population have a two-tier structure, often a falling block tariff where the management objective is cost recovery. However, it was anticipated there would be a shift to a rising block tariff. Jones (2003) notes an OECD trend of substituting flat rates by a rising block structure. Other changes noted by the OECD (2003) between 1998 to 2002:

- Australia's reduced use of decreasing block tariffs
- Canada's 6 per cent increase in increasing block tariffs
- Italy's phasing out of minimum charges
- England and Wales where metered households doubled from 1998 to 2002
- USA where volume rates rose and flat-rate tariffs declined.

temperature increases residential water demand between 0.02 and 2.8 per cent whereas a one percent decrease in precipitation increases demand between 0.02 and 0.31 per cent. It is reported that demand for domestic water is also influenced by the costs of other goods and services required by the household, but comparatively little work has been done because the price of water is often normalised against an index of consumer prices. Equally, little work has been done on industrial water demands because there are fewer consumers compared to residential customers, and there is great complexity, for example comparing commercial users such as car washes to hospitals or restaurants. Nevertheless, the examples cited above and in Box 6.1 demonstrate the potential use of price to influence consumption in some sectors and for some households.

Martinez-Espineira (2002: 162) reviewed a number of USA and European studies that illustrated 'water demand appears inelastic but not perfectly inelastic'. This study found values of –0.12 to –0.17 for North West Spain, comparable to other European examples, and showing only a modest reduction in demand relative to price increase. The study also suggested that in northern Spain climate (season) influences demand, but not as much as in drier areas of the Mediterranean. It also suggested a complex relationship of water demand against socio-demographic variables, with income being significant but not family size. Renzetti (2002: 157, 158) also concludes, 'In general, the demand for water is a function of its price, the prices for other goods, the scale of activity ... and the nature of the preferences or technology of the decision maker. ... except in very unusual circumstances, the value of water is neither zero nor infinite', but goes on to note that economic theory does not provide guidance on the actual decision making by the consumer. Box 6.2 and Figure 6.6 summarises the principle of water pricing mechanisms, and in particular the different goals they may serve: 'flat rate' primarily serves cost-recovery objectives, while volumetric,

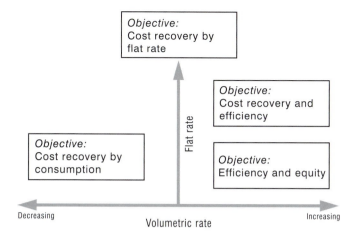

Figure 6.6 Relationship between pricing structure and management objectives
Source: after Montginoul, 2007: 864; permission to use granted by Springer
(*Water Resources Management*)

and particularly 'rising block', tariffs may also be aimed at reducing water use (and conserving environmental resources), particularly among those who consume most water.

One obvious observation on the data in Box 6.1 is that calculated values of water price elasticity vary across a very wide range. The variety in prices and structures also begs the question, 'Which is best?' Ayoo and Horbulyk (2008) suggest that this can only be answered in terms of the purpose of water pricing. This can vary widely: from cost-recovery and economic-efficiency considerations, to revenue or profit generation, to equity or social welfare or environmental protection.

It needs also to be acknowledged that unless the consumer understands the pricing regime, and has real-time information about their water consumption rates, then their behaviour is unlikely to be modified in the manner expected. Thus, complex pricing structures may not be effective. Kolokytha and Mylopoulos (2004) examined residential water consumption in Velos, Greece using questionnaires and three month water consumption data. In 1985 an increasing three-tier block rate structure (later four-tier) was implemented for residential customers with a fixed service charge. They found that 64.8 per cent of residents did not know the price of water, yet 48 per cent claim it is expensive. With only one out of six residents changing consumption in light of two-monthly information, they concluded that water pricing is not effective in achieving conservation. The need for greater public awareness of

pricing structures is also noted by Kenny et al. (2008) for the USA and
Martinez-Espineira (2002) in an analysis of water demand in North West
Spain using monthly data. Kenny et al. (2008) refer to technological
improvements through the Water Smart Reader that provides real-time
water consumption data.

There are concerns that responses to water pricing may be short-lived
and that, after a small decrease in consumption, previous levels are
re-established. In contrast Bar-Shira et al. (2006) found over a five-year
period that irrigation farmers in Israel (1992–97) responded immediately
to price rises and consumption remained price sensitive with an average
elasticity of −0.46 over the study period. The problem with including
discussions of agricultural water use alongside those of domestic
households is exemplified by Jones (2003), for the OECD, where only
Austria and the Netherlands make charges that are comparable. Chen
et al. (2005), in a study of a rural area of China, also argued that all water
pricing should follow the same patterns as for domestic water supplies,
with the expectation of irrigation efficiency gains. Burt (2007) argues the
need for volumetric charges for irrigation water, on grounds of financing
infrastructural improvements. Elsewhere, Jones reports a marked subsidy
for agriculture, with households paying more than 100 times as much as
agricultural users.

There is debate over which is the most appropriate pricing system for
agriculture as well as domestic supply. Bar-Shira et al. (2006) observe a
trend for adopting block rate tariffs for commercial and agricultural users
justified as aiding small producers and businesses and meeting social and
welfare objectives. It is, also criticised however, for supporting inter-
farm water inefficiencies because of their heterogeneity. In their study of
Israel they found that a block tariff structure lowered water use by 7 per
cent, and small farmers would pay a lower average price and use more
while larger farmers would use less and pay more. Russell et al. (2007)
looked at the opportunities for using economic instruments to manage
water in the Middle East and North Africa where agriculture accounts
for between 65 per cent (Algeria) to 95 per cent (Syria and Yemen) of
water withdrawals. As in Reisner's (1993) observations on irrigation in
the western US (Chapter 2), and Shah's (2007) account of northern India
(Chapter 4), they note that the cost of agricultural water is heavily
subsidised and it will be politically difficult to redress this under-pricing
of water. Gyau-Boakye and Ampomah (2003) contrasted approaches to
water pricing in Ghana which changed from heavy subsidies until 1985
to full cost recovery in urban areas with only operational and maintenance costs recovered for rural supplies. This was later changed to

strengthen rural community engagement plus a 5 per cent local contribution to capital costs but the authors note that greater private-sector involvement is required to finance these developments. Moreover, as noted in the case of the Office du Niger irrigation scheme in Chapter 4, a straightforward 'cost-recovery' approach raises the likelihood of a significant number (up to a third) of irrigators being unable to pay their water fees in the absence of a subsidy.

While there is now sufficient evidence that demand for water can be influenced by price, Kenny et al. (2008) point out that much still needs to be done to better understand behaviour, and hence provide more accurate forecasts of changes in consumption, while communication of pricing information also needs to be improved. Sharp (2006) echoes this perspective and criticises previous studies of demand management for their aggregation of data and consequent failure to understand household decisions sufficiently. Jansen and Schulz (2006) also note a lack of household (micro level) studies in the WDM literature and cite a household study from Sri Lanka which showed demand was price inelastic and income inelastic so price increases may not conserve water. Their study of demand in South Africa found that low-income groups demonstrated modest change to prices (-0.324), yet high-income groups proved to be highly price sensitive (-0.967) – as in the study we noted earlier, by Kenny et al. (2008) in Colorado, USA. They concluded that cheap water for poorer groups did not lead to greater wastage.

It is argued more needs to be done on developing the culture in which water demand is managed through informed choices which require real-time information on consumption and an easily understood tariff structure. The South African case is of particular interest because the 1997 Water Services Act explicitly recognised a 'right of access to basic water supply and sanitation' and stated that 'reasonable measures' must be taken by municipal water service authorities to realise these rights. Concerns about the affordability of water supply for the 29 million South Africans estimated to be living on an income of less than R1000/month (in 2005, data cited by Brown, 2005) prompted the government to proclaim a right to 'Free Basic Water' (FBW) of 6000 litres per household per month (equivalent to about 25 lcd). This subsidised water was to be funded in a variety of ways. In urban municipalities with substantial income from metered water users, a rising block tariff recouped the cost of the initial 'free' 6000 litres through higher tariffs on those who consumed more water. In less wealthy small towns where metered supply predominated, a system of targeted credits was advocated. In rural areas water supply was capped at a 'free basic' level by 'appropriate service levels' provided through standpipes with restricted flow rates. A feature of black people's resistance in apartheid South Africa was a boycott of payment for services, although Brown's (2005) study in Nelspruit led her to argue that:

> a culture of non-payment is not simply a legacy of the rates boycott, though this is still within living memory; it is made up of many contributing factors which have resulted in an environment where it is the norm not to pay service bills, regardless of ability to pay. There is evidence these factors do exist, and they may be socio-economic (lack of experience in prioritising, budgeting and managing household finances – a legacy of the deeply ingrained combined flat rate system), ideological (anti-privatisation sentiment), political (miscommunication of the FBW policy and a lack of political will to enforce cost recovery) as well as grievances (inability to understand meters or bills).
>
> (Brown, 2005: 20–21)

Gaudin (2006) also looked into the notion that lack of information on consumption and price i.e. billing information hampered the effectiveness of water pricing. It was found in the USA that price related information increases price elasticity by 30 per cent. This is clearly a possible reason for lower than expected elasticities in previous studies.

Water pricing as an alternative to supply enhancement?

Figure 6.6 shows that there are global differences in the amount of water privatisation across different regions, but there have been significant increases over the last ten years. The value of the privatised market is estimated to be US$45 billion (2005), although Anand (2007) notes that less than 5 per cent of the world's population are supplied water by the private sector.

In this recent period we see a trend toward treating water as an economic good rather than as a basic human need, a change many have argued was required in the face of increasing water scarcity (i.e. increasing cost of additional supplies). Kolokytha and Mylopoulos (2004: 263) are emphatic about the need for a change in direction, 'demand management is considered to be the best potential solution to meet future (water) needs' because, they argue, supply enhancement does not effectively deal with growing competition between consumers. Sharp (2006) notes that for England and Wales, where 40 per cent of abstracted water is

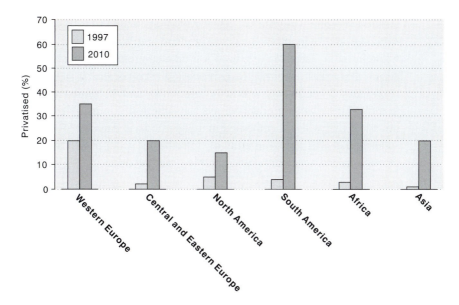

Figure 6.7 Predicted water supply and sanitation privatisation trend from 1997 to 2010

Source: after Swyngedouw, 2006

used for household consumption, the policy of government agencies, e.g. DEFRA and OFWAT is firmly in favour of managing demand. Many argue water demand management has been tested and shown to work, for example Kenny et al. (2008) found that, in Colorado, a range of water conservation measures in response to drought after 2002 reduced demand by 8 per cent in the first year, and by 26 per cent a year later. Widespread support for using pricing to manage water demand is also found by Page (2005) in the European Water Framework Directive, World Trade Organisation and the World Bank. However, the same author argues that it is contradictory to seek greater equity of access to water (e.g. in developing countries) while using economic means to control access.

While water pricing has widespread institutional support, therefore, there are many who question its effectiveness, while others question the right to place a value on a basic human right, i.e. water should be treated as a common good. Linked to this debate over the effectiveness of treating water as a commodity is a sometimes ideological discussion over the privatisation of water. Anand (2007) suggests that the notion of fairness is paramount in attitudes to privatisation and whether water should be treated as a public good. That is, how much they trust the government to ensure everyone is treated in a similar manner. Anand uses this as the explanation of why privatisation in England and Wales was accepted while in India, Tanzania or Ghana it produced protests. Page (2005) argues that water cannot be easily commoditised because it is different from many other objects: it is heavy and bulky and therefore tends not to move far between production and consumption. It is also the case, however, that water is embodied in many goods, and a visit to a supermarket will reveal an array of 'inter-basin' water transfers through traded agricultural and potable water products.

It has been shown that within water-demand management (WDM), 'water pricing' is not a simple matter of water 'supply' and 'demand', but a complex array of measures which may be deployed to achieve different goals, from promoting social equity to more efficient use of water resources. However, it is by no means clear that attempting to balance different policy objectives can be achieved by more complex tariff schemes. Ayoo and Horbulyk (2008) examine pricing schemes in Canada, where high demand is in the south (of Canada) and most water resources are to the north. They found water pricing for domestic consumers varied widely, such that 37 per cent of households have a flat rate, 39 per cent constant volumetric rate, 13 per cent declining block tariff and 10 per cent increasing block tariff. Yet, in some cases, the amount charged is less than 50 per cent of the supply cost and there is growing concern that water pricing is not meeting policy objectives. Similar concerns are also raised for Australia, and yet institutions such as the World Bank, OFWAT and EC continue to support pricing as the principal element of WDM. One issue, therefore, is the gap between simple models upon which water pricing is based (see Box 6.1) and the multiplicity of factors that apply in any specific context.

WDM has become synonymous with accepting water as a global commodity, embedded in the growth of water privatisation and increasingly criticised for not delivering the expected benefits. As Swyngedouw (2006: 61) states, 'The recent shift towards turning H_2O into a global commodity has profound implications on the social and political meaning and cultural valuation of water' and goes on to explain that supplying water increasingly serves economic aims of growth and profit maximisation. From this perspective the broader public interest in the environmental need to reduce consumption is contested by the private need to recover costs, including commercial criteria of returns on investment.

We are, then, at an interesting turning point in the management of water where the need to raise capital for investment and the adoption of economic principles for management have promoted private, non-state involvement in water management. Swyngedouw (2006) cites estimates that the annual investment required to meet the 'water for all' vision of Earth Summit 1992 and the MDGs is US$1.8 trillion by 2025 of which 70 per cent is expected from the private sector, including some US$550 billion in dams. This is taking place in a context where there is promotion of greater public engagement with sustainable development and public participation, in particular with water management (e.g. EU Water Framework Directive). Dungumaro (2007) cites public participation as a core of Integrated Water Resources Management. Hence questions of water pricing and private investment incentives intersect with those of governance, as stakeholder involvement has come to the fore. We will expand upon these in Chapter 7. However, it has become clear that water pricing needs to be implemented with a detailed understanding of how consumption behaviour can be affected.

Infrastructure and water conservation

Metering

A range of technical elements are prominent within most WDM strategy documents. For example, Jordan's National Water Demand Management Policy specifies, in addition to water pricing and raising public awareness, adoption of universal metering, leakage detection and reduction, and improving plumbing standards (Abdel-Khaleq and Dziegielewski, 2006). Magiera et al. (2006) found that across the Middle East and North Africa strategic management of water included water-efficiency measures. These typically include reducing leakage losses and reusing water, and, during periods of particular scarcity, water rationing. Both publications illustrate that conservation of water resources is achievable with more effective and efficient use through a combination of technical and behavioural responses. In this section we examine the first of these, the technical means available to conserve existing supplies. These typically include: metering; improving delivery infrastructure (leakage, retrofitting, pressure reduction); and wastewater recycling. We shall briefly consider each in turn.

Previously we have seen that, for demand to respond to water pricing, there needs to be timely monitoring of consumption. Thus OFWAT in England and Wales have supported the installation of water meters on new household properties, but Sharp (2006) reports that 74 per cent of households continue to pay a flat rate for water, based upon the notional value of their property, and not their actual consumption. Jones (2003: 11), writing for the OECD, states, 'The metering of water consumption is a prerequisite for the application of efficient water pricing policies', with two-thirds of OECD countries having at least 90 per cent of single family dwellings metered.

There are costs associated with the installation of meters, notably in multiple household dwellings. Meters may also suffer damage as they provide a focus for public opposition over water-pricing policies. Chambouleyron (2004) notes that water-cost estimates tend to ignore metering costs, since they focus on water production (abstraction, pumping, storage, etc), distribution, treatment and disposal and on marginal user and environmental costs. The result, it is argued, is that calls for universal metering seldom are based on an economic analysis that takes adequate account of metering costs. As a consequence, prices of metered water can end up being more than the 'socially optimal' marginal cost of delivering water. There is, therefore, an argument that the 'socially optimal' level of metering may be less than universal metering for all water consumers. In this case metering should focus on those consumers for whom 'savings in water costs more than offset the fall in consumer surplus (due to the increase in the volumetric charge) plus metering costs' (Chambouleyron, 2004: 306). This is evidently going to apply to those water consumers who will significantly reduce their water use in response to higher charges, and we have seen earlier that this will tend to be those consuming larger amounts of water. Thus, there is in the concept of 'optimal metering' the possibility of metering only those most able to afford the expense of metering. However, in order to achieve this, Chambouleyron advocates a fully commercial approach, relying on water users themselves to decide whether they will gain from meter installation.

In practice, water metering (and reduction in consumption) may present conflicts of interest for water companies. Swyngedouw (2006) argues the slow rate of adoption of water saving technologies is linked to the profit margins required by privatisation. Bakker (2000) also reported in her analysis of drought in Yorkshire (UK) that there had been relatively little investment, especially on leakage detection, and insufficient installation of meters. Meter installation and volumetric charging are associated with the idea of water as a commodity and water consumers may therefore view their introduction with suspicion that profit, not conservation or social welfare, is the prime motivation. And yet, without metering, it is difficult to manage demand and more difficult to achieve water conservation through leakage detection. It is also possible that the effect of metering on water consumption will only be significant when water charges are relatively high. Sharp (2006) reviews literature on the behavioural responses to water meters and notes that water consumption is often habitual hence difficult to change especially when, as in the UK, water costs are a small part of the household budget. She also considers that it is possible information on water consumption has a greater influence upon behaviour than the charges, called an 'information effect'. Nonetheless, OFWAT (2007) report no trend of decreasing consumption (1995 to 2006) in the UK, where household consumption remains stable at 145–150 lcd, although unmetered households consume an average of 154 lcd compared to 133 for metered households. Thus, there is only modest encouragement to introduce water saving measures via metering.

Infrastructure

Another potential function of metering is to support the management of water-supply infrastructure. Page (2005: 302) notes, 'It is imperative that more capital should be fixed in the form of pipes, taps and reservoirs in the developing world, but there is good reason to be sceptical of the claim that a single policy of commodifying water is likely to provide a universal solution.' The growing water infrastructure has led to concerns over the staggering amounts of water lost through leakage and associated transmission losses alongside evaporation. Nixon et al. (2006) state that pipe leakage is a worldwide problem and cite an average loss of 36 per cent for the Asia-Pacific region, within a range of 8 to 62 per cent, and 25 per cent being common in the USA for *new* systems. Lahlou (2001) suggests losses of 10 to 20 per cent can be considered 'normal' although not all losses are due to leaks, as theft and poor monitoring may also be factors. These are obvious financial losses as well as waste of an important environmental resource, while leaks can reduce water pressure so that higher energy, i.e. pumping costs are incurred.

The problem of water losses is not only encountered in urban areas; as Chapter 4 illustrated many irrigation systems have efficiencies well below 50 per cent. Continual maintenance of infrastructure is an obvious requirement but this has not always been the priority. During periods of acute water shortage greater focus may be applied to the water infrastructure. However, Bakker's (2000) study during the Yorkshire drought of 1995 found that leakage rates actually increased, becoming roughly equal to domestic consumption at the peak of the drought.

Leakage detection is a demanding strategy and not without difficulties, requiring a water-supply infrastructure that enables consumption to be monitored and charges levied. The section above explained the problems of monitoring consumption through meters. Even at a crude scale, monitoring can indicate a problem area that might then require more detailed local investigation. Lahlou (2001) describes the use of geo-phones to detect leaks through sound to pinpoint the location of a leak. The ability to save water resources has encouraged a growing industry for leakage detection, as an internet search will reveal.

In combination with leakage reduction, OFWAT (UK) requires water companies to promote water-saving devices such as reductions of toilet-cistern flushes while also increasing water efficiencies. Sharp's (2006) analysis of websites notes the following as being heavily pro-moted: rainwater storage; toilet-cistern displacement; water audits; replacement of household appliances with more water-efficient models. Retrofitting water-saving devices is one of the most common methods used to decrease consumption. For example, Deoreo et al. (2001) noted that in the USA household consumption could be significantly reduced to around 150 lcd (see Chapter 1 for comparable rates). Examples include the use of artificial turf in gardens, toilets that flush effectively using lower volumes, installing more efficient showers that consume less water and more efficient household appliances (Maddaus et al., 2008). Kolokytha and Mylopoulos (2004) cite examples of 20 per cent decrease in consumption through residential retrofitting, while using 'grey water' (e.g. rainwater) can save up to 39 per cent, although Kenny et al. (2008) caution that lower rates of reduction (around 10 per cent) may be more realistic. Beyond changes within the household there is also a need to recognise that water conservation can be aided by changing the ways in which urban areas are planned and managed (see also Chapter 5). White (2008) argues for a more 'absorbent' city to better manage water resources, while Gill et al. (2007) present ideas on green infrastructure that can aid urban adaptation to climate change impacts.

Recycling

We saw in Chapter 2 that re-use of wastewater is a key factor in con-serving water in more industrial and urbanised economies. It clearly hinges on the growing twentieth-century emphasis upon safe sanitation. We need to recognise, however, that this issue of wastewater recycling has been created by the decision to use water as a means of disposing of waste. The ancient Romans are often credited as having introduced sewerage systems (pipe and channel infrastructure) for the disposal of human wastes based upon water transport (Tebbutt, 1998). Current systems, however, owe much to the rapid development of waste-treatment techniques during the 'sanitary age' in the latter half of the nineteenth century where sewage-treatment works were built to harness natural purification processes (Newsom, 1994). It is important to remember that these water-using systems replaced 'earth closets' for waste disposal, and the latter can still be found in rural parts and where water conservation is an imperative.

In Chapter 1 (Box 1.1), we reviewed the classification used for different types of waste waters. Contaminated water can be treated through desalination (see Chapter 5), but there are more traditional and less energy-demanding techniques. Systems for removing organic materials and sediment, as found in sewage effluent, are well established as primary, secondary and even tertiary treatment (Viessman and Hammer, 1998; Abbassi and Al Baz, 2008) based upon techniques such as primary settling through to activated sludge and use of bio filters. Tebbutt (1998) identifies the basic methods as:

> Physical (screening, sedimentation, filtration, gas transfer)
> Chemical (coagulation, precipitation, ion exchange)
> Biological (biological filtration, activated sludge, anaerobic oxidation)

Sanchez et al. (2006) reviewed the use of filtration methods in developing countries noting slow sand filtration tends to be problematic for waters with high turbidity. This leads to the use of coagulation, which incurs higher costs and is more suited to larger, urban, installations. For rural areas, where capacity is lower and operational costs need to be minimised, they advocate multi-stage filtration, using a combination of coarse- and fine-grained materials. Natural filtration can take place through groundwater recharge, although a subsurface water detention period of up to two years may necessary for pathogens to be removed, and hence chlorination is necessary for potable supplies. Corkal et al. (2004) suggest several techniques suitable for rural Canada but note that wells less than 40m deep need to be considered the same as surface water because of the risk of contamination. Filtration and recharge can be enhanced through artificial percolation lagoons or even recharge dams, but this technique is best where stream or lake beds are highly permeable.

In Chapter 2, we noted that these water-treatment methods have been widely adopted since early in the nineteenth century and have arguably been a major reason for falling death rates in industrialised countries during the twentieth century. In non-industrialised countries the commitment to reusing domestic and agricultural wastewaters is less clear. The FAO–Aquastat (2009) report that in the Middle East total wastewater treatment is 2,663 Mm^3y^{-1} or 7.3 Mm^3d^{-1}, which is equivalent to the total amount of desalinated water in the region (see Chapter 5). Turkey dominates the region with 38 per cent of wastewater treated, 21 per cent for Syria, and 10 per cent for Israel. Given the amount of irrigation in the Middle East it might be expected that reuse of agricultural drainage waters was also a priority. Data on this are scarce, although the FAO-Aquastat reports substantial amounts are reused in Iraq (1,500 Mm^3) and Syria (2,246 Mm^3). This source has little data on waste water re-use elsewhere, but notes that in Latin America and the Caribbean only marginal amounts were treated while in Asia, with the exception of China and Japan, 'treatment plants for wastewater are inexistent'. Mara and Broome (2008) argue that it is possible to provide low-cost sanitation services in developing countries. They provide various reasons for the many failed previous projects, and they suggest the need for a move away from the conventional 'self-cleaning' approach through better understanding of the hydraulic principles.

It is also important to recognise the limitations in conventional approaches to wastewater treatment. Once treated water has been put back into the aquatic system where there may be need for further dilution by natural waters to ensure there is no further risk to the environment or those who subsequently extract it for potable water treatment. There may be cultural problems over the drinking of recycled waters, or even its use for growing crops and vegetables for human consumption, although Wilson and Pfaff's (2008) analysis suggested that this is overstated. Mara and Kramer (2008) caution that conventional wastewater treatment, with its focus on removal of nutrients and organic matter, may require further disinfection, while, in developing countries, stabilisation ponds can prove effective. We also noted in Chapter 2 that conventional water treatment is relatively ineffective in removing 'micro-pollutants', relatively complex, and often biologically-active, organic compounds arising from disposal of pharmaceutical and cosmetic products in waste water. The importance of re-using wastewater has recently been reviewed for water-scarce countries by Al Baz et al. (2008) who note that, for much of the Middle East and North Africa region, there has been a significant deterioration in the availability of renewable water resources over the last thirty years. They argue, however, that the approaches developed over the last hundred years in comparatively water rich areas may not be entirely suitable as they:

- are based upon sewers that require relatively high volumes of water flow derived from potable supplies,
- have a centralised and expensive infrastructure (capital and maintenance costs)
- only partly remove nutrients
- create problems of sludge disposal.

There are, therefore, reasons to consider that wastewater treatment as it has developed for the past two hundred years may not serve as effectively the needs of the future, and this is clearly a key field for future research.

Social approaches

Scaling–up

We have been concerned thus far with strategies that presuppose that either the state has the leading role in managing water demand (e.g. through establishing infrastructure, water charges and subsidies, etc), or that private-sector operators are contracted by government to take over such functions. In this section we consider the potential for demand management 'from below', in particular in contexts where investment is scarce. Chapters 2 and 5 showed that rural water supply and sanitation (RWSS) has historically lagged behind developments found in urban areas. The technologies for water treatment and supply are much the same but urban areas are more able to adopt such measures because they can secure the investments necessary for capital improvements and have greater access to resources for maintenance. There may also be other differences, for example in the degree of trust in government. Pumphrey et al. (2008), for example, report evidence that urban dwellers in the USA are more likely to trust government regulation than those in rural areas, although attitudes to water conservation were less differentiated and a combination of water pricing and regulation was felt to be effective in rural areas. The urban/rural dichotomy may, then, be overstated.

Although, as we have seen, the means to treat wastewaters have been well established since the mid-nineteenth century rural provision is still inadequate and the rapid growth of urban populations in the South means that many urban dwellers now live in squalor without appropriate disposal of wastes (de Bruijne et al., 2007). Ryan (2004) characterises the constraints as: limited resources (funds, human capital, institutional capacity); lack of knowledge or shared understanding; and 'resistance to change'. Yet, it is also the case that substantial improvements in sanitation, water treatment and water conservation do not require very sophisticated equipment.

A key question concerns the scaling-up of very local initiatives to achieve larger impact. Community-driven approaches, such as the Community Led Total Sanitation (CLTS) described by Deak (2008, see below), offer some promise, but there is no 'magic bullet', nor consensus, over the best approach. Moreover, the management of water resources does not happen in isolation, but will be affected by other policy priorities such as alleviating poverty. It seems unlikely that rural water management can depend on water projects requiring major infrastructure and international funding. Rather, the expectation is that rural water development will be focused upon household and village provision. At some stage the attention paid to the behaviour of households needs to be scaled up to community and even national levels for full impact to be realised. During the past two decades there has been a great deal of interest in the possibility of community-based natural-resource management, inspired by empirical studies of 'self-governing institutions' managing resources as commons (Ostrom, 1990).

Ryan (2004: 57) recently reviewed the literature on scaling-up, taken to be the 'accelerated implementation of safe sustainable water supplies in a timescale that might be judged acceptable'. Deak (2008) further distinguishes between vertical scaling (institutional scale) and horizontal scaling (spatial in scale). A difficulty in judging past efforts has been an over-emphasis upon anecdotal reporting of achievements rather than having clearly defined community-supported objectives. Nevertheless the depressing conclusion reached by Ryan is that up-scaling is unlikely to be successful within existing institutional conditions. The role of NGOs emerged as a clear issue, in particular how they are able to coordinate between similar organisations, and with government agencies. Ryan (2004) describes the following key roles and a 'check list' of questions for different stakeholders:

- Local community: managing the process and in control
- Central government: institutional support through financial and legal provision
- NGOs: facilitation, support and training

- INGOs: advocacy and support
- Private sector: supply-chain development
- International community: long-term commitment without intervention

On a more positive note, Deak (2008) draws upon experiences in Bangladesh on Community Led Total Sanitation (CLTS) projects which have now spread across the community. The emphasis of CLTS is that of fostering innovation and community commitment, that is social organisation rather than technology. CLTS has developed from the widely known Participatory Rural Appraisal (PRA) (Chambers, 1994; Cooke and Kothari, 2001; Hickey and Mohan, 2004), while also accepting that the state has a key role to play because 'grass roots' initiatives alone are insufficient. Deak lists the following observations for scaling up:

- Consideration together of coverage (spatial) and sustainability (timescale)
- Engagement of community and 'enabling environment', i.e. institutions and policy support
- Shift from projects (implementation) towards service delivery
- Vital role of partnerships
- Financial sustainability
- Capacity building and strengthening.

The dichotomy between a top–down or bottom–up approach is rejected in favour of a community-based approach that receives support in various forms from NGOs and the state. In trying to understand why successful spread occurs, Deak (2008) notes the importance of 'self spread'. This is a decentralised process through enthused champions and institutions. However, the latter may also sometimes create obstacles. A more thorough discussion of the shifting institutional configuration of agencies of state, market and 'civil society' in development, and particularly the tensions associated with the role of 'non-government' organisations, is found in Bebbington et al. (2008).

Moriarty and Butterworth (2003) report the use of the catchment as a natural boundary for an integrated assessment of water development priorities and strategies. Participatory watershed management has widely been applied as a means of conserving water resources. It seeks the engagement of small-scale 'land managers' in locally negotiated changes in land use that can improve water quality, groundwater recharge and other 'watershed services' (Hope et al., 2007). Such watershed approaches to scaling up water management take the catchment as the logical planning unit. However, this may not always translate into 'social catchments', while in many parts of the world which are reliant upon groundwater resources, a topographically defined surface catchment has limited relevance.

As with all collective management of natural resources, socio-economic heterogeneity among stakeholders presents problems for consistent engagement of individual resource users across the catchment. Brown (2007) found that stakeholder capabilities and preferences varied so greatly within catchments in South Africa that it was difficult to generate consensus as a basis for decision making. In the Mesta/Nestos river catchment between Bulgaria and Greece, water-management decisions are informed by activities in which different scenarios are presented to stakeholders and graded. This limits the range of options, and assumes similar understanding of each option. However, Ganoulis et al. (2008) suggest this may not matter since the goals of the process are less to do with up-scaling and community-led decision-making, but more to resolve disputes and inform trans-boundary negotiations by gauging local opinion. Integrated Catchment Management (see Chapter 8) is presented by Fenemor et al. (2008) as a strategy to enhance community resilience in New Zealand through developing active partnerships across communities and stakeholders in the catchment. They present this as a two-dimensional process involving the knowledge base of resource users and stakeholder involvement in management, a process of 'social learning', which we will next consider in more detail.

Social learning

Writing about hydropower projects, Hjort-as-Ornas (2008) recognises the potential adverse social impacts of such schemes, but, more significantly, describes (p. 4) growing stakeholder interaction in projects from Angola, Namibia and Vietnam, which goes beyond just engaging in 'mitigation' of negative effects of HEP. Relevant to such observations is the concept of 'social learning', described by Van Slobbe et al. (2006) as a move away from linear planning, with its emphasis upon expert knowledge. Blackmore et al. (2004: 6) explain social learning to mean: 'a process of knowing based upon experience and practice that is developed collectively and interactively among multiple interdependent stakeholders'. In their analysis of UK experiences of stakeholder engagement through an EC-funded project: Social Learning for the Integrated Management of Sustainable Use of Water at Catchment Scale (SLIM), they found a lack of evaluation of effectiveness but reported a number of promising case studies:

- Stakeholder groups and the Mersey Basin River Valley Initiatives
- West Country Rivers Trust work with farmers to tackle diffuse (agricultural) pollution
- Environment Agency and WFD.

They concluded that obstacles to stakeholder participation include the complex institutional context of water management in the UK (see Chapter 7) and lack of experience with the social-learning approach. On a more optimistic note, Van Slobbe et al. (2008) cite an example from North Brabant in the Benelux region of Europe. Faced with over-use of groundwater, the local authority and water boards actively engaged with some 3000 farmers and glasshouse owners to secure improvements in water conservation. This represented a break with the traditional approach of engaging with professionals and specialist agencies, although the authors also cite an example from Sri Lanka where bottom–up approaches have led to too much participatory engagement, and, as a consequence, inefficient water management through failure of regulatory instruments. They conclude that participatory practices require invest-ment of time for interactions and negotiation, that social spaces for this to take place are required, and there is a need for patience and tolerance as results are not instantaneous.

Brown (2007) has also cautioned about the limitations that participatory approaches may offer in practice, and the frequent absence of evaluation of the outcome of stakeholder involvement. Her study in South Africa suggested that, where stakeholders have strongly competing interests in water use and where there are major asymmetries in information, it is unlikely that consensus will be reached through consultation among 'local stakeholders'. Another South African study (Brown, 2009) broadly supports this conclusion. It also questions whether 'social learning' raises too many expectations with regard to stakeholders' collective interest, in collaborative management, being greater than their individual benefits from non-collective water use. These findings suggest that ideas of water users' engagement in water management need to be understood within a broader political and economic context that signifi-cantly constrains the possible outcomes from local participatory process.

Gender and water development

The key role for women in development has been widely recognised and was enshrined in MDG3 to promote gender equality and empower women. The UNDP's (2010) website for women's empowerment states forcibly that 'Development cannot be achieved if 50 percent of the population are excluded'. There are further examples of widespread support for this position, such as the Women's Environment Development Organisation (WEDO, 2010), USAID (2010) and the World Bank (2010b), while the volumes written on this topic have been summarised in an online bibliography by the Global Development Research Centre (GDRC, 2010). Here we have a more narrow focus with our principal concern being the roles for women specifically in water-resources development, both as beneficiaries of improved water supplies and as key players shaping demand in the water-development process.

In the past, water-resources development assumed that households would alter their practices in a predictable manner to take advantage of improvements in water supply, but this often neglected local customs and habits. Nyong and Kanaroglou (2001: 387) argue, 'They (the water planners) fail to recognize that newly available water may not be used optimally because it does not conform to the existing norms of such social groups'. Hunter (2006) also argues time allocation studies, seeking to explain household decision-making based upon labour supply, assume predictable responses to economic incentives, but thereby ignore the role of culture and intra-household power relations that mean changes tend to benefit men at the expense of women. From a household perspective, it might be expected that whoever manages the household's water supply has the capacity to influence water-resources consumption and sanitation improvement. Many authors report a primary role for women in securing household water. Were et al. (2008), for example, found 80 per cent of household water in rural Kenya was collected by women, a pattern widely reported for other rural areas in Africa and Asia. Perceptions of water use are similarly highly gendered, as Makoni et al. (2004) observed in Zimbabwe, where men regarded water as being mainly for drinking, while women's concerns covered also the health and hygiene benefits of clean water.

As a consequence of such perceptions of women's crucial role in household water development, many have argued that women remain an under-utilised resource in water-development approaches that are often dominated by external agencies and men. The UN WWDR (2006: 15) states, 'Of the 1.3 billion people living in abject poverty, the majority are women and children. They also happen to be the largest group systematically under-represented in water management arrangements'. It goes on to identify women's roles in household hygiene, food and water supply. Fonjong (2008) also explains the desire to engage women in water-resources development as informed by perceptions that often women have a closer relationship to nature, and water in particular, than men. However, there are risks of oversimplification and stereotyping associated with such 'essentialised' interpretations of women's role in water management (Jackson, 1993). Women are not everywhere central to household water collection, as Nyong and Kanaroglou (2001), working in north-eastern Nigeria, found that women were not always the primary adult collectors of water. In areas which were predominantly Hausa, for example, cultural requirements of gender segregation meant that less than 30 per cent of water collections were by women while 52 per cent were by adult males. There is also evidence that 'cultural' gender divisions of labour are subject to change, as Thompson et al. (2001) report in East Africa an increasing involvement of men in water collection, using carts or bicycles to transport water in plastic jerrycans. This is associated with the growth of informal water markets, especially in urban areas. More generally, then, there are risks in only focusing consultation and engagement with one group while ignoring male roles, children's contributions and the needs of the elderly.

It is interesting that so many studies on women and water development focus upon rural areas and yet as we have already seen urban populations now exceed rural while much of the future growth in urban populations will be amongst the urban poor in Africa and Asia, where some of the worst living conditions are to be found. The same lessons of empower-ment with greater ownership and governance of resources from rural areas can be translated to urban poor. It can even be argued that women in urban households have greater access to assets than their rural coun-terparts, hence more opportunities exist (Moser and Dani, 2008). Considering the urban poor in the 'south' Mitlin (2003: 404) identifies three significant development strategies:

- Strengthen social capital within low-income neighbourhoods, and political capital between low-income neighbourhoods and relevant state authorities
- Build financial capital enabling communities to invest in their own neighbourhoods and homes
- Influence the approach to urban development so that it is affordable.

Fonjong (2008: 473) looks beyond empowerment and suggests that further steps are also required to 'engender' all those involved in water-resources development, from local extension staff through to those taking decisions on policy formulation and implementation.

While women can directly contribute to water development, they can also benefit directly from improvements in water supply and sanitation. The provision of clean locally available water has a key labour-saving benefit, as collection and storage of household water are tasks which can consume significant amounts of household labour, averaging between two and three hours a day in East Africa (Thompson et al., 2001). Fisher (2008) notes as further benefits:

- Health and well-being (enhanced through safe and timely access to water and sanitation, more hygienic pregnancy and childbirth, the cause of 15 per cent of all maternal deaths).
- Education (reduced sickness and less requirement to care for the infirm enhances educational opportunities for girls)
- Income generation (Asian and African women walk on average 6km for each collection of water thus time is freed up for other activities while water itself is a commodity that can be sold).

It is also important to recall that Chapter 2 identified simple improvements in hygiene, such as hand washing, that could substantially reduce the risk of diarrhoeal diseases (Metwally et al., 2006), so we should not underestimate the positive contributions that can be made by relatively simple and inexpensive measures.

Engaging women in water development is more than simply reducing the time spent on walking to springs and standpipes, it demonstrates acceptance of empowering local peoples and engaging them in the decision-making process for water projects. Past experiences in water development suggest that men are already directly involved but there is still a need to engage more widely with women. Fonjong (2008), for example, observes in practice a 'gender blindness' (p. 467) in north-west Cameroon, where, although 'community development' draws upon knowledge contained in village development associations, yet women are hardly ever 'leading actors of change', even though they form 52 per cent of the population. Despite a 1979 convention to eliminate gender discrimination, it remains evident in lack of opportunities for ownership and inheritance of natural resources, lack of education and training and weak support services. For water in particular, Fonjong notes that in some villages women are prevented from:

- collecting water at night
- access to some streams and springs if pregnant or if these are designated as being inhabited by spirits
- fishing in rivers or owning fish ponds
- sitting on community water committees.

Thus, women's spatial mobility is limited in these 'public spaces'. Were et al. (2008) also note that in Kenya men restrict women's abilities to manage water, as men control land, finances, government and access to water. Fonjong argues that ownership or greater control over assets such as water resources is required to raise the esteem in which women are held, to increase their engagement in decision making and to create access to training and capacity building. It may be that such opportunities are presented through new water-supply provision, but this alone will not succeed without understanding the restrictions placed on women through the cultural barriers placed for moving from private (household) spaces and into the public realm. Sultana (2009) observes that, following discovery of arsenic contamination in many shallow boreholes in rural Bangladesh, restrictions on women's movement prevented them travelling beyond their local water supply in order to obtain non-contaminated water, and, since men do not traditionally collect water, this meant that some households continued to use arsenic polluted water. In contrast, Sultana also notes that, where poor families have access to a safe (green) borehole, women's empowerment may be enhanced, while recognition of the harmful effects of water from contaminated (red) boreholes has led, in some instances, greater engagement of males in water collection from further afield. There are, then, a range of responses in this one case study. This suggests care is needed to fully understand not only what is regarded as 'traditional' (and by whom) and how this may be modified by change. The introduction of irrigation has provided classic studies of the erosion of women's control over rice growing in West Africa as project administrators register irrigated plots in the name of (male) 'heads of households' (Carney, 1988). More recent study of water management in Swaziland (Peter, 2006) also provides evidence that water developments such as irrigation may reinforce gendered contestation over control of crop growing.

Were et al. (2008) investigated the role of women in three water community projects in the Kenya Highlands. Kenya has a long history of promoting self-help water groups, such that, by 2000, 30 per cent of rural populations were supplied by community-managed schemes and the Water Act of 2002 established the role of Water User Associations. Their study revealed very low engagement of women in community groups with decision-making powers, even, in one example, where women had been responsible for the creation of the group to improve water supplies. Many of the reasons provided above apply to the Kenyan study: men control land, access to water and public spaces. They further note that meetings were invariably organised in the evenings, thus excluding women, who were at this time responsible for cooking and looking after children. Registration of land primarily to male family heads means men are at the forefront of discussions over access to, and development of, any water resources. In contrast, women's roles tended to increase once the water source was operational. The study did reveal that, in practice, women gained some engagement in the management of the water supplies through informal means, but this was of a limited nature. Were et al. (2008) argue that the formation of water community groups needs to recognise the importance of gender relations through better guidelines on women's participation in the implementation of the Water Act. Peter (2006) notes the (then) draft Swaziland water policy emphasised the important role of women and increased representation of women at all levels.

Whether altering policy guidelines and appointing women to committees will bring about the cultural shift required is debatable. Many of the studies cited above have described how women are excluded from water resources projects while emphasising their central roles in water collection and consumption. There are fewer examples of how to empower women in decision making beyond formal institutional arrangements. Sultana's (2009) review of the gender and water literature finds an emphasis, as revealed above, upon distinguishing between ownership, control of productive assets and the different ways men and women are able to control natural resources. She advocates analyses of 'embodied subjectivities' (p. 428):

> Rather than seeing women as rational resource users or victims of environ-
> mental degradation, I argue that greater attention needs to be given to the
> ways that gendered and embodied subjectivities are produced and negotiated
> through (water) resource management practices.

Hunter (2006: 171) concluded from a study of household strategies in Ghana that

> stakeholder perceptions of development needs should be of major concern
> to policymakers since decisions must be made with regard to policy
> priorities ... equitable stakeholder involvement in developing contexts
> has been problematic as a result of social stratification that hampers
> meaningful participation across groups. Specifically, fewer contributions
> are made by less empowered individuals, particularly rural women and
> those of lower class.

This leads to broader questions of representation and governance which we will consider in Chapter 7.

Summary

During the twentieth century global fresh-water consumption increased sixfold (Hunter, 2006), more than twice the population growth. Lundqvist (2000: 259) notes that the standard response to increasing water demands has been 'to withdraw and convey water. ... In some parts of the world there is simply no more fresh water to develop, in other areas there is no investment capital'. Thus, at the same time it is recognised that water resources are becoming scarce in some parts of the world, huge investments are identified as necessary to meet the UN Millennium Development Goals to satisfy unmet water demand. As we saw in Chapter 1, debates about water resources are strongly shaped by questions of how this investment cost is to be allocated. To the extent that this investment is sought from commercial capital markets, pressures for 'full cost recovery' will reflect the need to provide a commercial return on capital. For countries whose governments lack a tax base adequate to fund universal water services, new water strategies need to be developed that do not simply rely upon more capital expenditure to increase supply, but require demand management that treats water as a finite resource.

This chapter has considered three main approaches to water-demand management (WDM). Water pricing is one mechanism of WDM, which serves not only to influence consumption, but also to raise revenue. It has proved the most controversial, as it is often linked to privatisation of water supply and waste management. As Swyngedouw (2006: 61) puts it: 'The recent shift towards turning H_2O into a global commodity has profound implications on the social and political meaning and cultural valuation of water.' We have seen that the rate of water privatisation has not met expectations but it remains a key strategy. Ultimately, outcome of water pricing, in terms of water demand and conservation, will depend less on the simple models of water 'scarcity' and 'price elasticity of demand', on which it was based, than on more context-specific political factors that influence individual water-consumption behaviour.

A similar conclusion appears relevant to the second element of WDM we have considered, that involves technological measures such as water-saving devices and water recycling. These, too, were found to offer possibilities, but also limitations contingent on the behaviour of water users. We have then focused upon WDM approaches that focus on engaging water users directly in water management, in the belief that this is fundamental to ensure changes are completed in a sustainable manner. Thus, we have moved in this chapter from neo-liberal strategies to neo-institutional approaches (see Chapter 1), where the latter lay more stress upon governance and social learning through greater stakeholder engagement. Neither has proved wholly successful, however. The experience of the neo-liberal approach and water privatisation suggests that wider issues of ownership and equity need to be addressed alongside the capacity for private investment. Conversely, the efforts to implement the institutional principle have exposed the limitations of local participation in bringing about change, particularly where asymmetries of power and conflicts of interest present obstacles to agreed courses of action. In this sense, the widely promoted advantages to be gained from local-level, participatory and social-learning approaches to water management have, so far, proved over-optimistic, particularly where they have failed to take account of discriminatory social values entrenched in local custom, or of the underlying influence of state agency. This prompts a need to re-evaluate the role of the state and the political processes through which development goals and priorities are determined and implemented. We turn in the final two chapters to the issues of institutions and the control of water, first through discussion of boundaries of power in catchment conflicts, and second the governance of water.

Further reading

It is helpful to understand the context for placing greater emphasis upon water consumption in contrast to the long-held strategy of supply enhancement, which was discussed in Chapter 5. There are two key areas for further reading: the literature that addresses the wider issues of equity and access to water; and the more specific literature on water-demand management. For an overview see:

Anand, P. (2007). *Scarcity, Entitlements and the Economics of Water in Developing Countries*. Edward Elgar, Cheltenham.

DfID 2001. Addressing the water crisis: healthier and more productive lives for poor people. UK Department for International Development: www.dfid.gov.uk/Documents/publications/tspwater.pdf

Gandy, M. 2008. Landscapes of disaster: water, modernity, and urban fragmentation in Mumbai. *Environment and Planning A* 40: 108–30.

Lundqvist, J. 2000. A global perspective on water and the environment. *Physics and Chemistry of the Earth* 25(3): 259–64.

Page, B. 2005. Paying for water and the geography of commodities. *Transactions of the Institute of British Geographers* 30: 293–306.

Swyngedouw, E. (2006). Power, water and money: exploring the nexus. Human Development Report 2006 Occasional Paper. New York: United Nations Development Programme.

WWDR 2006. UN World Water Development Report, 2nd Report: Water: a shared responsibility. UNESCO, Paris: http://www.unesco.org/water/wwap/wwdr/wwdr2/

The following sources provide a background on water-demand management and the use of pricing mechanisms:

Butler, D. and Memon, F.A. (eds) 2005. *Water Demand Management*. IWA, London.

OECD 2003. *Social Issues in the Provision and Pricing of Water Services.* Organisation for Economic Cooperation and Development. OECD Publishing, Paris.

OECD 2009. *Managing Water for All: An OECD perspective on pricing and financing.* Organisation for Economic Cooperation and Development. OECD Publishing, Paris.

Renzetti, S. 2002. *The Economics of Water Demands.* Kluwer, Dordrecht.

To read about water conservation and waste management, see:

Al Baz, I. Otterphohl, R. and Wendland, C. (eds) 2008. *Efficient Management of Wastewater.* Springer-Verlag, Berlin.

Deak, A. 2008. Taking community led total sanitation to scale: movement, spread and adaptation. IDS Working Paper 298, University of Sussex.

De Bruijne, G., Geurts, M. and Appleton, B. 2007. Sanitation for all? IRC International Water and Sanitation Centre (www.irc.nl).

7 Catchments and conflicts

Water catchments as commons

We concluded Chapter 6 by stressing the importance of social criteria for managing water demand and the role of water governance. We identified the need to better understand how control over water is exercised and emphasised efforts to promote stakeholder and community engagement. The discussion considered the particular challenge of vertical integration, through up-scaling from the individual to the institution. Here in Chapter 7 we move onto consideration of horizontal scaling, or organisation of water control over the spatial scale, a key element of the 'holistic management' identified with the ecological principle (see Chapter 1, fourth section The Emergence of an International Agenda for Water-resource Management, p. 27).

Conceptualising water as organised and subdivided by hydrological catchments introduces two important features of its management. First, it establishes the inter-linkage of water resources via hydrological flows either of surface waters or of sub-surface aquifers. As a consequence, a specific water use(r) within a catchment will potentially affect other water use(r)s. Water within a catchment may thus be considered as constituting a commons, whose use is 'rival'. Second, the idea of hydrological catchments establishes boundaries to the water resource, and hence the notion of a finite resource, increasing use of which will generate scarcity and increasing competition among users. The idea of water catchments is thus an important means of reconceptualising a mobile, renewable, and (in aggregate) abundant resource into one which is finite, scarce, and, if demand is not managed, threatened by over-exploitation and degradation: a 'tragedy of the commons' (Hardin, 1968).

Such a discourse of resource degradation could be perceived as consistent with the widespread evidence of human capacity to radically alter the distribution of water through large infrastructure projects, in many cases creating large numbers of losers as well as winners (Chapters 2 and 4). Losers would include those negatively affected by deterioration of water quality through industrial pollution, altered sediment loads in rivers or saline intrusions into aquifers, as well as those affected by altered timing (e.g. seasonality) or quantity of river flows. Many would regard the case of the Senegal River valley, considered in Chapter 4, as exemplifying such a situation. Even more dramatic cases widely cited are those of the Aral Sea in Central Asia, whose level dropped by 15 metres in 30 years as a result of diversion of some 90 per cent of upstream river flows for cotton irrigation in the 1960s, and Lake Chad in West Africa, whose area is now less than 10 per cent of that in the 1960s due to rising irrigation demands on the rivers that feed it (HDR, 2006: 211–12).

A significant element in the diagnosis of such 'tragedies of the commons' is the observation that the water catchments in question straddle international boundaries. The negative social and ecological consequences are therefore attributable to failures in international cooperation. Considerable attention thus focuses on the determinants of international collaboration or conflict over trans-boundary water resources – river catchments, lakes, or aquifers that extend across international frontiers – both from the point of view of sustaining productive and equitable use and, particularly with regard to the latter, avoiding the prospect of 'water wars' between states contesting control of water resources (Wolf, 2006). Such trans-boundary water resources include 60 per cent of all river flows and some 40 per cent of the world's population (HDR, 2006) and are more common where national entities are small, notably in post-colonial Africa and, more recently, post-Soviet Eastern Europe. In this chapter, we review case studies of the behaviour of governments in contexts where it is supposed that the potential for conflict over water is high, the Middle East and North Africa, where water use is highest relative to availability of renewable supply (see Chapter 5), and where the 'geopolitics' of water are likely to be acute. First, however, we will consider the role of the hydrological catchment as a framework for water-resources allocation and regulation before exploring the linkages between the exercise of power and the distribution of water.

Catchments and regulation

The prevailing acceptance of market approaches to water management in the last three decades has promoted privatisation of the water resource, notably in the UK (Box 7.1), but also in a number of other countries among which Chile is the most unequivocal example. Elsewhere, recent water reforms have tended to assert water is a public good held in trust for the nation by the state. In some instances, notably South Africa, this has effected a nationalisation of what was previously regarded as private property. It is important to recall from Chapter 6 that privatisation is not necessary for the operation of a pricing system for water, since volumetric water charges are as easily applied by public water providers or management agencies. Also, state water-management agencies as well as commercial water companies may issue water users with permits to use defined amounts of water (or contracts to supply water) which may themselves be tradable. What is essential to the operation of such a system is an effective means of assessing the size of the water resource and of monitoring the amounts being used by different users. The failure to achieve this in South Africa (Brown and Woodhouse, 2004; Woodhouse, 2008) has proved a barrier to the implementation of a licensing system which was intended to allow re-allocation of water to black farmers who have been historically excluded from irrigation use.

More broadly, processes of institutional and regulatory reform have tended to move in one of two directions. A 'sectoral' approach where the emphasis is upon each economic sector, i.e. land drainage, irrigation, urban supply and so on. Where surface water does not dominate management this approach is more likely to be found. For example, Ahmed (2005) lists for Jordan the following lead institutions and roles:

- Ministry for Water and Irrigation (water- and wastewater-development programmes)
- Water Authority of Jordan (municipal water)
- Jordan Valley Water Authority (catchment and agricultural water).

In addition there are water responsibilities for:

- Ministry of Agriculture
- National Centre for Agricultural Research and Technology Transfer
- Ministry of Municipal and Rural Affairs and the Environment
- Water and Environment Research and Study Centre.

Box 7.1

Water Management in England and Wales

1945 Water Act enabled the amalgamation of previous water-supply companies and myriad of river boards such that between 1920 and 1963 over 2000 water 'undertakings' had reduced to 100 Water Boards, plus 50 Local Authorities and 29 privately owned water supply companies.

1973 Water Act created 10 Water Authorities with responsibility to plan and control all water uses including water supply and sewerage. The Chairman of each WA was appointed by the Secretary of State and local government was represented in the membership.

1983 Water Act reduced the role of local government in the Water Authorities with additional measures to reduce public engagement. To compensate, Consumer Consultative Committees were established by the WAs.

1989 Water Act (privatisation) created 10 Water Service Companies with responsibilities for water supply and sewage disposal based on the previous WAs. Environmental regulatory (environmental protection and monitoring) responsibility was separated and given to the newly created National Rivers Authority through the development of catchment management plans. The formulation of the latter had some degree of public consultation. An industry economic regulator was also created, OFWAT, with duties of setting prices and protecting customers while also recognising the needs for capital investment and environmental improvements. Ten regional Customer Service Committees (renamed Water Voice Committees since 2002) were created to represent customers.

1995: The Environment Agency (EA) replaced the NRA and merged with previous institutions for air quality control and waste management for greater integration of environmental management and policy formulation. Local Environment Action Plans replaced catchment management plans.

1996 Water Act introduced the requirement for the EA to contribute to sustainable development.

2000 EC Water Framework Directive (adopted in 2003 through UK Water Environment Regulations) requiring by 2008 for public participation to be included in water planning.

2003 Water Act links water abstraction to local water-resource availability.

Source: after Blackmore et al., 2004

Elsewhere, reform and re-organisation have tended to follow the trajectory towards management units focused upon large river basins or catchments. Chapter 2 charted the changing role of the state in water-resources development, using the USA as an example, where from the nineteenth century onwards the scale of operations meant that individuals were quickly replaced by private water companies which in turn were replaced by municipal authorities and from the 1920s US Federal agencies such as the Bureau of Land Reclamation and the TVA. An increasingly important driver of these processes has been the perception of advantages of managing water resources in a holistic or integrated fashion. In theory this would allow for a full inventory of all resources, above and below ground level, for competing demands to be assessed against resources both in a geographic framework, the catchment, and with modelling through time. Thus management and allocation decisions would be better informed, environmental considerations could be included and future demand requirements and environmental changes anticipated. Unfortunately, in reality, river-basins leak and prediction is challenging, while institutions operate in a legal, financial and political context. As a consequence 'Integrated Water Resource Management' (IWRM) has become the subject of considerable dispute (Biswas, 2004).

The evolution of water-management agencies in England and Wales exemplify both the move towards an integrated or catchment approach, but also a retention of regulatory function by national government. Cook's (1998) examination of UK water regulation notes the plethora of legislation from the early nineteenth century commencing with the 1830 Lighting and Watching Act, though land-drainage and flood-prevention measures can be traced to thirteenth century at least. Over the last 100 years in the UK water has been managed through river boards based around individual catchments. These were scaled up to reflect the need for larger spatial units that could better integrate competing water demands and different sources of water. Today there are 12 large-scale water (supply) and sewerage operators covering the UK, which are privately owned in England and Wales (see map at http://www.water.org.uk/home/resources-and-links/links/water-operators/supply-only/a4-supply-only-companies-map-0609.pdf). There is a close correspondence between the major geographical areas covering the water and sewerage operators and the boundaries of the Environment Agency which has key roles in promoting sustainable development and monitoring water, land and atmospheric environments in an integrated fashion (see map of EA regions at http://www.environment-agency.gov.uk/aboutus/organisation/35673.aspx). In practice there are differences, not least because while there are ten large privatised water supply and sewage companies covering England and Wales, there are also an additional 14 smaller, water-supply-only, companies. In many areas the surface catchment structure is not ideal as river basins vary in size. Whereas in East Anglia groundwater is significant, surface supplies dominate in the North-West where rivers flow East to West, but water resources need to be transferred from North (Lake District) to South to meet demand (Greater Manchester and Merseyside). Furthermore, abstraction (licences) is managed through CAMS (Catchment Abstraction Management Strategies, EA, 2002) using 129 smaller geographical areas. There are, then, a variety of catchments and regulatory boundaries in England alone, despite the attempts to create an integrated framework.

This pattern of the reform of water institutions during the twentieth century has been taking place across the globe but in such an array of local contexts, geographical and political, that it is difficult to generalise. Centralised control and expert-led decision making still dominates. However, as we have seen in Chapter 6, the move to greater participation by water users and other stakeholders in water governance has challenged this position since the 1990s, just as privatisation has added further complexity.

In South Africa, the 1998 National Water Act retained a national government agency (DWAF) as responsible for national water strategy, including definitions of environmental requirements (e.g. minimum flows to be maintained in rivers), questions of inter-basin water transfers and international agreements on shared river basins. It also divided the country into 19 'water-management areas' each to be managed by a Catchment Management Agency (CMA) funded by a water-management charge levied on water users within the catchment. Specific responsibilities were to be devolved to CMAs as their competence and capacity increased. In Brazil, a similar structure was established in 1997, with a federal agency (ANA) responsible for strategic planning and, of more immediate impact, for facilitating inter-state agreement on the use of waters that crossed state boundaries (in effect, most of the country's major rivers). The Brazilian water policy is also to promote the development of water-basin committees. As in the South African case, these are to be representative of water users in the catchment and are to serve as a means to reconcile competing interests among water users. The early implementation of this approach in the context of extreme drought in the state of Ceará was noted in Chapter 4 (see p.158). We may note that there is a significant difference between the South African case, in which the catchment management agency is supervised by a national water authority, and the Brazilian case where it is subordinate to the department of water resources in the relevant state. There is, in particular, a marked difference in potential connection between 'water management' and questions of local political and economic development, which has relevance for effectiveness of local water-management agencies in 'IWRM' as discussed in Chapter 8.

Other models include that of New Zealand, where large integrated institutions called Regional Councils have responsibility for not just water but also land, air and coastal environments (Fenemor et al., 2008). These were created in 1991 and replaced Regional Water Boards, thus representing a broader integration of water with other activity. Gyau-Boakye and Ampomah (2003) describe the organisation of water management in Ghana as being arranged around the GWSC (Ghana Water and Sewage Corporation) created in 1965 by combining previous responsibilities for urban, rural and industrial water supply and sewerage held by the Public Works Department and the Department of Rural Water Development. In 1994 the CWSD (Community Water and Sanitation Division of the GWSC) was created with specific responsibility for the rural sector. This alternative approach to the integrated catchments which is further demonstrated by another institution created in 1977, the GIDA (Ghana Irrigation Development Authority) revealing the country's specific needs. Recognising the need for an integrated and strategic approach the WRC (Water Resources Commission) was created in 1996 to formulate water policies on the regulation and management of water resources.

Water allocation as an exercise in power

While uneven spatial distribution of water plays a role in access to water (notably in terms of cost), it is important to recognise that the inequalities of access to water are a reflection of differences of power embedded in social relations. Thus, in a study of a community-managed water supply system in a village of some 5000 people in northern Tanzania, Cleaver and Toner (2006) found that the local Water User Association (UWUA) had successfully recuperated and extended a government-constructed piped water supply system that had fallen into disrepair. The UWUA organised voluntary communal labour from villagers to dig trenches for pipes, and managed a charging system for water delivered from private and public taps. While able to support its running costs, including the salary of a professional manager, and serving the whole village – thus eliminating the need for women to walk 12km to fetch water in the dry season – the UWUA consists of a relatively small number of residents willing and able to pay the membership fee, and is seen by some as a private company rather than a representative village organisation. 'Many felt that communal labour is for the "have-nots", whilst membership of the UWUA is for the wealthier people in the village' (Cleaver and Toner, 2006: 214). There is also evidence that the financial sustainability of the system is given higher priority, through increased prices charged for water, at the expense of affordability for the poorer villagers, who are directed towards the elected Village Council for assistance it has no funds to provide. Meanwhile, in parts of the village where public taps are few, wealthier villagers are able to sell water from their private taps at higher prices than those charged at public taps. For these members of the UWUA there may even be, therefore, a disincentive to extend the provision of public taps.

This sketch of community-managed water supply illustrates that, while local initiative and organisation can bring about an overall improvement in water availability, it cannot be assumed that this precludes the emergence of unequal access to water that reflects differences of wealth and influence. It suggests, on a very small scale, an association between water and power reflected, on a much larger scale, in extremely unequal water use that is characteristic of major cities in developing countries (Swyngedouw, 2004, 2006). Similar inequalities in water use are evident in societies subject to gross discrimination, such as South Africa, where Gini coefficients suggest that the considerable level of income inequality (Gini 0.64) is dwarfed by the disparity in direct water use (Gini 0.95) in rural areas (Cullis and van Koppen, 2007), reflecting historical exclusion of the black-majority population from property rights in water under the apartheid regime (Woodhouse, 1995). In another extreme example, discriminatory regulation of permitted well depth by Israeli authorities results in Israeli settlers using 53 per cent of the total groundwater extracted in the West Bank despite having only 13 per cent of the wells, so that annual per capita water consumption of settlers averages 620 m^3 per person compared to the Palestinian average of < 100 m^3 per person (HDR, 2006: 215–16). In each case, it is not scarcity of water that creates such inequality of access, but the social processes of unequal economic and political power through which scarcity is mediated.

The aggressive use of power in its diverse forms, including asymmetries of knowledge, in the assertion of unequal access to water is a cultural theme with international resonance (cf. the films *Chinatown* and *Jean de Florette*) and explains, perhaps, the durability of the scenario of 'water wars' on an international stage. Others observe that such conflicts have not, in fact, developed over the past 4000 years, pointing to the high cost of warfare relative to the low value of water as the main reason for this. Conversely, the considerable and additional mutual benefits achievable through cooperative strategies of developing shared (trans-boundary) water resources, notably through infrastructure investments, are argued to constitute a strong incentive to choose cooperation over conflict (Wolf, 2006). The achievement of such agreements will depend, however, upon the capacity of the different parties to support the cost and gain the benefit of the cross-boundary water agreements. As a consequence, the negotiation of such agreements may typically take many decades until agreement is triggered by the need for one or more of the parties to secure new or increased services from the shared resource. This can be seen in the 45-year negotiations between India, Bangladesh (previously East Pakistan), Bhutan and Nepal over the agreements for joint management and investment on the Ganges and its tributaries (Crow and Singh, 2000). The signing of five agreements within two years (1995–96) was partly a reflection on political change within India, including a greater influence of private industrial capital on government policy, but also of the growing pressure for expanded electricity-generating capacity to meet India's industrial and irrigation pumping needs. Of the five agreements signed, three were for joint development of hydroelectric power schemes in which the bulk of the electricity was to be purchased by Indian power companies.

Case studies of catchment geopolitics

As indicated in the previous section, the exercise of power permeates all water allocation, whether at the local level of a village or city, or at the level of international agreements on the sharing of waters of major river basins. Growing recognition that water-resource management involves reconciling competing demands for water underlies the concern of current best practice (see Chapter 1) with 'integrated' water-resource management. In this chapter we focus particularly on international contexts in which water occupies the position of a strategic resource, and thus forms the object of geopolitics. In the first of the cases examined here, that of the Nile, perceptions of competing water use among the principal riparians is relatively recent, and it is a concern with future change and uncertainty – not least that arising from climate change – that engages countries and regions that have historically had little political or economic interaction. A second case, that of the Jordan valley, concerns relations between international parties with a history of intimate and protracted struggle that has been characterised, by some at least, as being motivated by competition over water.

Both cases are located in the Middle East and North Africa (MENA) region which is widely regarded as facing the most acute shortage of water relative to supply (Allan, 2001; SIWI, 2006), due to a population increasing rapidly, at between 3 and 4 per cent per year, from 100 million in 1960 to 311 million in 2005 and projected to reach 430 million by 2025 (SIWI, 2006) and a low level of economic diversification in many of the countries of the region. As a consequence, about 87 per cent of water continues to be used for irrigation although agriculture contributes only about 13 per cent to the regional GDP, ranging for individual countries from 2.4 per cent (Jordan) to 25 per cent (Syria).

The Nile

Despite playing a leading role for more than four thousand years in the development of agriculture and civilisation, the Nile has been conceptualised as an integrated hydrological basin for only a matter of decades. For much of its human history the hydrology of the Nile in Egypt has been more linked to the development of the Mediterranean than with the highlands of East and Central Africa where its waters are generated. Since 1999, however, the Nile Basin Initiative, promoted by the World Bank and supported by multilateral and bilateral funders, has provided a mechanism through which a Council of Ministers of Water Affairs of the Nile Basin States (Nile–COM) have elaborated 'a shared vision and a set of agreed policy guidelines to provide a basinwide framework for cooperative action' (NBI, 2008). The great length (6670 km) of the river, and its interruption, by non-navigable rapids (cataracts) south of Aswan and by the Sudd swamps of southern Sudan, were undoubtedly factors that disconnected the exploitation of the waters in the arid lower basin states of Egypt and Sudan from any consideration of the hydrology of the river in the upper basin riparian states of Ethiopia, Eritrea, Kenya, Uganda, Tanzania, Rwanda, Burundi and the Democratic Republic of Congo. Together, the ten riparian states make up nearly a third of the African continent, and the Nile catchment itself covers 12 per cent of the continent. An overriding feature of the hydrological system is that all the water is generated by rainfall in the relatively poorer riparian states of the upper basin, while none is contributed by the two principal

Table 7.1 *Schematic summary of annual water flow in the Nile River basin*

	Flows (km³/yr)
Lake Victoria outlet	42
Upper White Nile (Mongalla)	52.6
Sudd evapotranspiration	26
Blue Nile mouth	48
Atbara mouth	11
Aswan inflow	84
Aswan evaporation	10
Aswan Outflow: Egypt	55.5
Aswan Outflow: Sudan	18.5

Source: based upon sources in Howell and Allan, 1994; see Allan, 2001: 65 for more details of Nile flows

water-using states of the lower basin, Egypt and Sudan. Egypt by far out-weighs all the other riparians in terms of economic and technical resources.

Flood levels were systematically recorded in Egypt from around 3000 BC, but current understanding of the hydrology of the Nile catchment relies on flow records first established at Aswan in the 1860s and extended progressively during the twentieth century through the establishment of flow gauges on the tributaries of the Nile in the upper catchment. The overall pattern that has emerged is summarised in Table 7.1, and Box 7.2 (see also Figure 7.1). The length and complexity of the Nile River basin, and the high variability of the rainfall that feeds it, make a precise measurement of system-wide water balances unlikely, but the figures in Table 7.1 highlight in a schematic way some important components that have been identified from data recorded during the past 150 years.

Box 7.2

Determinants of flow in the Nile

Total run-off of about 120 km³/year is delivered by two principal sets of headwaters: the Blue Nile draining the Ethiopian highlands, and the White Nile flowing through the great lakes of equatorial Africa, principally Lake Victoria, Lake Kyoga and Lake Albert. Without human interference, it is estimated that some 80 km³ would flow annually from the Nile into the Mediterranean, whereas in the 1990s only 17 km³/year reached the sea, suggesting that three quarters of the 'natural' flow was being used (Chesworth, 1994).

The Blue Nile, originating at Lake Tana, and the Atbara River, both fed by rainfall on the Ethiopian highlands, together provide about 70 per cent of the flow of the Nile reaching Egypt from Sudan, and, since 1966, feeding the reservoir of the Aswan High Dam. The White Nile flow emerges from Lake Albert as the Bahr el Jebel, crossing the frontier from Uganda to Sudan to spread across Africa's second largest wetland, the Sudd, from which it is estimated that at least 50 per cent of the inflow is lost as evapotranspiration (Allan, 2001). The White Nile that drains northwards from the Sudd is augmented by the Sobat before it joins the Blue Nile at Khartoum. The Nile waters stored by the Aswan High Dam (AHD) are divided under the terms of the Nile Waters Agreement (1959): 55.5 km³/year to Egypt and 18.5 km³/year to Sudan.

To this simplified picture it is necessary to add the element of rainfall variability, which caused the annual flow of the Blue Nile to vary between 20.6 and 79 km³ over the period 1961–1990 (Conway, 2005). In addition to inter-annual variation the rainfall and river flow records suggest a cycle in which a greater likelihood of low flows alternates with a greater frequency of high flows. Such 'decadal' variations in flow recall biblical references to seven years of famine followed by seven years of feast, and record the particular severity of the Ethiopia droughts of the mid-1970s and mid-1980s.

The determinants of the flow of the White Nile are more complex, but dominated by the level of Lake Victoria. Figure 7.2 shows that, from the 1960s onwards, a succession of years with heavy rainfall in the catchment draining into Lake Victoria resulted in a rise of more than 2 metres in the level of the Lake that has persisted for more than 40 years, though with decadal variations. As a consequence, the average rainfall for the Lake Victoria region was some 8 per cent higher for the period 1961–90 than for the previous thirty years (1931–60), but outflows from the Lake roughly doubled, from 20.6 km³/yr in 1961 to 50.5 km³/yr in 1964, as a result of the higher level of the Lake (Conway, 2005). A further effect of this greatly increased outflow from Lake Victoria was rapid flooding in southern Sudan, with the loss of tens of thousands of lives and hundreds of thousands of livestock (Collins, 1994: 124), and a major expansion in the area of the Sudd wetland, from 13,000 km² to about 30,000 km².

The accumulation of flow and rainfall data for the Nile river basin over the course of the twentieth century have enabled the identification of a relationship between rainfall intensity in East Africa and atmospheric circulation patterns resulting from changes in ocean surface temperature differentials in the Indian Ocean (Conway, 2005). While this suggests mechanisms through which future rainfall, and hence flows in the Nile, may be affected by rising ocean temperatures associated with global warming, the specific direction of effects on rainfall is as yet unclear.

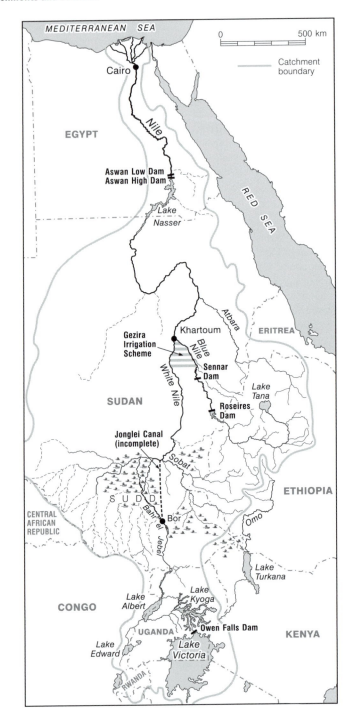

Figure 7.1 The Nile River catchment

Source: after Agnew and Anderson, 1992: 232; permission to use granted by Taylor & Francis (Routledge)

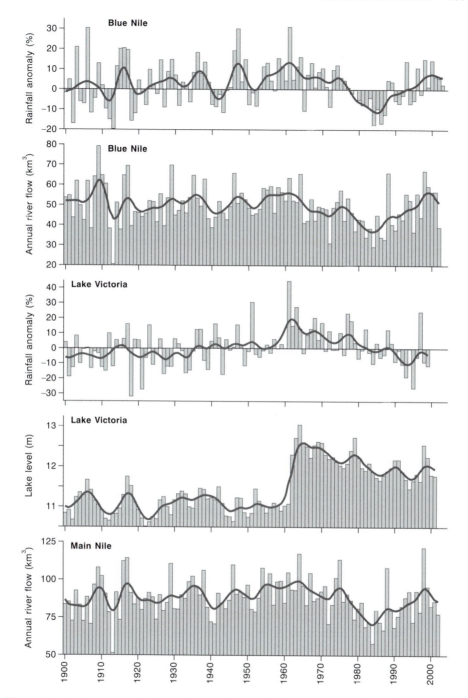

Figure 7.2 Time series data of annual rainfall anomalies and hydrology in the Nile basin

Sources: Conway (2005: 103 and 2005: 104); permission to use granted by Elsevier (*Global Environmental Change*)

More recent research has linked rainfall variability in the upper Nile catchments with changing circulation patterns in the Indian Ocean. Although capacity to predict local rainfall effects remains limited (see Chapter 3), this emerging picture of the determinants of rainfall variability in the upper catchment and its impacts on the flow of the Nile 5000km downstream can only enhance the interest of the major users of the Nile in the arid north in ensuring an integrated approach to managing the river basin's water resources. This interest is also heightened by recent proposals in Uganda and Ethiopia to develop use of the water of the Nile's upper catchment for irrigation and hydroelectric power, which for most of the twentieth century had been undertaken exclusively in Egypt and Sudan.

Egyptian agriculture's dependence on the Nile waters is often portrayed as constituting an absolute constraint on the development of the Egyptian economy, as, for example, in a reduced capacity of Egyptian agriculture to meet the food needs of the country's population (Allan, 2001). The growth of the Egyptian population has indeed accompanied increasing investment in works to control the volume and timing of the Nile's flow, to enable ever more intensive use of the land. Yet the forces driving Egypt's economic and population growth have derived from Egypt's economic and political relationship with the international economy, rather than changing water availability from the Nile. This is illustrated by expansion of irrigation in the nineteenth century.

The first major investment in infrastructure was for barrages across the Rosetta and Damietta branches of the Nile delta, instigated by Mohammed Ali Pasha, who ruled Egypt from 1805 to 1848. Prior to this, Egypt's population had remained fairly stable at between 2.5 and 4 million for 1900 years since a peak of 5 million at the time of Roman occupation (Chesworth, 1994: 67–71). Over the same period, water management for irrigation remained at essentially a pre-Roman level: water lifted using the *shaduf* and the *saqia* ('Persian wheel') to supply canals distributing water to fields. It is important to note that the main purpose of Ali Pasha's barrages was to raise the minimum level of the river, so as to irrigate crops with longer growing periods, notably cotton and sugar cane, to supply international markets emerging with the industrial revolution. The barrages were completed in 1861, enabling Egyptian producers to take advantage of the shortage of cotton for European textile industry during the American Civil War (1860–64). Population growth followed the expansion of cultivated area, rising from around 5 million in the 1840s to 8 million in the 1880s and 11 million by 1900.

Subsequent geopolitics of the water resources of the Nile can perhaps be most easily discussed in terms of four historical periods: an early colonial period (1880–1920); a late-colonial period (1922–1963); a post-independence period dominated by the Cold War and development in Egypt and Sudan (1963–1990); a post-Cold War era in which development by the states of the upper catchment has come more to the fore (1990 to the present).

The early-colonial period (1880–1920)

British expeditions to the equatorial lakes in East Africa in the mid-nineteenth century had identified the main components of the Nile River basin. Subsequent establishment of British rule in Uganda and Kenya, coupled with the earlier British occupation of Egypt (from 1882) and the Sudan (from 1898), prompted plans to manage the Nile basin as a hydrological unit. Sir William Garstin, Hydrological Adviser to the British Government in Cairo, investigated the upper Nile, and in 1904, he proposed raising the levels of the equatorial lakes (Victoria, Kyoga and Albert) in order to store water for release when required for agriculture in Egypt and Sudan. The proposals included construction of a canal in southern Sudan to divert water past the Sudd swamps, so reducing the evapotranspiration (equivalent to 50 per cent of the flow upstream of the Sudd) 'lost' as the water spread across the vast wetland. At the same moment, the British administration in Egypt was completing construction of a dam at Aswan to store part of the annual Nile flood to provide year-round irrigation. The earlier barrages in the Nile delta were also re-built at the same time and the Aswan Dam was further raised in 1912. The First World War halted this period of strategic hydrological planning by the British administration in Egypt. It was not to be renewed in the post-war years, which were marked by interdepartmental struggles and preparations to hand over to Egyptian government in 1922.

The late-colonial period (1922–1963)

Egyptian independence introduced the need for international agreement on the use of Nile water by Egypt and (British-administered) Sudan. The development, in 1925, of 400,000 ha of irrigation at Gezira, at the confluence of the Blue and White Nile and supplied by a dam on the Blue Nile at Sennar, greatly increased Sudan's use of Nile water. The Nile Waters Agreement of 1929 made minimum annual allocations of 48 km^3 and 4 km^3 of Nile water to Egypt and Sudan respectively, but also asserted Egyptian rights over Nile waters more generally: 'Save with the previous agreement of the Egyptian Government, no irrigation or power works are to be constructed ... on the River Nile or its branches or on the lakes from which it flows so far as these are in the Sudan or in countries under British administration, which would. ... either reduce the quantity of water arriving in Egypt, or modify the date of its arrival, or lower its level' (quoted in Howell and Allan, 1994: 85). This 1929 Agreement established a framework for development and management of the Nile for the next 25 years. In it, the British administration in Sudan had agreed an Egyptian veto over any water development in British-administered Uganda, Kenya and Tanganyika, although British administrations of those territories had not, in fact, been consulted on the matter.

During the 1930s the height of the Aswan Dam was raised again, to further increase the area of year-round irrigation in Egypt. In parallel, two sets of British hydrologists, working at the Physical Department of the Ministry of Public Works in Egypt and at the Sudan Irrigation Department, developed a number of alternative and competing plans to increase Nile storage of water in the equatorial lakes – the 'Equatorial Nile Project' – and to construct a canal to bypass the Sudd – the 'Jonglei Canal'. All shared the same basic goals: to store water in the equatorial lakes, where evaporative losses were relatively low, and transmit the stored water to Egypt as speedily as possible. The logic appeared incontrovertible, and the 'Equatorial Nile Project' was adopted by the Egyptian Government in 1947 (Collins, 1994).

In the aftermath of the Second World War, facing increasing unrest and political opposition, the bankrupt British jettisoned their colonial territories, formally ceding independence to Sudan in 1956, and to the three East African territories of Uganda, Tanganyika and Kenya between 1961 and 1963. A disastrous British attempt in 1956 to prevent militarily Egypt's President Nasser from nationalising the Suez Canal resulted in cessation of all diplomatic links between Britain and Egypt for five years. The geopolitics of the Nile River Basin were now an arena for the Cold War, and the 'hydrological unity' of the 'Equatorial Nile Project' was replaced by national interests whose competitive implications and 'zero sum' consequences were exacerbated by support from one or other of the two principal Cold War protagonists, the Soviet Union and the United States. This is exemplified in the 1959 Nile Waters Agreement (NWA – see Box 7.3), through which Egypt and Sudan effectively asserted prior claims on the entire annual flow of the Nile and its tributaries. As in the earlier 1929 agreement, upstream states in East Africa were to refrain from using Nile water except where it is 'in excess' of that required to maintain the specified storage at the Aswan High Dam (AHD), to be funded by the Soviet Union. As with the earlier agreement,

Box 7.3

The 1959 Nile Waters Agreement

In Nasser's Egypt, the loss of trust in the British following the Suez Crisis, coupled with deepening uncertainty about the security of water infrastructure developed on the Nile in upstream states – of which the civil war in southern Sudan since the early 1950s was evidence – suggested a radical alternative to increase the availability of water for Egyptian agriculture.

The Aswan High Dam (AHD) was designed to hold the entire annual flood of the Nile in a reservoir of some 162 km^3, of which 109 km^3 would be usable to support the agricultural needs of Egypt's population, which then stood at about 30 million. The reservoir's capacity was larger than the average annual flow of 84 km^3 (1905–1961) at Aswan, thus offering the prospect of smoothing out the year-to-year variation in the Nile flow. On the negative side, evaporation losses from the reservoir were high, at 10 km^3 per year. Furthermore, the reservoir would extend across the frontier into Sudan, displacing 70,000 Sudanese people from their homes.

Sudanese agreement was achieved through an increased share in the use of Aswan storage (from 4.5 to 18.5 km^3 per year), a 50 per cent share for Sudan in any 'new' water developed (e.g. through swamp drainage, or the Jonglei Canal), Egyptian funding for the resettlement of displaced Sudanese and agreement to Sudan's proposed construction of the Roseires Dam (3.35 km^3 capacity) on the Blue Nile to allow a doubling of the size of the Gezira irrigation scheme.

This agreement on the AHD formed part of the 1959 Nile Waters Agreement (NWA), which, in guaranteeing annual minimum flows of 55.5 km^3 and 18.5 km^3 to Egypt and Sudan respectively, effectively asserted prior claims on the entire annual flow of the Nile and its tributaries, thus – in the spirit of the earlier 1929 agreement – requiring upstream states in East Africa to refrain from using Nile water except where it is 'in excess' of that required to maintain these minimum volumes at Aswan. The 1959 NWA also established a Permanent Joint Technical Commission (PJTC) through which Egypt and Sudan would plan and implement further development of the Nile. As with the earlier agreement, neither Britain, nor its soon-to-be independent East African territories, nor Ethiopia were consulted on the 1959 Nile Waters Agreement, but the Soviet Union undertook to fund the Aswan high Dam.

neither Britain, nor its East African territories, nor Ethiopia were consulted on the 1959 Nile Waters Agreement.

The 1959 NWA established a Permanent Joint Technical Commission (PJTC) through which Egypt and Sudan would plan and implement development of the Nile. British efforts in the 1960s to negotiate with the PJTC on the development of irrigation, notably for sugar-cane plantations, from the Nile in Uganda proved ineffectual. However, in 1958 the US Bureau of Reclamation made a study of the Blue Nile in Ethiopia and produced a 'Blue Nile Development Plan' including proposals for four major dams totalling 51 km³ of storage, equivalent to the mean annual flow of the Blue Nile, and hydroelectric generating capacity three times that of AHD, together with irrigation requiring 6 km³ annually, equivalent to 17 per cent of Egypt's irrigated area. Collins (1994) argues that 'if properly managed' the Blue Nile Development Plan would result in only a small loss (2.5 km³ per year) of water for Egypt, since the steady release of water from the HEP plants in Ethiopia would allow the AHD reservoir to operate using less of its full capacity and thus lower evaporation losses due to reduced surface area. However, in the context of the 1960s Cold War, the Blue Nile Development Plan offered a prospect of upstream development over which Egypt would have no control.

Post-independence (1963–1990)

In the event, the early 1960s were marked not by water shortages but by successive years of heavy rains. Water levels in the equatorial lakes rose by over 2 m, and flows in the White Nile doubled (Figure 7.2). Tanganyika's President Nyerere was quick to repudiate the 1929 Nile Waters Agreement as having been made by a colonial power with no jurisdiction in his newly independent state. However, political turmoil meant there was little immediate pressure for increased water use in the upper catchment for a quarter of a century. In Uganda, after 1971 a series of coups and economic collapse culminated in a civil war that ended with National Resistance Movement taking power in 1986. In Ethiopia, revolution in 1974 was followed by secessionist wars that ended when an alliance of secessionist movements (EPRDF) took power in Addis Ababa in 1991. In contrast, the 1970s saw a cessation in the civil war that had raged since the 1950s and the Sudanese government proceeded with the excavation of the 'Jonglei Canal' (Box 7.4). By 1983, however the project was suspended by renewed conflict in the region and remains incomplete.

Post-Cold War (1990–present)

Two decades of political stability in Uganda and Ethiopia have brought proposals for water development in the upper Nile basin, notably for irrigation in Ethiopia, and for hydroelectric power generation in Uganda.

Box 7.4

The Jonglei Canal

The civil war that had raged in southern Sudan since before independence in 1956 was halted by President Nimeiri's government, which in 1972 negotiated an accord with the southern Sudanese rebels based on ceding a degree of self-government through a Southern Regional Assembly. The peace accord presented an opportunity for the Egyptian–Sudanese PJTC to revive the 'Jonglei Canal' project in order to generate up to 5 km³ 'new water' per year in the White Nile downstream from the Sudd wetland. This 'new water' was to be gained by reducing the evaporation losses from the Sudd, estimated to be about 26 km³y⁻¹, or about 50 per cent of the flow entering the swamps from the White Nile.

The project was perceived to be more viable than when originally proposed by the British in the 1940s as the availability of more powerful excavation machinery offered the prospect of completing the 360 km canal, from the town of Bor to the Sobat River at its confluence with the White Nile, in under five years, compared to 20 years in the original proposal. A contract for construction of the canal, 28–50 m wide and 4–7 m deep, and with a raised all-weather road on its east bank, was awarded to the Compagnie de Constructions Internationals in 1976, and excavation began in 1978.

The project quickly became a target of opposition leaders in the Southern Regional Assembly and civil unrest ensued, leading the government to promise local economic development and health and education services to people in the areas affected by the canal. By the early 1980s international environmental organisations, including the United Nations Environment Programme, had voiced concerns about the environmental impact of the canal. Collins (1994) has argued that many of the environmental concerns, on fish stocks, aquifer recharge, and on nutrient supply to Sudd vegetation, were groundless, and that the reduction in the extent of the permanent swamp (by 35 per cent) and seasonal swamp (by 22 per cent) would not have a major effect on livelihoods of people living in the area. However, he and others (Howell and Lock, 1994) identified major problems arising from the impact of the canal on the seasonal movement of livestock by Nuer and Dinka herders. More specifically, the particular line selected for excavation of the canal separates herders' villages and cultivated areas from *toich* dry-season grazing areas that are flooded during the rainy season. To reach these seasonal pastures some 700,000 animals and 250,000 people would have to cross the canal annually, most likely by swimming since none of the planned bridges had been constructed.

The failure to deliver promised health and education services, and the technical obstacles to constructing an all-weather road (heavy clay soils, and the unavailability in the entire Sudd of any stone for construction) meant that the canal became a focus for renewal of southern Sudanese perceptions of their marginalisation by the national government, and by 1983 the civil war had broken out again, leaving the Jonglei Canal incomplete (260 out of 360 km).

In the case of the latter, the Owen Falls 180 MW power station (renamed Nalubaale) built in 1954 at the outlet of Lake Victoria was extended in 2000 by a 200 MW station at Kira, downstream, both stations being operated by a Ugandan subsidiary of South Africa's Eskom electricity utility. After repeated delays, from 1994 to 2002 to 2005, and with World Bank financial support (US$250 million) being withdrawn in 2003 and then reinstated (US$360 million) in 2007, construction of a further 250 MW hydroelectric station at Bujagali has begun, despite environmental and tourist industry opposition. Hydrolectric-power-generation potential from the Nile in Uganda is estimated at 2000 MW (USDA, 2005), and plans to develop this resource are not limited to the country's domestic demand (already larger than existing generating capacity), but are premised on selling electricity to an East African regional market.

A decade of low rainfall in the late 1970s and early 1980s reduced storage in the AHD reservoir below that needed to supply Egypt the 55.5 km^3 guaranteed under the 1959 Nile Waters Agreement. Heavy rains in 1988 avoided the need to cut water supply to Egyptian agriculture, but the AHD – and the strategy of water-resource management based upon it – were shown to be vulnerable, not to annual variability of rainfall and river flow, but to 'decadal' variation where available water is reduced by the cumulative effect of years of low rainfall following one another (Conway, 2005). Subsequent periods of high flows, notably in the late 1990s, have been used to develop two major areas of new irrigation in Egypt, 90,000 ha in the Sinai requiring 1.7 km^3, and 210,000 ha in Western Desert (New Valley Project) requiring 5 km^3. The Jonglei Canal seems increasingly unlikely to overcome objections to its social and environmental impact, and such projects will have an assured supply of water only if demand elsewhere in Egypt is reduced, for example by diversifying the economy away from irrigated agriculture, or if supply is managed in collaboration with upstream states. The World Bank-sponsored Nile Basin Initiative, launched in 1999 is clearly an opportunity for discussion of such issues, although thus far it has focused primarily on data collection. The World Bank has also funded a study (Blackmore and Wittington, 2009) of possible 'Joint Multipurpose Projects' on the Blue Nile for the 'Eastern Nile Committee of Ministers' (ENCOM) of Egypt, Ethiopia and Sudan. This has produced recommendations that recall those of the Blue Nile Development Plan of the 1950s, noted above. The report warns, 'Continued unilateral development, high system losses, growing demand, climate change … would soon result in stress, putting the eastern Nile water resources at serious risk' (p. ix).

A key question is whether sufficient mutual benefits can be identified to promote joint action by states that share the Nile River basin.

The Jordan valley

Water resources and the roots of conflict

Competition for water in the Jordan River basin is perhaps the most prominent of all cases in which water 'scarcity' has been linked to conflict and an inevitability of 'water wars' in the future. While the existence of armed conflict is indisputable, writers with a variety of analytical perspectives (Wolf, 1995; Allan, 2001, Selby, 2003) argue that, while control of water has been perceived as strategic to achieve political and economic goals, it was neither the determinant of those goals, nor, therefore, the cause of conflict. The broad contours of conflict in the Jordan River basin reflect the nineteenth-century Zionist project to settle Jewish immigrants in Palestine with a view to establishing a Jewish state, and the advance of that project in the twentieth century through armed conflict to overcome resistance, initially from British Mandate administration and displaced non-Jewish 'Arab' Palestinians, and subsequently from a coalition of Arab governments of neighbouring states (see Box 7.5).

Selby (2003: 64) draws on Shafir's (1989) historical study of early Jewish immigration to argue that the settlement model of exclusively Jewish farming communities emerged from a need to protect Jewish immigrants from the local market for cheaper, Arab, agricultural labour. He argues that this separatist nature of Jewish settlement, and the institutions that were established to support it, lies at the root of the 'Arab–Israeli problem'. A rapid rise in Jewish immigration and settlement provoked hostility among Arab Palestinians alleging that Jewish land purchases from absentee landlords were making Palestinian tenants landless. Jewish arguments in favour of expanded immigration were based on the possibility of greatly increasing land productivity – and hence the population that could be supported from it – by the use of irrigation. Seen from this perspective, access to water was a strategic requirement of Jewish immigration.

It is important to note, however, that the ensuing conflict has taken particular forms and directions at specific historical moments as a result of changing interactions between the Zionist idea and other political processes. First, opportunities and constraints have altered according to geopolitical priorities in the Middle East of other powers, notably the British and French following the collapse of the Ottoman Empire during the First World War, the United States and Soviet Union during the Cold War, and subsequently the United States and its allies and their opponents in radical Islamic movements. Second, the nature and outcome of conflict has been shaped by the shifting balance of power between political and economic groups within the evolving state of Israel and elsewhere. Analysis of conflict in the Jordan River basin thus cannot be resolved into simple competition between 'nations', but must recognise a more disaggregated definition of 'national interest' that results in changing priorities, reversals of policy, and even shifting definitions of what constitutes an 'international boundary'.

The water resources of the Jordan River basin

The principal water resources of the Jordan River basin are summarised in Table 7.2, together with estimates of quantities used at the end of the twentieth century. It is important to note that these estimates are from published sources but are widely contested, particularly in the case of groundwater extraction rates. Allan (2001: 83), for example, quotes a figure of 700 Mm^3y^{-1} flowing westwards into Israel through the Northern and Western aquifers – a much higher figure than the 507 Mm^3y^{-1} Israeli hydrologists claimed in 1995 was the total annual recharge of these aquifers. Such data uncertainty reflects the deeply controversial status of water resources in the region, and is in addition to the expected variability due to annual fluctuation in rainfall in the north of the Jordan Basin, which feeds the Jordan headwaters and in the mountains of the West Bank, which supplies the aquifers. Despite these reasons for caution, the data nonetheless suggest that water resources in the Jordan basin are being exploited at close to their limit, and, indeed, the flow in the Jordan south of the confluence with one of its principal tributaries is commonly referred to as a trickle (Allan, 2001), fed chiefly by sewage effluent and saline springs that have been diverted away from points where water is extracted for irrigation or urban use. As a consequence, the Dead Sea, the final destination of the Jordan River, has been reduced in both volume and extent.

From 1948 until 1967 the initial Jewish settlement strategy was continued: strategic areas were secured by agricultural settlements whose viability depended upon the ability of the Israeli state to secure and deliver water for irrigation. The development of water infrastructure and technology was therefore intimately connected to national security strategy. Israel's 1959 Water Law brought state control of both agriculture and water and established the institutional capacity with which to exercise that control: a Water Commission responsible for planning and licensing water use;

Table 7.2 Water resources and usage in the Jordan River basin

	Estimated average annual flow (Mm^3y^{-1}) or total storage (Mm^3)	Users	Estimated annual usage (Mm^3y^{-1})
Surface water[1]			
North Jordan			
Hasbani River (Lebanon)	125		
Dan Spring (Israel)	250		
Banias River (Golan Heights)	125		
Local run-off	140		
		Israel	100
Total inflow to Lake Tiberius	540		
Precipitation over Lake	65		
Local run-off	70		
Inflow from Yarmuk	100		
Evaporation from Lake	−270		
		Israel Nat. Water Carrier	500
Outflow from Lake Tiberius to Jordan River + saline springs diverted from upstream of Lake	70		
Yarmuk River	400–495		
		Syria (net irrigation)	200
		Jordan (East Ghor Canal)	150
		Israel	80–100
Al Wehda Dam (Yarmuk, from 2007)	110		
Net flow in Jordan below Yarmuk confluence	72		
Lower Jordan springs	185		
Zarqa River	322		
King Talal Dam (Zarqa river)	75		
East Ghor return flow	32		
Inflow (saline) to Dead Sea	611		
Total surface flow	1477		1050
Groundwater[2]			
North-east aquifer	145	Israel and West Bank[3]	145–184
Western aquifer	362	Israel and West Bank[3]	362–621
Eastern aquifer	172	Israel and West Bank[3]	94–204
Total	679		601–1009

Sources: 1 Surface water estimates from Kolars, quoted in Wolf (1995); 2 Groundwater recharge figures are those presented to Oslo Peace Negotiations, 1995 (quoted by Selby, 2003); 3 Data on groundwater use are contested: Lower figure is that presented to Oslo Peace Negotiations. Upper figure is from Palestinian Water Authority, 2000, quoted by Klawitter, 2006. See also Table 7.3

a National Water Authority, Mekorot, charged with construction and operation of water infrastructure; Tahal, responsible for estimates of water demand and strategic planning; and the Israeli Hydrological Service, responsible for monitoring and modelling water resources.

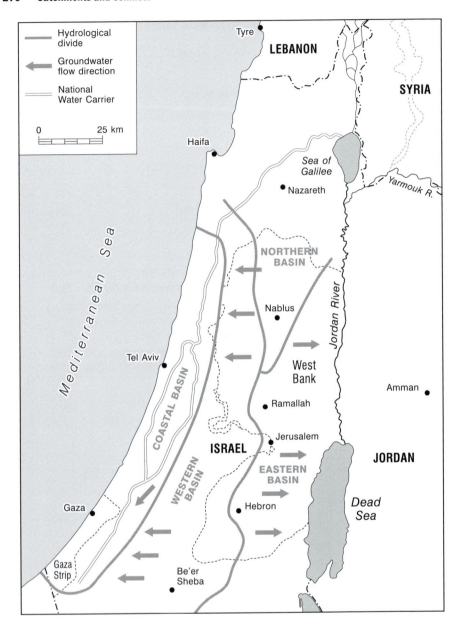

Figure 7.3 Map of Jordan River basin

Source: after Feitelson, 2002 and HDR, 2006

Over this period armed conflict between Israel and Syria focused on Israel's efforts to divert water from the Jordan headwaters to irrigate the Negev Desert to the south, and Syria's efforts to prevent this by constructing a diversion canal upstream on the Hasbani or Banias tributary southwards to a dam on the Yarmuk River, to be shared between Syria and Jordan. Amid heightened tension, in 1953 the United States sent special envoy Eric Johnston to devise a comprehensive allocation of the Jordan waters between the riparian states (Lebanon, Syria, Jordan and Israel). Over the subsequent two years, a series of proposals and counter-proposals involved both Israel and representatives of Lebanon, Syria and Jordan under the leadership of Egypt, with Johnston acting as intermediary. A compromise plan involving increasing the storage of water in the Sea of Galilee/Lake Tiberius and a dam at Maqarin on the Yarmuk (175 Mm^3) was agreed by the governments taking part – with the proviso that both construction and operation of the scheme were under international control. The 'Johnston Plan' proposed allocations of the Jordan waters according to availability of irrigable land: 400 Mm^3 to Israel, 720 Mm^3 to Jordan, 132 Mm^3 to Syria, and 35 Mm^3 to Lebanon (Wolf, 1995: 48). The plan was accepted by the Israeli government and President Nasser of Egypt but ultimately rejected by the Arab League in 1955. However, the plan appears to have set an important 'objective' yardstick for allocation of the Jordan waters to which future negotiators could refer, and, in that respect, it is most unfortunate that it did not include any consideration of groundwater.

Box 7.5

Early conflict (1890–1948)

Although the creation of a Jewish State in Palestine had been discussed since 1897 (Wolf, 1995: 16), opportunity to advance only arose as the Ottoman Empire collapsed in 1917. The emerging nationalist claims of the populations of the Middle East were incorporated into negotiations at the 1919 Peace Conference between the British and French who were mandated to take over administration of the newly divided territories as protectorates: Lebanon and Syria (French), Palestine (British) and Mesopotamia (British).

Zionist claims for the headwaters of the Jordan to be included in the territory of Palestine were subordinated to the two colonial powers' strategic interests in key infrastructure (roads, railways, oil pipelines etc.), so that the frontier divided the Jordan headwaters between Syria and Palestine. Critically, the purpose of the British mandate in Palestine was compromised by the ambiguities of the notorious Balfour Declaration of 1917, which stated 'His Majesty's Government view with favour the establishment in Palestine of a national home for the Jewish people, and will use their best endeavours to facilitate the achievement of this object, it being clearly understood that nothing shall be done which may prejudice the civil and religious rights of existing non-Jewish communities in Palestine, or the rights and political status enjoyed by Jews in any other country' (quoted in Wolf, 1995: 19). Unable to resolve the contradictions inherent in this Declaration, by the late 1930s, the British responded to a rising tide of violence by considering partitioning the territory between Jewish and Arab 'nations'.

In the decade that followed, Jewish settlement efforts intensified so as to occupy key boundaries and resources, particularly near the Jordan headwaters in northern Palestine. The UN Partition Plan of 1947 was accepted by the Jewish leadership but rejected by Arabs in Palestine and neighbouring territories. In the violence that followed British withdrawal in 1948 the Jewish settlers were able not only to withstand attacks from neighbouring Arab countries but also to displace some 700,000 non-Jewish residents from the territory they controlled, creating a state with a Jewish majority population. The remaining territory of Palestine was annexed by the Hashemite monarchy of Jordan, which had been established as a separate political entity east of the Jordan River by the British in 1922. Some 450,000 Palestinians displaced from the territory of Israel in 1948 went to Jordan (Wolf, 1995).

With the failure of the 'Johnston Plan', Israel pressed ahead with its plans to divert water from Lake Tiberius into an aqueduct – the 'National Water Carrier' – to the coastal cities and irrigation in the Negev Desert. Costs were greatly increased by continuing conflict with Syria, which meant moving the intake from the preferred site north of the Lake, to Eshed Kinrot on the north-western lakeshore. As a result, water needed to be pumped to a height of 250m, rather than simply flowing to the Negev by gravity. In 1965 the Arab League renewed construction of a diversion canal upstream of the intake of Israel's National Water Carrier, prompting Israeli military attacks on the construction site in Syria. This set in train an escalating series of border clashes that ended in 1967 with Egyptian blockade of the Israeli port of Eilat and the pre-emptive air-raids with which Israel began the Six-Day War and gained control of the Jordan headwaters in the Golan Heights, Sinai and the entire West Bank of the Jordan. Although Israeli military hegemony was subsequently tempered by Egypt's successful action to regain Sinai in 1973, and by fruitless Israeli incursions into Lebanon in the 1980s and in 2007, the control over water had shifted decisively in Israel's favour, as it was now an upstream user not only of the Jordan flow, but also of the West Bank mountain aquifers from which it had previously been pumping via wells 'downstream' within Israel. The former remained subject to trans-boundary water management with Syria and Jordan. The latter became subject to Israel's internal water management (see Box 7.6).

Box 7.6

1967–1994 water allocations across 'internal' boundaries: the West Bank

With the occupation of the West Bank and Gaza in 1967, Israel acquired an additional non-Jewish population of 1.5 million, which it placed under a military administration. Local municipal government was largely left alone, but economic activity was closely regulated by the Israeli authorities, initially through restrictions on the development of new industry or agriculture, and on access to export markets, resulting in a heavy dependence of the West Bank and Gaza economies on employment in Israel. Shortly after the start of the occupation, water in the West Bank was declared property of the Israeli state, to be managed by the Israeli water utility, Mekorot. All water installations required new permits from the Israeli administration, and in 25 years only 23 permits were issued for wells, only three of which were for agriculture (Isaac, quoted by Selby, 2003: 42).

During the first decade of occupation, Jewish settlements were limited to strategic frontier zones, particularly along the Jordan River valley. Growing influence of political parties favouring Jewish settlement of the West Bank resulted in a shift in control of Israeli government. From the early 1980s there was a major expansion of land acquisition and construction of dormitory towns in the West Bank for Jewish settlers employed in Israel, with a total of 280,000 Jewish settlers resident in the West Bank by 1993 (Selby, 2003). The spread of Jewish settlements was accompanied by the construction of new water infrastructure that effectively integrated the West Bank water-supply system, fed mainly by wells sunk into the aquifers, into the Israeli national water network. Selby's (2003: 87–88) research in the southern West Bank has documented how the detailed design of this system both integrated but also discriminated between Jewish and non-Jewish communities, through differences in the diameters of water pipes, the heights of off-take pipes in water reservoirs and so on, resulting in continuous water supply in Jewish settlements but intermittent supply for neighbouring non-Jewish communities.

Deteriorating income and living standards among Palestinians produced growing hostility to the occupation, manifest from 1987 in the Intifada campaign of non-cooperation with Israeli authorities, and in suicide attacks on Israeli cities.

Trans-boundary relations 1967–1994

Following defeat in 1967, the Arab League supported the Palestinian Liberation Organization's efforts to continue an armed resistance against Israel from exile. However, Israel's military supremacy enabled it to attack neighbouring Arab states, including their vulnerable water infrastructure. Israel's destruction of the East Ghor Canal, in 1969 is believed to have prompted the Jordanian army's action to expel the PLO from Jordan in 1970. Expelled also from Lebanon following the Israeli invasion in 1982, the PLO was subsequently based in Tunis and increasingly marginal to developments in Palestine, particularly after the Soviet Union collapsed in 1989, and PLO support for the Iraqi invasion of Kuwait in 1990. A shift in Cold War allegiance of some of Israel's Arab opponents, notably Egypt following disillusion with Nasser's socialist republic, was followed by the first peace accord, in 1978, between Israel and an Arab country.

While the Hashemite Monarchy did not formally sign a peace accord with Israel until 1994, a number of published accounts (Wolf, 1995; Sosland, 2007) have documented a long-term informal cooperation – epitomised by 'picnic talks' two or three times a year – between Israeli and Jordanian water engineers in managing water resources, notably those of the Yarmuk River, where the Israelis pump water from downstream of the intake to the Jordanian East Ghor Canal that delivers water to irrigate the east bank of the Jordan. In these accounts, Jordan's long-term allegiance to Britain and the United States appears an important asset, enabling the Jordanian leadership to seek US pressure on Israel to reach compromises on water sharing. Following the 1994 peace accords between Israel and Jordan, cooperation on managing the Jordan has been increased further to include joint use of Lake Tiberius to store water from the winter flow of the Yarmuk, for subsequent release to Jordan via an aqueduct (the KAC), and agreement on the construction of the Al-Wehda Dam (110 Mm^3 capacity, completed in 2007) at Maqarin on the Yarmuk river, as originally proposed by the Johnston Plan in 1955.

Re-partition: new trans-boundary water conflict (1994–present)

In contrast to the settlement of trans-boundary conflict over surface waters of the Jordan, the aquifer system remains a source of acute tension, fuelled by a history of discriminatory water access for the population under military occupation (Box 7.6). In the wake of the 1991 Gulf War, the United States government sought to check the deteriorating relations between the Israeli government and Palestinians with proposals to establish a degree of Palestinian self-government as part of a 'peace process' towards the creation of a two-state 'solution': a non-Jewish state in Gaza and the West Bank co-existing with Israel. The PLO leadership was recruited from exile in Tunis to head a new Palestinian Authority to administer the West Bank and Gaza. Negotiations over water, and particularly Palestinian rights to groundwater, formed an important part of the 'peace process'. Box 7.7, which draws on the work

Box 7.7

Water negotiations in the Oslo Accords

During negotiations in 1993 and 1995, the issue of water shortages experienced by Palestinians was prominent and the 1995 ('Oslo II') Interim Agreement stated explicitly that the water rights of Palestinians would be recognised, although the details of what those 'rights' were was left to 'final status' negotiations in the future. In the interim agreement, water allocations were to be made to meet the 'needs' of the Palestinians: 23.6 Mm^3y^{-1} from the West Bank aquifers to meet immediate needs, and 70–80 Mm^3y^{-1} to meet future needs. Given that total water use by West Bank Palestinians in 1995 amounted to only 118 Mm^3y^{-1}, these allocations offered a significant improvement in water availability for the Palestinians living on the West Bank (Selby, 2003: 104). The 1995 agreement also stipulated the formation of a Joint Water Committee charged with surveying and protecting water sources and developing and maintaining infrastructure, and Joint Supervision and Enforcement Teams for monitoring and policing the West Bank's water resources and systems. Day-to-day management was to remain unchanged, however, with Mekorot responsible for supplies to Israeli settlements and the West Bank Water Department managing the billing and connection of Palestinian users on behalf of Mekorot. Finally, the agreement provided for the transfer to Palestinian Authority control water systems that 'related solely to Palestinians'.

Although the agreement was heavily criticised from within Israel as compromising the security of Israel's water supplies, the great asymmetry in hydrological information available to the Israeli and Palestinian negotiators undermined Palestinian gains. Selby's detailed study of the operation of water supply on the West Bank concluded that the 'joint water management' arrangements after 1995 simply formalised the day-today operations of water supply that existed under the military occupation. Most significantly, of the 'new water' for Palestinians less than 10 per cent (3.1 Mm^3y^{-1}) was to be re-allocated from Israel's National Water System. The remainder was to be developed, at the Palestinians' expense, from the eastern Aquifer in which Israeli hydrologists said 78 Mm^3y^{-1} were as yet 'unused'. Selby (2003: 191–131) argues that the available data gives widely varying estimates of the extent of unused water in the eastern aquifer, and that the impact of more intensive water extraction on salinity in the aquifer is not well understood, posing significant hazards of degrading the quality of water supplied from the aquifer. Certainly, Palestinian data on water extraction from West Bank aquifers (Table 7.3) suggest that in the eastern aquifer, as in the northern and western aquifers, rates of water withdrawal greatly exceed the average recharge levels on which the Oslo II negotiations were based.

The outcome of perhaps greatest significance from the 1995 agreement was that Israel successfully prevented any increase in 'upstream' Palestinian water use from the 'transboundary' (northern and western) aquifers, thus maintaining its own use of more than 85 per cent of the water extracted from them. By thus avoiding any Palestinian claims on these aquifers, the Israeli negotiators ensured that they would not be liable to any payments for 'Palestinian' water, since the 1995 agreement requires that 'in the case of purchase of water by one side from the other, the purchaser shall pay the full real cost incurred by the supplier ... all the way to the point of delivery' (quoted in Selby, 2003: 108). On the contrary, the Israeli control of supply from the transboundary aquifers ensured that they were always 'suppliers', able to charge full cost for water used by Palestinians, while free to subsidise water supplied to Jewish settlements. The final Israeli gain from the joint water management agreed in 1995 was that the liability for non-payment of water bills by Palestinians was transferred to the Palestinian Authority and its international funders. As Selby (2003: 108) concludes: 'The formalisation of Israeli–Palestinian co-operation had enabled Israel to divest itself of some of its most onerous burdens of occupation, without losing control of either water resources or supplies to Israeli settlements, and without having to forgo its discriminatory pricing policy.'

of Selby, suggests a classic case of asymmetric information about the water resources producing inequitable outcomes that have done little to reduce Palestinian grievances or to regulate over-exploitation of the aquifers (Table 7.3).

In practice the Oslo 'peace process' quickly foundered. The Palestinian Authority did not deliver the 'security' Israel expected, and the Palestinians saw little improvement in their economic situation but further expansion of Jewish settlement in the West Bank. The Israeli army re-occupied the West Bank in 2000 and in the following years a deepening political crisis was marked by increasingly invasive and repressive control of Palestinian movement including about 600 Palestinians killed each year,[1] the construction of a wall to physically

Table 7.3 Annual recharge and water use (Mm³y⁻¹) from West Bank aquifers

Aquifer	Recharge	Use from wells			Use from springs			Total	Use/recharge
		Israel	West Bank		Israel	West Bank			
Location Control			Israeli	Palestinian		Israeli	Palestinian		
Western	362	542	3	27	47	–	3	621	1.7
Northern	145	59	13	19	75	–	18	184	1.27
Eastern	172	2	34	26	8	88	45	204	1.18
Total	679							1009	

Source: after Selby, 2003

seal off Palestinian territory from Israel, a second Palestinian *intifada*, the refusal by Israel and by Palestine's international funders to recognise the elected Palestinian Authority, and factional violence among Palestinians in Gaza. In 2006, the populations of Israel and the West Bank with Gaza were estimated at 7 and 4 million, respectively (World Bank, 2007). The same source lists per capita gross national income as US$18,580 in Israel but only US$1230 in the West Bank and Gaza. Aside from the obvious humanitarian concerns raised by such impoverishment of Palestinians, the continuation of such inequality must call into question the future stability of separate development under a 'two-state' model.

The role of water in the current context is potentially quite different from that of the past. The Israeli economy has become highly diversified, with agriculture accounting for only 3 per cent of GDP. Since the remaining 97 per cent of GDP is achieved from only 5 per cent of the total water used (Allan, 2001: 79) there is pressure on Israeli agriculture to cut its water use. More significantly, industrialisation has greatly increased Israeli capacity to produce water, notably through recycling of urban wastewater for use in agriculture, and through desalination. Allan (2001: 91) states that by 2000 recycled wastewater provided 15 per cent of Israel's irrigation, and 10 per cent of overall water use. In contrast, he argues, although the technology has long been available, investment in desalination plant in Israel has been delayed by Israeli government perceptions that the 'peace process' provided an opportunity to obtain international assistance for investment in desalination. This was to be achieved, first, by insisting negotiations on water with the Palestinians should discuss only 'new water' (as distinct from existing waters in aquifers), and, second, by presenting investment in desalination as the source of this new water. The position was stated succinctly by the Director of Water Research at the Israel Institute of Technology, Uri Shamir, in 2004, when proposing a desalination plant at Hadera-Caesaria on the Mediterranian coast to supply water to 250 Palestinian villages the northern West Bank: 'The plant will be funded by the world for the Palestinians. Israel will not be willing to carry this burden, and the Palestinians are not able to' (quoted by Pearce, 2004: 6).

It is possible, however, that the changing nature of the Israeli economy may introduce other priorities into water policy, as argued by Feitelson (2002). The launch of Israel's Desalination Master Plan in 2000, envisaging 400 Mm³y⁻¹ output by 2005 and 750 Mm³y⁻¹ by 2020, provided a platform for Israeli desalination companies to develop as competitors in the international market for desalination expertise. IDE is the Israeli desalination company partnering the multinational Vivendi in constructing and operating the world's largest desalination plant, which started producing 100 Mm³ of fresh water per year from seawater at Ashkelon on the coast of Israel in 2006. By 2008 IDE had announced contracts to build desalination plants in China and Australia (*Business Week*, 25 April 2008, *Arutz Sheva*, 23 August 2008). Such developments may bring about a change in Israeli perceptions of the geopolitics of water resources. Whether this will bring an end to the Israeli-Palestinian conflict seems unlikely.

Conflict and cooperation in catchment management

Management of trans-boundary catchments inevitably tends to confer advantages on upstream users' ability to achieve absolute control over the resource, although their ability to justify such *de facto* advantage on grounds of principle, in terms of 'sovereign rights' over resources within their territory, depends on their ability to dominate downstream water users. In the case of the Jordan headwaters, Syria's attempt to impose its control was overturned by Israel. In the case of the Yarmuk, Israel's ability to impose its superior military power over upstream water user Jordan was constrained by Jordan's role in the wider political strategies of Israel's main sponsor, the United States. In the West Bank, Israel's acquisition of 'upstream' control of the westward-flowing aquifers enabled it to direct water towards Israel and towards Jewish settlements in the West Bank.

With the prospect of losing upstream control of the aquifers to an internationally recognised Palestinian state, Israel invoked a principle of 'prior appropriation' – that the aquifers were already fully exploited – and thus effectively excluded the trans-boundary aquifers from all negotiations until those of 'final status' – now indefinitely postponed. The same principle of 'prior appropriation' underpinned the Egyptian and Sudanese Nile Waters Agreements of 1929 and 1959 that sought to restrict Nile water use in the riparian countries of the upper catchment. The much greater distances involved, and the corresponding difficulty in projecting military or economic action on upstream water users will make it relatively difficult for Egyptian governments to impose their claims on the Nile.

Claims to water based on absolute principle, whether of (upstream) sovereign rights, or (downstream) prior appropriation, are likely to prevail only where they can be backed by coercive power, with the logical implication of 'water wars'. To avoid this, a number of international agreements and resolutions that include a water dimension have been drawn up since the beginning of the twentieth century (Box 7.8).

Box 7.8

International agreements and resolutions relating to water management

International Agreements:

- UN Convention on the Law of the Non-Navigational Uses of International Watercourses (1997)
- United Nations Convention to Combat Desertification (1994)
- Ramsar Convention on Wetlands of International Importance especially as Waterfowl Habitat (1971)
- Convention Relating to the Development of the Hydraulic Power affecting more than one state, and Protocol of Signature (1923)
- Convention and Statute on the Regime of Navigable Waterways of International Concern (1921)

International Resolutions:

- International Decade for Action, 'Water for Life', 2005–2015 (2003)
- International Year of Freshwater (2000)
- International Law Commission, Resolution on Confined Transboundary Groundwater (1994)
- Resolution 3129 (XXVIII) on Cooperation in the Field of the Environment Concerning Natural Resources Shared by Two or More States (1973)
- Resolution Concerning the Utilization of Non-Maritime Waters for Purposes Other than Navigation (1961)
- Madrid Declaration on International Regulations Regarding the Use of International Watercourses for Purposes other than Navigation (1911).

Source: IWLP (International Water Law Project, 2009: http://www.internationalwaterlaw.org)

Efforts to formulate an explicit framework of principles for sharing trans-boundary water resources were initially drawn up as the Helsinki Rules in 1966. These were subsequently codified as the 1997 United Nations Convention for the Non-Navigational Use of Shared Waterways. Three core principles of the convention relate to 'equitable and reasonable utilisation', 'no significant harm' and 'prior notification of works'. The key distinction between these principles and those of absolute principle are that they engage negotiations with the 'needs', rather than 'rights', of different water users. This does not make negotiation of shares any simpler, however. Nor is there a mechanism for ensuring compliance with agreements. Symptomatic of these problems is that, despite the existence of 27 trans-boundary water-management organisations covering the world's major water resources including the Amazon Cooperation Treaty Organization, the Mekong River Commission, the Nile Basin Initiative and the Zambesi River Authority, only 14 countries have joined the Convention (HDR, 2006: 218). This struggle to gain support, particularly from upstream states within large international river basins, has been contrasted by Hunt (2004) with the widespread engagement with the Ramsar convention, established in 1971 for the protection of wetland environments, and which had grown by 2002 to include 133 participating countries with over 1000 wetlands (see also the FAO's (1998) legislative review on international water law and the Water and Law Standards project of the UNFAO and WHO: http://www.waterlawandstandards.org/).

Two elements that seem important in negotiating water shares that are considered fair can be identified in the cases explored in this chapter. The first is the importance of transparency and the role of agreed data on water resources and water use as a prerequisite for achieving it. The second is the need for an arbitrator who has the trust of the different parties to the negotiation. In the case of the Jordan River basin, the Johnston Plan for sharing water between riparian countries according to the amount of irrigable land appeared to satisfy both requirements, and provided a reference point for decades. Conversely, the large and continuing asymmetry in information about aquifer resources and their use has enabled Israel to maintain its discriminatory use of aquifer water, but at the cost of continuing illegitimacy (among Palestinians) that feeds its sense of insecurity. It is significant that the first major action of the Nile Basin Initiative is to establish a data-recording and processing system to provide the riparian governments with hydrological data for the basin as a whole.

Summary

We noted at the start of this chapter that inherent hydrological characteristics of water have the effect of linking different users drawing water from the same catchment or aquifer, creating the possibility of competition and conflict when the water resource becomes scarce relative to the demands placed upon it. However, the discussion we have developed has not led us to conclude that scarcity of water will lead inexorably to water wars. This is because the evidence that we have reviewed has suggested that questions of water allocation need to be informed by wider understanding of economic and political (and military) power. In that sense, water is more likely to be an instrument than a cause of war.

This is most clearly the case in instances where systematic discrimination by governments against particular social groups is manifest in gross inequity of access to water. However, we have also seen that at a local village scale disparities in wealth and power may be played out through a differentiation between 'haves' and 'have nots' in relation to water. As we saw in Chapter 6, efforts to manage demand for water have tended to focus on relatively 'technical' approaches, such as water pricing but also 'social engineering' of water consumption behaviour. The discussion in this chapter takes us into the terrain of broader political values, that is: what economic or political goals should water management achieve? This type of question is evident in the efforts made in many countries over the past two decades to reform management of water catchments in ways that enable representation of the views of different water users. A theme here that we shall return to in Chapter 8 is the critical political role of the state in establishing 'ground rules' in terms of political values that should govern such negotiated approaches to water management. The South African government's slogan for water reform – 'Some, for all, forever' – is perhaps a neat exemplar of this.

At a geopolitical level, of course, there is no vertical state authority, but only horizontal relations of power and mutual interest. It seems clear from cases like the Nile catchment that large asymmetries of power will be inimical to the pursuit of mutual interest. Yet it also appears that international conventions have yet to gain purchase on the regulation of trans-boundary water agreements. In this context, the major challenge is to identify the potential for mutual benefit to be achieved from water developments as a means of fostering collaborative, rather than competitive initiatives.

Further reading

Allan, J.A. 2001. *The Middle East Water Question: Hydropolitics and the global economy*. I.B. Tauris, London.

Cleaver, F. and Toner, A. (2006). The evolution of community water governance in Uchira, Tanzania: The implications for equality of access, sustainability and effectiveness. *Natural Resources Forum* 30: 207–18.

HDR 2006. *Beyond Scarcity* Human Development Report 2006. United Nations Development Programme, New York.

Lenton, R. and Muller, M. (2009). *Integrated Water Resources Management in Practice*. Earthscan, London.

Wolf, A. (1995). *Hydropolitics along the Jordan River*. United Nations University Press, Tokyo.

8 Water resources and development conclusions

Much has previously been written on water resources and their development, often framed by the inescapable necessity of water for life. However, as we noted in the opening chapter of this book, explaining our interest in water through reference to water being 'essential' is too simplistic. It is undoubtedly the case that 'not having access to water and sanitation' is widely understood as a peculiarly severe form of deprivation, due to the centrality of water to enhancing hygiene and maintaining human health. However, in this book we have sought a broader scope, to recognise that water as an essential resource (and therefore constraint) in the development of complex society ('civilisation') both historically and currently. Water can empower development, enabling higher levels of industrial and agricultural production, and the demands of water management have profoundly influenced human society, from ancient Egypt to the Tennessee Valley Authority to the Office du Niger. Equally, we have argued, water is 'produced' in different ways as a consequence of human economic and political development. Many of the world's rivers have borne the effects of development for so long that data to indicate their 'natural' condition are scarce. Even where 're-engineering' is taking place to improve the ecological conditions of rivers, such as on the Rhine, the 'restored' river is a product of human values and goals, such as re-establishment of salmon migration and spawning grounds.

A consequence of this stance is that, while recognising the negative impacts of development on water resources, through wider and more intense exploitation and depletion, and pollution from agricultural, urban and industrial waste, we have also emphasised the relative abundance of water, at least at an aggregate level. In this respect we have taken a different course from writing on water that has emphasised its scarcity, and particularly scarcity in terms of a 'global water crisis'. Rather, we have emphasised that water scarcity is, above all, a result of local water use exceeding local supply. Responses to scarcity are to make further investment to 'produce' more water (from distant sources or using new technology), or to take measures to reduce demand. In this book we explored these alternatives in Chapters 5 and 6, respectively. One important implication of this line of argument is access to water is a consequence of investment of capital and/or labour, and that inequality of access to water is a reflection of the relations of political and economic power that determine patterns of social invest-ment (Swyngedouw, 2004; Gandy, 2008). This suggests that 'scarcity' of water needs to be questioned as justification for *inequality* of access to water (at least for the basic needs of drinking, cooking, hygiene), lest it allow us to ignore political and economic processes that underlie it (Mehta, 2001).

In Chapter 1 we reviewed the growth of international concern over the management of natural resources, fuelled by two perceptions. On the one hand human activity was perceived to be causing major, and likely undesirable, impacts on hydrological systems. On the other hand lack of access to water and sanitation was perceived to be a key indicator of the increasing number of people living in poverty. The era of international initiatives to address environmental problems, following the Earth Summit in 1992, saw the future of water management debated in a series of high-profile international meetings, of which the World Water Forums were perhaps the largest. Two key features of these debates were, first, a general acknowledgement that water resources needed to be managed in a more holistic way. This involved both tempering water-engineering design with environmental considerations, and also reconciling different water uses within water catchments. Second, and linked to the first, was an overall shift from strictly engineering supply-enhancement approaches to a greater emphasis on managing the demand for water. These two elements were relatively uncontroversial, and accompanied a parallel shift in emphasis from centralised and techno-cratic to decentralised and democratic models for managing water resources. The evolving debate also exposed major controversies over ways in which these water-management models should be governed, specifically in relation to the role of markets in determining water allocation. This raised questions of governance, and in particular ideas of 'integrated water resource management', to which we will turn later in this chapter.

In this book we have viewed 'development' as an economic or political imperative for which increasing utilisation of water is fundamental. Water is therefore a key resource for human aspirations and challenges in the twenty-first century. In this book we focused particularly on:

- Water and economic development
- Water and growing (climate) uncertainty
- Water and poverty reduction
- Water and conflict.

Chapter 2 explored the critical role of investment in water infrastructure as an instrument of economic development. A model that has endured since its inception in the US during the 1930s is that of state investment in large water-engineering projects, such as the 'multipurpose' large dam supplying HEP, irrigated agriculture, flood control and flow stabilisation for navigation. We suggested that to these early priorities a wealthier, more urbanised society had added new goals of ecological and water-quality improvement. This set in train not only a process of 're-engineering' rivers, but also moved the emphasis in water management from one relying exclusively on engineering to one that seeks political and economic regulation of individual and collective water use. In many instances (as exemplified by the Rhine), this requires new approaches to planning and governing the use of water resources that cross administrative or national boundaries.

In reviewing the impact of climate change on water resources, in Chapter 3, the inevitability of growing uncertainty was evident. Although considerable strides have been made in understanding the science of the atmosphere and its impact on energy and water balances, there remain challenges in translating predicted effects between different spatial scales. This means that great uncertainty attaches to the effect of climate change on rainfall patterns at any but very large, i.e. regional scales (small cartographic scale). Given the highly variable nature of existing rainfall patterns in many areas prone to water deficits, such as the semi-arid tropics, the challenges to decision making may be best addressed by 'risk management' approaches.

In Chapter 4 we considered the use of water to reduce poverty in economies where industrial growth does not provide higher-paid employment, specifically the use of irrigation to protect poor people from famine and provide a base for 'rural development'. As in the case of urban and industrial economies, we observed a shift from engineering equality of opportunity through state-managed 'protective irrigation' to an emphasis on 'social engineering' through decentralised organisations such as water users' associations to ensure that water use was both productive and equitable. We found evidence that irrigation has indeed enabled growth in aggregate agricultural output (with broad social benefits), but with the hazard that the benefits are unevenly shared, creating increasing inequality among rural populations.

'Winners' were able to exploit irrigation to achieve commercial levels of output, often increasing their share of water use in the process, while 'losers' struggled to produce enough to cover the costs of water. To the extent that the aggregate outcome has been to improve national 'food self-sufficiency', governments may more or less actively foster this widening divergence of socio-economic conditions among the rural population, as in the case of India's green revolution. Does this mean that irrigation is not an effective way to reduce poverty? We would rather draw the conclusion that wherever water enables better productivity, then it will have an impact on social relations. Population will increase through immigration, and competition for land that in the past may have been of marginal interest may become acute (Woodhouse et al., 2000; Woodhouse, 2003). The outcome of such processes, in terms of who wins and who loses and by how much, will depend on the state's political goals and regulatory capacity. Again, early focus on engineering design has become tempered by a recognition that outcomes are as much, if not more, a function of 'political' factors underpinned by power relations within state and society.

Chapters 5 and 6 considered ways of redressing local imbalances between water supply and demand through, respectively, options of increasing supply or moderating demand for water. These options manifest a range extending from investments in engineering and technology (e.g. desalination) to a reliance on incentives to change the behaviour of water users (e.g. water pricing). Neither course is unproblematic, however. The relative costs, in economic (capital investment and operation and maintenance costs) and political (legitimacy of regulation and distribution of costs and benefits) terms have a major influence on the relative feasibility of different options. In each case the question of how to decide water allocations between competing uses of water, or whether to increase supply or curb demand, has been left open, and it is to this matter of 'governance' of water resources that we turn in this final chapter.

In Chapter 7, the idea of water as a commons was explored, and extended to consider cases of trans-boundary management of shared water resources in river basins or aquifers. This suggested that, while a shared water resource may imply a common interest in collaborative management, it may also become an instrument of conflict. Questions of 'scarcity' and consequent competition between users may accentuate the potential for conflict, but do not determine it. A key role was identified for institutions that generate transparency (notably through assessing and disseminating data on the water resource) and offer arbitration in cases of dispute, returning us again to a consideration of governance.

Issues of water governance have been evident since communities established alongside the Tigris, Euphrates and Nile started to organise their efforts to dam and divert flows and to compete for the water that thereby became available. We noted in Chapter 1 that in the last 100 years a change in the character of water management, from being an entrepreneurial activity with private ownership to a state-run nationalised industry to gain an emphasis upon water as an economic good. We also charted the emergence of an international consensus on 'best practice' in the last two decades of the twentieth century. This brought together three separate perspectives on water management. First, the nature of water is that it connects different activities within a given hydrological unit (catchment or aquifer) such that water use by one may have an impact on water availability to another. This suggests a need for 'holistic' or 'integrated' management of water in which individual use is subordinated to a perception of the wider social or ecological 'common good'. Second, the idea that increasing the scale of water use could induce local scarcity, promoted the logic of treating water as an economic good and using water pricing as a means of allocation between competing users. Third, the need to allocate water between different users in ways that were perceived as fair and legitimate underpinned the need for decision making that was informed by different users' views. We observed that these three strands of thinking, formulated as a set of (ICWE or 'Dublin') principles and embedded in the UN's Agenda 21, actually represent distinct twentieth-century views of how society should be governed:

- Weberian: engineering/science as arbiter of water management: Integrated Catchment Management (the ecological principle)
- Neo-liberal: water as an economic good and allocation through water pricing (the instrument principle)
- Neo-institutional: subsidiarity and self-government by water users (the institutional principle).

Given that the three principles are based on markedly different conceptions of how decisions should be made, it is legitimate to question whether they contain inherent contradictions that prevent their equal implementation under all circumstances. For example, Page (2005: 294) quotes the European Commission (2000) as stating 'water is not a commercial product like any other but rather a heritage which must be protected, defended and treated as such', while also advocating 'full cost' pricing of water. We traced the ideological struggle over the commodification of water evidenced at World Water Forums following the publication of the 'World Water Vision' in The Hague in 2000. One consequence of these debates is that the past decade has witnessed growing appreciation of the importance of the governance of water, as WWDR (2006: 47) state: 'water is power, and those who control the flow of water in time and space can exercise this power in various ways'. Figure 8.1 shows dimensions of water governance, as presented by the WWDR (2006). It may alternatively be argued that there are really only three dimensions, i.e. environmental, economic and social/political, in keeping with the three principles identified above. The important point is that governance cannot be divorced from these constituent elements.

By the 1970s the assumptions of state-led development that had dominated the previous four decades were being questioned as being unaccountable and unresponsive to the changing and complex needs of society, giving rise to inappropriate patterns of development.

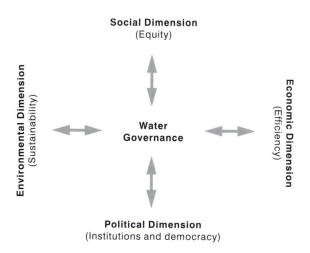

Figure 8.1 Dimensions of water governance

Source: after WWDR, 2006: 46, and based upon Tropp, 2005

The increasing signs of environmental damage were frequently used as evidence to support this perception. Two lines of thinking proposed alternatives to 'top–down' government planning. One, associated with Jürgen Habermas, promoted ideas of a more active 'participatory' citizenship, with decisions being subjected to 'deliberative democracy' in which technical experts would be engaged in discussion by ordinary 'lay' citizens to examine the rationale and consequences of new developments. The second approach sought to make resource allocation decisions subject to the 'discipline' of markets as understood in classical economics. From this viewpoint, markets were the aggregate of decisions made by 'utility-maximising' individuals. Hence, because each individual was making the best decision to satisfy their own needs, then society as a whole would satisfy its needs. In the case of scarce resources, markets would ensure that resources (in this case water) would be allocated to those able to gain the greatest marginal economic benefit from its use, to the aggregate benefit of society. In terms of Figure 8.2 (based on attitudes to wetlands (Van Slobbe et al., 2006), the ideas of participatory citizenship are consistent with the upper two quartiles, whereas 'neo-liberal' ideas that emphasise the primacy of market allocation are more congruent with the lower-left-hand quartile, while 'top–down' development might be regarded as identified with the lower-right-hand quartile.

It should also be clear that the different approaches to decision making are reflected in the ICWE water-management principles identified above. Of these, in the 1980s it was the neo-liberal approach that made the greater impact. The development of environmental economics provided the tools

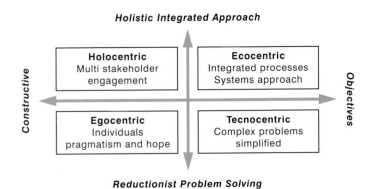

Figure 8.2 Paradigms for water-resource management

Source: after Van Slobbe et al., 2006: 274

for allocating economic values to natural resources, including – through devices such as 'contingent valuation' of public 'willingness to pay' – 'future' or 'existence' values of conserving resources such as rivers or landscapes in their 'natural' condition. The approach enabled the development of a framework for allocating costs of improving water quality, for example through policies of charging polluters for their use of rivers to dispose of waste. It provides a means not only to generate funding to compensate or mitigate damage (e.g. by investing in water treatment), but provides economic incentives for individual water users to change their behaviour and produce more socially optimal outcomes. An example from the Rhine is the proposal to operate basin-wide funding of water management that would include compensating (through reduced charges) upstream riparian authorities for investing in flood-control measures that also benefit those downstream (Hooijer et al., 2004). The wide application of valuation approaches has been termed the 'neoliberalisation of nature' (Castree, 2005). This has seen valuation initially undertaken to emphasise the importance of conserving resources such as wetlands or forests because of the 'environmental services' they provide in terms of regulating water quality and quantity. Subsequently, the recognition of such valuation has been translated into a market for environmental services. The critical step in such a transition is the establishment of private-property rights that enable resources to be traded. The regulation of markets for such environmental services to ensure that they really deliver the desired environmental benefits remains problematic, however (Robertson, 2004).

Initial moves to privatise water resources, notably in the UK, were directed at divesting the state of responsibility for environmental improvement and to provide incentives to commercial companies to undertake the task. The creation of commercial providers of water and sanitation services also created expectations of international contracts to provide such services in developing countries, as promoted by the 2000 World Water Vision (see Chapter 1). In practice, these expectations have not been realised. Not only has there been a widespread rejection in developing countries of the idea of treating water as a simple commodity, as we noted in Chapter 1, but there have been difficulties in reconciling notions of pricing water for 'full-cost' recovery with the need to provide subsidised water to poor people (Chapter 6). Both Swyngedouw (2006) and Page (2005) report most recently that privatisation is not accelerating, with difficulties (even prior to the 2008 financial crisis) of raising sufficient capital, and reports of recent privatisation in Africa being 'at best modest'.

There is evidence, therefore, of a turning away from private-sector engagement with water.

Part of the resistance to water privatisation is based on a perception of increased prices being used to generate private profit, as in the case of the Cochabamba riots in Bolivia (see Chapter 1). Similarly, Page (2005) commences his analysis of water as a commodity by describing public opposition in Cameroon to the (new) water company's requirement that people abandon community standpipes and instead buy their water from local entrepreneurs. A further limitation to a straightforward market approach to water management is suggested by Bakker's (2000) study in Yorkshire (UK) following privatisation of water in 1989 illustrating public opposition to treating water as a commodity. She notes that in five out of the seven years following privatisation Yorkshire Water Company applied hosepipe bans, whereas the expected (pre-privatisation) frequency had been only one year in ten. When standpipes were introduced for household supplies during the1995 drought, there was public resistance to curb consumption, in contrast to the pre-privatisation experience of the 1976 drought. Bakker notes (2000: 16): 'public perception of company performance played an important role in some consumers' refusals to respond to appeals to conserve water'. This is consistent with Sharp's (2006) argument that people often see water as a communal good whereas much of demand management is aimed at the individual, prompting a negative reaction to those who claim ownership over the resource.

Such 'perverse' behaviour is often regarded as a matter of inadequate information or 'awareness' of the need to conserve water, to be remedied by better communication, or 'education', to ensure consumer engagement with demand management. Page (2005: 298) summarises the danger of this 'top–down' stance: 'The engineers and politicians who now manage water supplies in Cameroon claim that most consumers are reluctant to pay for their water because they are ignorant; their attitudes need to be modified.' It is thus perhaps unsurprising that water companies' media messages lack real influence on water users' behaviour (Sharp 2006). Kenny et al. (2008) also conclude education programmes are only modestly beneficial. There is growing recognition that public perceptions of the fairness of water-pricing policy determine whether education will be met by a willingness to respond. Bakker (2008: 22) argues for the need to shift focus from 'management of unproblematised "natural" resources to socio-political struggle within and reregulation of, a dynamic naturalized resource landscape'.

Such a 'socio-political struggle' involves regarding people primarily as 'citizens' involved in collective activity rather than as 'consumers' acting as individuals. Efforts to involve people as 'stakeholders' in planning activities have been fostered by many reforms to water-management institutions since the 1990s and by efforts to use spatial planning (Howe and White, 2004) to contribute to sustainable development. Carter (2007) argues that spatial planning has become a key discipline in new strategies designed to manage water resources through participatory measures, a notion supported by researchers and institutions, most notably the World Summit (1992), Dublin (1992), the WFD (EC 2000) and the UNECE's Aarhus Convention (1998, http://www.unece.org/env/pp/). The benefits of stakeholder engagement are listed as:

- Increasing democratic legitimacy
- Building consensus
- Strengthening decision making.

Although some countries, such as South Africa, have enshrined stakeholder engagement in water legislation (Brown, 2007), in practice, opportunities may vary. Furthermore, the experience may not always be beneficial. Carter (2007) notes procedural barriers with, for example, stakeholders being involved too late in the planning process while 'consultation' may be favoured rather than active engagement. Therefore, as we concluded in Chapter 6, there are challenges to the institutional approach. In particular, it is evident that the institutional principle assumes a context in which water-management consultation and decision making will be consistent with broader political goals, such as equity and sustainability in water use, which is not always the case (Brown, 2007; Few et al., 2007).

This brings us back to the first, or 'ecological', principle of water management, which treats water as a holistic resource requiring an integrated (basin) approach. Gregory (2000: 414) notes that the UN first promoted 'Integrated River Basin Management' (IRBM) in 1958 and describes it as 'The process of managing water resources within the drainage unit in a manner which optimises water use throughout the basin and minimizes deleterious effects for water, river channels and landuse.' The notion of integrated water resource management (IWRM) and development was promoted again at the earth Summit (UNCED, 1992) as a goal of Chapter 18 on fresh-water resources. WWDR (2006: 185) note that, unlike the common sectoral approach to water management, IWRM 'makes the link between water resources and human activities throughout the hydrological cycle and allows ecological and socio-economic issues to be considered within an eco-system approach'. As with IRBM, IWRM uses a topographically defined river basin or catchment as the unit of management. The use of the watershed to demarcate hydrological boundaries is thus well established. The same notion has been advocated by the WWF and called the 'ecoregion' based on a geographically distinct ecology in which many large rivers and basins can be delineated (Pittock et al., 2006).

The approach is not without its critics. Not all water resources are uniquely defined by surface watersheds. Areas, including many cities, that are largely dependent upon groundwater resources, or those – typically in arid regions – where exogenous supplies dominate, may reject the IBRM approach. Calder (2005) notes that, after three decades of promoting the IWRM approach in more than 20 major conferences, the UN have established it as a guiding principle for the management of water resources, but concludes that implementation falls far short of a 'blue revolution'. The principle that integration is necessary for effective management is, then, widely accepted but the notion is beset with difficulties over terminology and implementation. Lenton and Muller (2009: 209) have argued that part of the problem is that many critics of IWRM appear to consider it a rigid framework that (inevitably) fails to be implemented in the widely varying contexts of the real world. They counter that IWRM is constituted by basic principles:

- The unitary nature of the resource
- The physical interventions need to manage it
- The limits of those interventions
- The need for an institutional framework that:
  - brings stakeholders together in an equitable manner that gives voice to the weak as well as the powerful
  - seeks a balance of interests between them, recognising the value and importance of the waters concerned
  - identifies the environmental dimension of water management
  - develops appropriate organisations to promote the overall approach

More generally, they promote IWRM as an exercise in 'adaptive man-agement' and, perhaps most significantly, they conclude that 'successful IWRM efforts adopt an integrated approach in order to address specific development problems; they never have an integrated approach as their objective' (Lenton and Muller, 2009: 208). This seems consistent with the argument adopted in this book: that water-resource management is not an end in itself, but a means to achieve development goals. A similar argument is developed by Mollinga et al. (2007) who state that the hydrological catchment imposes a false boundary and that 'problemsheds' which are socio-political constructs are more appropriate as a unit of management, and hence regulatatory organisation. They suggest (p. 702) that: 'attempts to impose particular models of river basin organisations in developing countries, especially those from the experiences of rich countries, are not likely to succeed because the objectives and institutional contexts are so different'. Rather than a prescriptive institutional model, in which 'social engineering' is likely to be needed to fit 'people' into predefined technological or organisational models, Mollinga et al. (2007) suggest three 'structural features' of policy reform: the state is the central promoter of reform; information and capacity building is a key factor in ensuring accountable and transparent decisions; and implementation needs coalitions of 'empowered stakeholders'. This returns the discussion to a further iteration of how to achieve better decisions about using water for development. It is quite explicit about the exercise of power, and recognises a role for both 'top–down' and 'bottom–up' elements to the achievement of change. More broadly, it suggests that wider political agendas for development are critical to the outcomes of 'participatory' approaches.

To summarise, then, there is an array of technical measures that have been shown to tackle water scarcity, but at a cost, which means they sometimes work to the benefit of a select few rather than the whole community. Technology alone is not enough. The experience of the neo-liberal approach through water pricing and privatisation suggests that wider issues of ownership and equity need to be addressed alongside the capacity for private investment. Conversely, the efforts to implement the institutional principle have exposed the limitations of local participation in bringing about change, particularly where asymmetries of power and conflicts of interest present obstacles. This prompts a need to re-evaluate the role of the state and the political processes through which development goals and priorities are determined and implemented.

In conclusion, we need to reiterate that, even though there are vast amounts of water across the globe, parts of the world are facing severe water stress, where there is insufficient natural supply to meet intense local demand. The past response of developing new supplies through technology has enabled industrial and agricultural revolutions along with the development of megacities. It is unlikely that this response alone will be sufficient in the future, and it is not surprising to read predictions, repeated over the last two decades, of an impending water crisis that requires immediate action – and yet continues unabated. We believe there is a growing consensus that strategies based upon the three principles outlined above may form the basis for sustainable development, in particular to embrace the institutional principle with an emphasis upon governance and stakeholder engagement while noting the significance of the instrument principle, i.e. that investments need to be financed in a sustainable manner. The key issue is to understand the tensions and interactions between these principles in specific development contexts.

 # Notes

4 Water resources in colonial and post-independence agricultural development

1 Consisting of Angola, Central African Republic, Congo, Democratic Republic of Congo, Equatorial Guinea, Gabon, São Tomé and Principe (FAO, 2008)
2 The most prominent of these movements in Brazil is the MST – *Movimento dos trabalhadores rurais Sem Terras*. In the northeastern state of Bahia, in addition to MST, mobilisation is coordinated through a regional movement, the CETA – *Movimento dos Trabalhadores Assentados, Acampados e Quilombolas do Estado da Bahia*.

7 Catchments and conflicts

1 From September 2000 to July 2007, 4228 Palestinians and 1024 Israelis were recorded as killed as a result of the Palestinian – Israeli conflict. (United Nations Office for the Coordination of Human Affairs, 2007.)

References

Note: websites accessed January 2010 unless otherwise indicated.

Abbassi, B. and Al Baz, I. 2008 Integrated waste water management, Chapter 3 in Al Baz, I. Otterphohl, R. and Wendland, C. (eds) *Efficient Management of Wastewater*. Springer-Verlag, Berlin. pp. 29–40

Abdel-Khaleq, R.A. and Dziegielewski, B. 2006 A national water demand management policy in Jordan. *Management of Environmental Quality* 17(2): 216–25

Abdulla, F.A. and Al-Shareef, A.W. 2006 Assessment of rainwater roof harvesting systems for household water supply in Jordan, in Hlavinek, P., Kukharchyk, T., Marsalek, J. and Mahrikova, I. (eds) *Integrated Urban Water Resources Management*. Springer, Dordrecht. pp. 291–312

Abu-Zreig, M. Attom, M. and Hamasha, N. 2000 Rainfall harvesting using sand ditches in Jordan. *Agricultural Water Management* 46: 183–92

Adams, A. 1981 The Senegal River Valley, in Heyer, J. et al. (eds) *Rural Development in Tropical Africa*. Macmillan, London

—— 2000 Social Impacts of an African Dam: Equity and Distributional Issues in the Senegal River Valley. World Commission on Dams. www.dams.org_docs_kbase_contrib_soc193.pdf

Adams, W. 1988 Rural protest, planning policy and the planning process on the Bakalori Project, Nigeria. *Africa* 58(3): 315–36

—— 1992 *Wasting the Rain*. Earthscan, London

Agnew, C.T. 2002 Drought, desertification and desiccation; the need for further analysis. *Geography Journal* 87(3): 256–67

Agnew, C.T. and Anderson, E. 1992 *Water Resources in the Arid Realm*. Routledge, London

Agnew, C.T. and Anderson, E.W. 1994a Integrated rural development in Uganda, Household water collection and use. Ministry of Energy, Minerals and Environmental Protection, Kampala

—— 1994b Integrated rural development in Uganda, Water quality: springs and households. Ministry of Energy, Minerals and Environmental Protection, Kampala

Agnew, C.T. and Anderson, E.W., Lancaster, W. and Lancaster, F. and Mahafir, A. 1995 Water harvesting system in the Eastern Jordan (Badia) desert. *Geo-Journal* 37(1): 69–80

Agnew, C.T., Clifford, N. and Haylett, S. 2000 Identifying and alleviating low flows in regulated rivers: the issue of the rivers Bulbourne and Gade, Hertfordshire, U.K. *Journal of Regulated Rivers: Research and Management* 16: 245–266

Agnew, C.T. 1998 Climate and vegetation in the Tropics. In Webster, C.C. and Wilson, P.N. *Agriculture in the Tropics*. Blackwell Science

Agnew, C.T. and Chappell, A. 1999 Drought in the Sahel. *GeoJournal* 48: 299–311

—— 2000 Desiccation in the Sahel. In McLaren, S. and Kniveton, D.R. (eds) *Linking Climate Change to Land Surface Change*. Kluwer Academic Publishers, Netherlands. pp. 27–48

Agnew, C.T. and Fennessey, S. 2001 Climate change and nature conservation, in Warren, A. and French, J.R. (eds) *Habitat Conservation: Managing the Physical Environment*. John Wiley & Sons, Chichester. pp. 273–304

Ahmad, I.H. and Opschoor, J.B. 2009 Reaching a climate deal in Copenhagen. UN-DESA Policy Brief No 17. Department of Economic and Social Affairs, New York

Ahmed, J. 2005 Assessing runoff potential for water harvesting generation in Jordan. PhD thesis, University of Manchester

Ahrens, B. and Beck, A. 2008 On upscaling of rain-gauge data for evaluating numerical weather forecasts. *Meteorology and Atmospheric Physics* 99(3–4): 155–67

Akgul, D., Cakmakci, M., Kayaalp, N. and Koyuncu, I. 2008 Cost analysis of seawater desalination with reverse osmosis in Turkey. *Desalination* 220(1–3): 123–31

Al Baz, I. Otterphohl, R. and Wendland, C. (eds) 2008 *Efficient Management of wastewater*, Springer-Verlag, Berlin

Ali, S. Ghosh, N.C. and Singh, R. 2008 Evaluating best evporation estimate model for water surface evaporation in semi-arid India. *Hydrological Processes* 22(8): 1093–1106

Allan, J.A. 2001 *The Middle East Water Question: Hydropolitics and the global economy*. I.B. Tauris, London

Allen, C.H. 1987 *Oman*. Westview Press, Boulder

Allen, M.R. and Igram, W.J. 2002 Constraints on future changes in climate and the hydrological cycle. *Nature* 419 (September): 224–32

Altinbilek, D. 2002 The role of dams in development. *Water Resources Development* 18(1): 9–24

—— 2004 Development and management of the Euphrates-Tigris basin. *Water Resources Development* 20(1): 15–33

Anand, P. 2007 *Scarcity, Entitlements and the Economics of Water in Developing Countries*. Edward Elgar, Cheltenham

Anderson, K. and Bows, A. 2008 Reframing the climate change challenge in the light of post-2000 emission trends. *Philosophical Transactions of the Royal Society A* 366: 3863–82

Andronova, N. and Schlesinger, M. 2004 Importance of sulfate aerosol in evaluating the relative contributions of regional emissions to the historical global temperature change. *Mitigation and Adaptation Strategies for Global Change* 9: 383–90

Arnell, N.W. 1992 Factors controlling the effects of climate change on river flow regimes in a humid temperate environment. *Journal of Hydrology* 132: 132–42

—— 2002 *Hydrology and Global Environmental Change*. Pearson, Harlow

—— 2003 Relative effects of multi-decadal climatic variability and changes in the mean and variability of climate due to global warming: future stream flows in Britain. *Journal of Hydrology* 270: 195–213

Arnell, N. Liu, C. Compagnucci, R. da Cunha, L., Hanaki, K. Howe, C., Mailu, G., Shiklomanov, I. and Stakhiv, E. 2001 Working Group II: Impacts, Adaptation and Vulnerability. Chapter 4: *Hydrology and water resources*. IPCC. http: //www.grida.no/publications/other/ipcc_tar/?src=/climate/ipcc_tar/wg2/180.htm

Arnfield, A.J. 2003 Two decades of urban climate research: a review of turbulence, exchanges of energy and water, and the urban heat island. *International Journal of Climatology* 23: 1–26

ASCE 2010 US Infrastructure report card: Dams. *American Society for Civil Engineers*. http://www.infrastructurereportcard.org/sites/default/files/RC2009_dams.pdf and http://www.infrastructurereportcard.org/fact-sheet/dams

Aw, D. and Diemer, G. 2005 *Making a Large Irrigation Scheme Work. A Case Study from Mali. Directions in Development*. Washington, DC, The World Bank

Ayoo, C. and Horbulyk, T.M. 2008 The potential and promise of water pricing. *Journal of International Affairs* 61(2): 91–104

Babel, M.S., Das Gupta, A. and Pradhan, P. 2007 A multivariate econometric approach for domestic water demand modeling: An application to Kathmandu, Nepal. *Water Resources Management* 21: 573–89

Bajard, Y. Draper, M. Viens, P. 1981 Rural water supply and related services in developing countries – comparative analysis of several approaches. *Journal of Hydrology* 51: 75–88

Bakker, K.J. 2000 Privatising water, producing scarcity: the Yorkshire drought of 1995. *Economic Geography* 76(1): 4–27

Bakker, K.J. 2008 The 'Commons' versus the 'Commodity': alter-globalisation, anti-privatisation, and the human right to water in the global south. In Mansfield, B. (ed) *Privatization*. Oxford: Blackwell Publishing

Balek, J. 1977 *Hydrology and Water Resources in Tropical Africa*, Elsevier, Oxford

Balek, J. 1983 *Hydrology and water resources in Tropical Africa*. Elsevier, Oxford

Bandyopadhyay, J. 2002 A critical look at the report of the World Commission on Dams in the context of the debate on large dams on the Himalayan rivers. *Water Resources Development* 18(1): 127–45

Bar-Shira, Z., Finkelshtain, I. And Simhon, A. 2006 Block rate versus uniform water pricing in agriculture: an empirical analysis. *American Journal of Agricultural Economics* 88(4): 986–99

Barron, J. and Okwach, G. 2005 Runoff water harvesting for dry spell mitigation in maize (Zea mays L.): results from on farm research in semi-arid Kenya. *Agricultural Water Management* 74: 1–21

Barrow, C.J. 1987 *Water Resources and Agricultural Development in the Tropics.* Longman Scientific, Harlow

—— 1999 *Alternative Irrigation: the Promise of Runoff Agriculture. Earthscan, London*

Bartlett, J.P. 1979 Environmental and legal consideration in weather modification activities in the Northern Sierra Nevada. *Water, Air and Soil Pollution* 12: 29–34

Bates, B.C. Kundzewicz, Z.W. Wu, S. and Palutikof, P.S. 2008 Climate change and water Technical paper VI. IPCC (http://www.ipcc.ch/publications_and_data/publications_and_data_technical_papers_climate_change_and_water.htm)

Batram, J. 1990 *Drinking Water Supply Surveillance.* Robens Institute, University of Surrey

Beaumont, P. 1989 *Environmental management and development in drylands.* Routledge, London

Bebbington, T., Hickey, S. And Mitlin, D. 2008 (eds) *Can NGOs Make a Difference? The Challenge of Development Alternatives.* London: Zed Books

Bell, J.N.B. and Treshaw, M. 2002 *Air Pollution and Plant life.* Wiley International

Bell, V.A. Kay, A.L. Jones, R.G. and Moore, R.J. 2007 Use of a grid-based hydrological model and regional climate model outputs to assess changing flood risk. *International Journal of Climatology* 27(12): 1657–71

Bemont, F. 1961 L irrigation en Iran. *Annales Geographie* 70: 587–620

Ben-Asher, J. and Berliner, P.R. 1994 Runoff irrigation. *Advances in Agricultural Sciences* 22: 126–53

Betts, A., Helms, S., Lancaster, W., Jones, E., Lupton, A., Martin, L. and Matsaert, F. 1990 The Burqu/Ruwaysid project: preliminary report on the 1988 field season. *LEVANT* XXII: 1–20

Bevan, K.J. 2003 *Rainfall Runoff Modelling.* J. Wiley & Sons.Chichester

Bicknell, J., Dodman, D. and Satterthwaite, D. (eds) 2009 *Adapting Cities to Climate Change,* London: Earthscan

Billington, D. and Jackson, D.C. 2006 *Big Dams of the New Deal Rra: A Confluence of Engineering and Politics.* University of Oklahoma Press

Bird, J. 2002 Nine months after the launch of the World Commission on Dams Report. *Water Resources Development* 18(1): 111–26

Biswas, A.K.1979 North American water transfers: an overview, in Golubev, G. and Biswas, A.K. (eds), *Interregional Water Transfers,* Pergamon Press, Oxford. pp. 79–90

—— 2002 Stockholm Water Symposium. Editorial. *Water Resources Development* 18(1): 5–7

—— 2004 Dams: cornucopia or disaster? *Water Resources Development* 20(1): 3–14

Blackmore, C., Collins, K., Furniss, P., Morris, D. and Reynolds, M. 2004 The UK policy context for water management I. The English and Welsh policy context, SLIM-Case study monograph 12A. http: //slim.open.ac.uk

Blackmore, D. and Wittington, D. 2009 Opportunities for cooperative water resources development on the Eastern Nile: Risks and rewards. An independent report of the scoping study team to the Eastern Nile Council of Ministers. Washington, DC: The World Bank

Blenkinsop, S. and Fowler, H.J. 2007 Changes in drought characteristics projected by the PRUDENCE regional climate models. *International Journal of Climatology* 27(12): 1595–1610

Bloschl G. and Sivaplan, M. 1995 Scale issues in hydrological modelling: a review. *Hydrological Processes* 9: 251–90

Bocock, R. 1993 *Consumption*. Routledge, London

Bohm, H., Haupter, B., Heiland, P. and Dapp, K. 2004 Implementation of flood risk management measures into spatial plans and policies. *River Research and Applications* 20: 255–67.

Bosart, L.F. 1985 Weather forecasting, in Houghton, D.D. (ed.) *Handbook of Applied Meteorology*. J. Wiley, New York. pp. 205–79

Botterill, L.C. 2005 Lessons for Australia and beyond. Chapter 12 in Botterill, L.C. and Wilhite, D.A. (eds) *From Disaster Response to Risk Management*. Springer-Verlag, Berlin. pp. 177–83

Braatne, J.H., Rood, S.B., Goater, L.A. and Blair, C.L. 2007 Analyzing the impacts of dams on riparian ecosystems: a review of research strategies and their relevance to the Snake River through Hells Canyon. *Environmental Management* (online 10.1007/s00267-007-9048-4)

Braham, Jr. R.R. 1986 Precipitation enhancement–a scientific challenge. *American Meteorological Society Meteorological Monographs* 21: 43

Bray, F. 1986, 1994 *The Rice Economies: Technology and Development in Asian Societies*, University of California Press

Breuer, G. 1980 *Weather Modification*. Cambridge University Press, Cambridge

Briggs, D. and Smithson, P. 1992 *Fundamentals of Physical Geography*. Routledge, London

Brimblecombe, P. 1987 *The Big Smoke*. Routledge, London

Broad, K., Pfaff, A., Taddei, R., Sankarasubramanian, A. and de Souza Filho, F. 2007 Climate, stream flow prediction and water management in northeast Brazil: societal trends and forecast value. *Climatic Change* 84: 217–39

Brooks, D.B. 2006 An operational definition of water demand management. *Water Resources Development* 22(4): 521–28

Brooks, N., Grist, N. and Brown, K. 2009 Development futures in the context of climate change: challenging the present and learning from the past. *Development Policy Review* 27(6): 741–65

Brookshire, D.S., Burness, H.S., Chermak, J.M. and Krause, K. 2002 Western urban water demand. *Natural Resources Journal* 42: 873–98

Brown, H. 2009 Social learning within participatory, catchment-based water management processes in South Africa and Namibia. PhD thesis, University of Sheffield, UK

Brown, J. 2005 Water services and the poor: A case study of the Greater Nelspruit Utility Company, Mbombela Municipality, South Africa. CRC Working Paper 12[1] (http://www.competition-regulation.org.uk/publications/working_papers/)

—— 2007 Participatory processes and outcomes of South Africa's water reforms. PhD thesis, School of Environment and Development, University of Manchester

Brown, J. and Woodhouse, P. 2004 Pioneering redistributive regulatory reform. A Study of implementation of a catchment management agency for the Inkomati Water Management Area, South Africa. Working Paper 89, Centre on Regulation and Competition. http://www.competition-regulation.org.uk/publications/working_papers/wp89.pdf

Bruintjes, R.T. 1999 A review of cloud seeding experiments to enhance precipitation and some new prospects. *Bulletin American Meteorological Society* 80(5): 805–20

Bryson, R.A. 1974 A perspective on climate change. *Science* 184: 753–60

Buonomo, E., Jones, R., Huntingford, C. and Hannaford, J. 2007 On the robustness of changes in extreme precipitation over Europe from two high resolution climate change simulations. *Quarterly Journal of the Royal Meteorological Society* 133: 65–81

Burbano, A.A., Adham, S.S. and Pearce, W.R. 2007 The state of full scale RO/NF desalination. The results of a worldwide survey. *American Water Works Association. Journal* 99(4): 116–127

Burgess, J., Collins, K., Harrison, C., Munton, R. and Murlis, J. 1999 An analytical and descriptive model of sustainable development: report for Environment Agency. University College London

Burt, C. 2007 Volumetric irrigation water pricing considerations. *Irrigation Drainage Systems* 21: 133–44

Burt, T.P. and Horton, B.P. 2007 Inter-decadal variability in daily rainfall at Durham since the 1850s. *International Journal of Climatology* 27: 945–56

Burt, T.P. and Shahgedanova, M. 1998 An historical record of evaporation losses since 1815 calculated using longterm observations from the Ratcliffe meteorological station. *Journal of Hydrology* 205: 101–11

Butler, D. and Memon, F.A. (eds) 2005 *Water Demand Management*. IWA, London

Byers, H.R. 1974 History of weather modification, in Hess, W.N. (ed.) *Weather and Climate Modification*. J. Wiley, London. pp. 3–44

Cabanes, C., Cazenave, A. and Le Provost, C. 2001 Sea level rise during the past 40 years determined from satellite and in situ observations. *Science* 294: 840–42

CADT 2008 Desalination: A national perspective. Committee on Advancing Desalination Technology, National Research Council. Water Science and Technology Board. The National Acadamies Press

Cairncross, S. 1987 The benefits of water supply. WEDC Conference: Developing World Water. Grosvenor Press, Hong Kong. pp. 30–34

Calder, I.R. 2005 *The Blue Revolution* 2nd ed. Earthscan, London.

Campling, P., Gobin, A., Beven, K. and Feyen, J. 2002 Rainfall runoff modelling of a humid tropical catchment: the TOPMODEL approach. *Hydrological Processes* 16: 231–53

Cancelliera, A., Mauro, Di. G. Bonaccorso, B. and Rossi, G. 2007 Drought forecasting using the standardised precipitation index. *Water Resources Management* 21: 801–19

Cantor, L.M. 1985 *Environmental Impact of Water Resources Projects*, Lewis Publishers, Michigan

Carney, J. (1988) Struggles over crop rights in contract farming households on a Gambian irrigated rice project. *Journal of Peasant Studies* 15(3): 334–49

Carney, D. 2002 *Sustainable Livelihoods Approaches: progress and possibilities for change*. DfID. http://www.eldis.org/vfile/upload/1/document/0812/SLA_Progress.pdf

Carruthers, I. and Clark, C. (eds) 1983 *The Economics of Irrigation*. Liverpool University Press

Carson, R. 1962 *Silent Spring*. Houghton Mifflin, Boston

Carter, J. 2007 Spatial planning, water and the Water Framework Directive: insights from theory and practice. *The Geogaphical Journal* 173(4): 330–42

Carter, J.G., White, I. and Richards, J. 2008 Sustainability appraisal and flood risk management. *Environmental Impact Assessment Review* 29: 7–14

CAS 1980 *Committee on Atmospheric Sciences: weather and climate modification, problems and progress*. Grand River Books, Detroit

Castree, N. 2005 *Nature*. Routledge, London

Caviedes, C.N. 1988 The effects of ENSO events in some key regions of the South American continent, in Gregory, S. (ed.) *Recent Climate Change*. Belhaven Press, London. pp. 252–66

CAWMA 2007 *Comprehensive assessment of water management in agriculture*. Colombo/London: International Water Management Institute and Earthscan. (see also Molden 2007)

CDFA 2010 California Department of Food and Agriculture. http://www.cdfa.ca.gov/

CDIAC 2008 Carbon Dioxide Information Analysis Centre. http://cdiac.ornl.gov/

CDWR 2009 California Drought: an update, June 2009. California Department of Water Resources. Sacramento. http://www.water.ca.gov/drought/docs/Drought_report_30june2009_web.pdf

Chambers, R. 1994 The origins and practice of participatory rural appraisal *World Development* 22(7): 953–69

Chambouleyron, A. 2004 Optimal water metering and pricing. *Water Resources Management* 18: 305–19

Chandimala, J. and Zubair, L. 2007 Predictability of stream flow and rainfall based on ENSO for water resources management in Sri Lanka. *Journal of Hydrology* 335: 303–12

Channabasappa, K.C. 1982 An overview of desalination systems for brackish water and their applicability to desert conditions, Proceedings of conference on Alternative strategies for desert development and management, Sacramento, Pergamon Press, New York

Chappell, A. and Agnew, C.T. 2001 Geostatistical analysis and numerical simulation of West African rainfall, in Conacher, A.J. (ed.) *Land Degradation*. Kluwer, Netherlands, pp. 19–35

—— 2004. Modelling climate change in West African Sahel rainfall (1931–90) as an artifact of changing station locations. *International Journal of Climatology* 24(5): 547–54

Chappell, A. and Agnew, C. 2008 How certain is desiccation in west African Sahel rainfall (1930–90)? *Journal of Geophysical Research-Atmospheres* 113 D07111, doi: 10.1029/2007JD009233

Chattanooga Times 2009 TVA critics step up pressure after 3 mishaps in 3 weeks *Chattanooga Times Free Press* 15/01/2009. http://epaper.timesfreepress.com/Repository/getFiles.asp?Style = OliveXLib (article accessed 16 April 2009)

Chen, Y. Zhang, D., Sun, Y., Liu, X., Wang, N. and Savenije, H.G. 2005 Water demand management: a case study of the Heihe River Basin in China. *Physics and Chemistry of the Earth* 30(6–7): 408–19

Chesworth, P. 1994 History of water use in the Sudan and Egypt, in Howell, P. and Allan, J.A. (eds) *The Nile: Sharing a Scarce Resource*. Cambridge University Press, Cambridge. pp. 65–80

Chilton, J. (ed.) 1999 *Groundwater in the Urban Environment*. A.A. Balkema, Rotterdam

Chisari, Omar, Antonio Estache and Catherine Waddams Price 2001 Access by the poor in Latin America's utility reform. WIDER discussion paper no. 2001/75. World Institute for Development Economics Research, United Nations University: Helsinki.

Christman, K. 1998 The history of chlorine. Accessed at, Water quality and health http: //www.waterandhealth.org/drinkingwater/history.html

Church, M. 1995 Geomorphic response to river flow regulation: case studies and timescales, *Regulated Rivers: Research and Management* 11, 3–22

Churchill, A. 1987 Rural water supply and sanitation: Time for a change. World Bank discussion papers 18. Washington, DC

CIMMYT 2009 International maize and wheat improvement centre. http://www.cimmyt.org/english/wps/about/index.htm and http://dtma.cimmyt.org/

Cissé, S. 1985 Land tenure and development in the Niger Delta of Mali, in Hill, A. (ed.) *Population, Health and Nutrition in the Sahel*. KPI, London

CityMayors. 2006 City Mayors Statistics: World largest cities and urban areas 2006. http: //www.citymayors.com/statistics/urban_2006_1.html

Clarke, A.J. 2008 *An Introduction to the Dynamics of El Niño & the Southern Oscillation*. Academic Press, London

Cleaver, F. and Toner, A. 2006 The evolution of community water governance in Uchira, Tanzania: The implications for equality of access, sustainability and effectiveness. *Natural Resources Forum* 30: 207–18

Cochran, J. and Ray, I. 2008 Equity re-examined: a study of community based rainwater harvesting in Rajasthan, India. *World Development* 37(2): 435–44

Collins, R. 1994 History, hydropolitics and the Nile: Myth or reality, in Howell, P. and Allan, J.A. (eds) *The Nile: Sharing a Scarce Resource*. Cambridge University Press, Cambridge. pp. 109–36

Conway, D. 2005 From headwater tributaries to international river: observing and adapting to climate variability and change in the Nile basin. *Global Environmental Change* 15: 99–114

Conway, D., Hanson, C., Doherty, R. and Persechino, A. 2007 GCM simulations of the Indian Ocean dipole influence on East African rainfall: Present and future. *Geophysical Research Letters* 34. L03705 (doi: 10.1029/2006GL027597)

Cook, H.F. 1998 *The Protection and Conservation of Water Resources: a British Perspective*. John Wiley & Sons, Chichester

Cooke, W. and Kothari, U. 2001 *Participation. The New Tyranny?* Zed Books, London

Corkal, D. Schutzman, W.C. Hilliard, C.R. 2004 Rural water safety from the source to the farm tap. *Journal of Toxicology and Environmental Health Part A* 67: 1619–42

Cosbey, A. 2009 Developing country interests in climate change action and the implications for a post-2012 climate change regime. UN Conference on Trade and Development UNCTAD/DITC/BCC/2009/2 New York

Cotton, W.R. 1986 Testing, implementation and evolution of seeding concepts; a review in Braham Jr. R.R. ed. Precipitation ehancement – A scientific challenge. *Meteorological Monographs of American Meteorological Society* 21:139–49

—— 2007 Basic cloud seeding concepts. Southwest Hydrology March/April. (online) http: //www.swhydro.arizona.edu/archive/V6_N2/feature1.pdf

—— 2008 Parallels and contrasts between deliberate cloud seeding and aerosol pollution effects, in Levin, Z. and Cotton, W.R. (eds) *Aerosol Pollution Impact on Precipitation*, Springer Science, New York. pp. 277–94

Cotton, W.R. and Pielke, P. 1995 *Human Impacts on Weather and Climate*, Cambridge University Press, Cambridge

—— 2007 *Human Impacts on Weather and Climate*, 2nd edn. Cambridge University Press, Cambridge

Cotula, L. 2006 Water rights, poverty, and inequality. The case of dryland Africa. Human Development Report 2006 Occasional Paper, New York, United Nations Development Programme

Countryside Commission 1995 Climate change, air pollution and the English countryside. Summary of potential impacts. UK Countryside Commission, Walgrave.

Cowden, J.R. Watlins, D.W. and Mihelcic, J.R. 2008 Stochastic rainfall modelling in West Africa. Parsimonious approaches for domestic rainwater harvesting assessment. *Journal of Hydrology* 361: 64–77

Crang, P. 2005 Consumption and its geographies. Chapter 16 in Daniels, P., Bradshaw, M., Shaw, D. and Sidaway, J. (eds) *Human Geography*, 2nd edn. Pearson Education, Harlow. pp. 359–82

Crosbie, R.S., Wilson, B., Hughes, J.D. and McCulloch, C. 2008 The upscaling of transpiration from individual trees to areal transpiration in tree belts. *Plant Soil* 305: 25–34

Crow, B. and Singh, N. 2000 Impediments and innovations in international rivers: the waters of South Asia. *World Development* 28(11): 1907–25

Crowley, E. 1991 *Resource Tenure in Mali: an anthropological analysis of local institutions.* Club du Sahel and OECD, Paris

Croxton, P.J. Huber, K. Collinson, N. and Sparks, T.H. 2006 How well do the Central England temperature and the England and Wales precipitation series represent the climate of the UK? *International Journal of Climatology* 26: 2287–92

CTGPC 2009 China Yangtze Development Corporation (CTGPC): China Three Gorges Corporation. http: //www.ctgpc.com.cn/en/

Cullis, J. and van Koppen, B. 2007 Applying the Gini Coefficient to Measure Inequality of Water Use in the Olifants Water Management area, South Africa. Research Report 113, International Water Management Institute. Colombo

Cummings, B.J. 1990 *Dam the Rivers, Damn the People*, Earthscan, London.

Dai, A., Fung, I.Y. and Del Genio, A.D. 1997 Surface observed global land precipitation variations during 1900–988. *Journal of Climate* 10, 2943–62

Dai, A., Lamb, P.J., Trenberth, K.E., Hulme, M. Jones, P.D. and Xie, P. 2004 The recent Sahel drought is real. *International Journal of Climatology* 24: 1323–31

Daniels, P. Bradshaw, M. Shaw, D. and Sidaway, J. 2005 *Human Geography*, 2nd edn. Pearson Education, Harlow

D'Almeida, C. Vorosmarty, C.J. et al. 2006 A water balance model to study the hydrological response to different scenarios of deforestation in Amazonia. *Journal of Hydrology* 331: 125–36

D'Amato, N. and Lebel, T. 1998 On the characteristics of the rainfall events in the Sahel with a view to the analysis of climatic variability. *International Journal of Climatology* 18: 955–74

D' Arrigo, R. Allan, R. Wilson, R. Palmer, J. Sakulich, J. Smerdon, J.E. Bijaksana, S. Ngkoimani, L.O. 2008 Pacific and Indian Ocean climate signals in a tree-ring record of Java monsoon drought. *International Journal of Climatology* 28(14): 1889–1901

Darwish, M.A. Al-wadhi, F.M. and Darwish, A.M. 2008 Energy and water in Kuwait Part 1: a sustainability view point. *Desalination* 225: 341–55

Davie, T. 2003 *Fundamentals of hydrology.* Routledge, London

Davies, S. 1996 *Adaptable Livelihoods. Coping with Food Insecurity in the Malian Sahel.* Macmillan Press, Basingstoke

Davis, R.J. 1974 Weather modification litigation and statues in Hess, W.N. (ed.) *Weather and Climate Modification.* J. Wiley, London. pp. 767–86

Davis, R.J. 1975 Legal response to environmental concerns about weather modification. *Journal of Applied Meteorology* 14: 681–85

Davis, R.J. and Grant, L.O. (eds) 1978 Weather modification and the law. *American Association for Advancement of Science* 20: Westview Press, Boulder,

Deak, A. 2008 Taking community led total sanitation to scale: movement, spread and adaptation. IDS Working Paper 298, University of Sussex.

de Bruijne, G., Geurts, M. and Appleton, B. 2007 Sanitation for all ?IRC International Water and Sanitation Centre (www.irc.nl)

de Georges, A. and Reilly, B. 2006 Dams and Large Scale Irrigation on the Senegal River. Impacts on Man and the Environment. Human Development Report 2006 Occasional Paper, New York, United Nations Development Programme

de Graaf, R. van Giesen, N. and van de Ven, F. 2007 Alternative water management options to reduce vulnerability for climate change in the Netherlands. *Journal of the International Society for the Prevention and Mitigation of Natural Hazards* (online 10.1007/s11069-007-9184-4)

de Groot, S. 1992 Decline and fall of the salmon fisheries in the Netherlands: is restocking the Rhine a reality? *Aquaculture and Fisheries Management* 23: 253–64

Delyannis, A. and Delyannis, E. 1980 *Seawater and Desalting*, Springer-Verlag, Heidelberg.

Dennis, A.S. 1980 *Weather Modification by Cloud Seeding*. Academic Press, London

Deoreo, W.D., Dietemann, A., Skeel, T. and Mayer, P.W. 2001 Retrofit realities. *American Water Works Association Journal* 93(3): 58–72

Dessler, A.E. and Parson, E.A. 2006 *The Science and Politics of Global Climate Change: a Guide to the Debate*. Cambridge University Press, Cambridge

De Wilde, J. 1967 *Experience with agricultural development in Tropical Africa*. John Hopkins University Press, Baltimore

DfID 2001 Addressing the Water Crisis: healthier and more productive lives for poor people. UK Department for International Development. www.dfid.gov.uk/Documents/publications/tspwater.pdf

—— 2010 Irrigation DfID-KAR-Water for Food website. http://www.dfid-Kar-water.net/W5outputs/index.html

Diaz, H.F. and Markgraf, V. (eds) 2000. *El Niño and the Southern Oscillation: Multiscale Variability, Global and Regional Impacts*. Cambridge University Press, Cambridge.

Dickey, G.B. 1961 *Filtration*. Van-Nostrand Rheinhold, New York

Diemer G. and van der Laan, E. (1987) *L'Irrigation au Sahel*. Paris, Editions Karthala.

Dinar, A. 2000 Political economic of water pricing reforms in Ariel Dinar (ed.) *The Political Economy of Water Pricing Reforms*, World Bank: Washington, pp. 1–26

Dogra, B. 1986 The Indian experience with large dams, in Goldsmith, E. and Hildyard, N. (eds) *The Social and Environmental Effects of Large Dams*, Vol. 2 Case Studies, Wadebridge Ecological Centre, Cornwall. pp. 201–8.

Donaldson, D. 1987 Community participation in rurban water supply and sanitation systems, Conference on Developing World Water, Grosvenor International, London. pp. 86–89

Doorenbos, J. and Pruitt, W.O. 1977 Guidelines for predicting crop water requirements. FAO Irrigation and Drainage Paper 24, Food and Agriculture Organization of the United Nations. Rome

DfID 2010 Irrigation.DfID-KAR-Water for Food website. http://www.dfid-kar-water.net/w5outputs/index.html

Dow, K. and Downing, T. 2006 *The Atlas of Climate Change: Mapping the World's Greatest Challenge*. Earthscan, London

Dreizin, Y. Tenne, A. and Hoffman, D. 2008 Integrating large scale seawater desalination plants within Israel's water supply system. *Desalination* 220: 132–49

Dresner, L. and Johnson, J. 1980 Hyperfiltration, in Spiegler, K.S. and Laird, A.D.K. (eds) *Principles of Desalination, Part B*, Academic Press, London

Dungumaro, E.W. 2007 Socioeconomic differentials and availability of domestic water in South Africa. *Physics and Chemistry of the Earth* 32: 1141–47

Ehlers, W. and Goss, M.J. 2003 *Water Dynamics in Plant Production*. CABI, Wallingford

Ehrlich, P. 1968 *The Population Bomb*. Ballantine Books, New York

Ekstrom, M., Jones, P.D., Fowler, H.J., Lenderink, G., Buishand, T.A. and Conway, D. 2007 Regional climate model data used within the SWURVE project 1: projected changes in seasonal patterns and estimation of PET. *Hydrology and Earth System Sciences* 11(3), 1069–1083.

Elhassadi, A. 2008 Horizons and future of water desalination in Libya. *Desalination* 220: 115–22

El-Sayed, Y.M. and Silver, R.S. 1980 Fundamentals of distillation, in Spiegler, K.S. and Laird, A.D.K. (eds) *Principles of Desalination, Part A*, Academic Press, London

Elshorbagy, W. and Elhakeem, A-B. 2008 Risk assessment maps of oil spills for major desalination plants in the United Arab Emirates. *Desalination* 228: 200–216

Emanuel, K. 2005 Increasing destructiveness of tropical cyclones over the past 30 years. *Nature* 436: 686–88

Engel, H. 1997 The flood events of 1993–94 and 1995 in the Rhine River Basin. Destructive water: water-caused natural disasters, their abatement and control. Proceedings IAHS conference held at Anaheim, California. IAHS publication No 239.

Engelhard, P., Ben Abdallah, T. And others 1986 *Enjeux del'Après-Barrage*. ENDA-Tiers Monde (Dakar) and Ministère de la Coopération (Paris)

Engels, F. 2005 (1845) *The Condition of the Working Class in England*. Penguin, London

Environment Agency (UK) 2002 *Managing Water Abstraction*. Environment Agency, Bristol

—— 2007 *Identifying Areas of Water Stress*. Environment Agency, Bristol

—— 2008 *Identifying Areas of Water Stress*. Environment Agency, Bristol
—— 2009 *Water for Life and Livelihoods: A consultation on the draft river basin management plan, Thames River basin district*. Environment Agency, Bristol
—— 2010a Water Quality Assessment. http://www.environment-agency.gov.uk/
—— 2010b Water Quality Report(2008) for England and Wales. http://www.environment-agency.gov.uk/research/library/data/34383.aspx

Environment Australia 2010 Department of the Environment, Water, Heritage and the Arts. http: //www.environment.gov.au/

EPA, 2000 *The History of Drinking Water Treatment*, Environmental Protection Agency, Office of Water (4606), Fact Sheet EPA-816-F-00-006, United States http://permanent.access.gpo.gov/websites/epagov/www.epa.gov/safewater/sdwa/trends.html

European Environment Agency 2007 Fourth Environmental Assessment Report. http: //www.eea.europa.eu/publications/state_of_environment_report_2007_1/chapter2.pdf

Evans, M.G., Burt, T.P., Holden, J. and Adamson, J.K. 1999 Runoff generation and water table fluctuations in blanket peat: evidence from UK data spanning the dry summer of 1995. *Journal of Hydrology* 221: 141–60

Evans, M. and Warburton, J. 2007 *Geomorphology of Upland Peat*. Blackwells, Oxford

FAO 1975 *Water available but mismanaged*. CERES 8(4).
—— 1994 *Water harvesting for improved agricultural production*. Water Reports No 3. FAO, Rome
—— 1998 Sources of international water law. FAO Legislative study 65, Rome

FAO–AQUASTAT 2008 Food and Agricultural Organisation, UN. http: //www.fao.org/nr/water/aquastat/data/factsheets/aquastat_fact_sheet_ind.pdf
—— 2009 Food and Agricultural Organisation, UN. AQUASTAT http: //www.fao.org/nr/water/aquastat/regions/

Fearnside, P.M. 2006 Dams in the Amazon, Belo Monte and Brazil's hydro-electric development of the Xingu river basin. *Environmental Management* 38(1): 16–27

Fedorov, A. and Philander, S.G. 2000 Is El Niño changing? *Science* 288(5473): 1997–2002

Feitelson, E. 2002 Implications of shifts in the Israeli water discourse for Israeli-Palestinian water negotiations. *Political Geography* 21: 293–318

Fenemor, A., Deans, N., Davie, T., Allen, W., Dymond, J., Kilvington, M., Phillips, C., Basher, L., Gillespie, P., Yong, R., Sinner, J., Harmsworth, G., Atkinson, M. and Smith, R. 2008 Collaboration and modelling: Tools for integration in the Motueka catchment, New Zealand. *Water South Africa* 34(4): 448–55

Few, R., Brown, K. and Tomkins, E. 2007 Public participation and climate change adaptation: avoiding the illusion of inclusion. *Climate Policy* 7: 46–59

Fisher, J. 2008 Women in water supply, sanitation and hygiene programmes. *Municipal Engineer* 161 (Dec.): 223–29

Flerchinger, GN and Cooley, K.R. 2000 A ten-year water balance of mountainous semi-arid watershed. *Journal of Hydrology* 237: 86–99

Fleskens, L., Stroonsnijder, L., Ouessar, M. and De Graaff, J. 2005 Evaluation of the on site impact of water harvesting in southern Tunisia. *Journal of Arid Environments* 62: 613–30

Folland, C.K. 1987 Sea surface temperatures predict African drought. *New Scientist* 116(1580): 25

Fonjong, L.N. 2008 Gender roles and practices in natural resource management in the North West Province of Cameroon. *Local Environment*, 13(5): 461–475

Foster, S. Morris, B., Lawrence, A. and Chilton, J. 1999 Groundwater impacts and issues in developing countires–an introductory review, in Chilton, J. (ed.) *Groundwater in the Urban Environment*. A.A. Balkema, Rotterdam, pp. 3–18

Fowler, A. 2002 Assessment of the validity of using mean potential evaporation in computations of long-term soil water balance. *Journal of Hydrology* 256: 248–63

Fowler, H.J., Ekstrom, M. Kilsby, C.G. and Jones, P.D. 2005 New estimates of future changes in extreme rainfall across the UK using regional climate model integrations. 1. Assessment of control climate. *Journal of Hydrology* 300: 212–33

Fowler, H.J., Kilsby, C.G. and Stunell, J. 2007 Modelling the impacts of projected future climate change on water resources in north-west England. *Hydrology and Earth System Sciences* 11(3): 1115–26

Flerchinger, G.N. and Cooley, K.R. 2000 A ten year water balance of a mountainous semi arid water-shed. *Journal of Hydrology* 237: 86–99

Franke F. and Chasin B., 1980: *Seeds of Famine*. Allanheld, Osman and Co., New Jersey.

Frost, D.H. 1987 Developing water resources in low rainfall areas, Conference on Developing World Water, Grosvenor International, London. pp. 75–77

Gandy, M. 2002 *Concrete and Clay: Reworking Nature in New York City*. MIT Press, Cambridge, MA

—— 2008 Landscapes of disaster: water, modernity, and urban fragmentation in Mumbai *Environment and Planning A* 40: 108–30

Ganoulis, J., Skoulikaris, H. Monget, J.M. 2008 Involving stakeholders in transboundary water resources management: the Mesta/Nestos HELP basin. *Water South Africa* 34(4): 461–67

Garrett, V., Ogutu, P., Mabonga, P., Ombeki, S., Mwaki, A., Aluoch, G., Phelan, N. and Quick, R.E. 2008 Diarrhoea prevention in a high risk rural Kenyan population through point of use chlorination, safe storage, sanitation and rainwater harvesting. *Epidemiology Infection* 136: 1463–71

Gaudin, S. 2000 Estimating evaporation and surface resistance from a wet grassland. *Physics and Chemistry of the Earth* 25(7–8): 599–603

—— 2003 Evaluating the reliability if point estimates of wetland reference evaporation. *Hydrology and Earth System Sciences* 7(1): 3–10

—— 2006 Effect of price information on residential water demand. *Applied Economics* 38: 383–93

Gavin, H. and Agnew, C.T. 2004 Modelling actual, reference and equilibrium evaporation from a temperate wet grassland. *Hydrological Processes* 18: 229–46

GDRC 2010 Urban water resources management: Global Development Research Center. http://www.gdrc.org/uem/water/index.html

Geilen, N. Jochens, H., Krebs, L., Muller, S., Pedroli, B., van der Sluis, T., van Looy, K. and van Rooij, S. 2004 Integration of ecological aspects in flood protection strategies. Defining an ecological minimum. *River Research and Applications* 20: 269–83

Gerik, R. 1973 Legal aspects of weather modification in Texas. *Baylor Law Review* XXV: 501–10

Germani, G., Olalde, A., de Oliveira, G., Quan, J., Coelho Neto, A., Santos C., Estrela, E., Chamo, L., Souza, M., Santos, T. and Menezes, W. 2007 Acesso a terra e desenvolvimento territorial no Medio São Francisco, Bahia. Land and Territory Research Paper No. 5. Greenwich, Natural Resources Institute

Giakoumakis, S. and Tsakiris, G. 2001 Experimental validation of a linearized kinemantic wave equation of micro-catchment water harvesting design. *Water Resources Management* 15: 235–46

Gill, S., Handley, J.F., Ennos, R. and Pauleit, S. 2007 Adapting cities for climate change: the role of green intrastructure. *Built Environment* 33(1): 97–115

Gine, R. and Perez-Foguet, A. 2008 Sustainability assessment of national rural water supply program in Tanzania. *Natural Resources Forum* 32: 327–42

Giorgi, F. 2006 Climate change hotspots. *Geophysical Research Letters* 33: L08707 doi: 10.1029/2006GL025734

Girard, M. and Stewart, R.A. 2007 Implementation of pressure and leakage management strategies on the Gold Coast, Australia: Case Study. *Journal of Water Resources Planning and Management* 133(3): 210–16

Gleick, P.H. 1988 Regional hydrological impacts of global climate changes, in Whitefield, E.E. et al. (eds) *Arid Lands Today and Tomorrow*. Westview Press, Boulder, pp. 43–59

—— 1993 Water in the 21st Century, in Gleick, P.H. (ed.) *Water in Crisis*. Oxford University Press, Oxford. pp. 105–13

Glantz, M.H. 1987 Drought in Africa. *Scientific American* 256(6): 34–40

—— 1994 *Drought Follows the Plow, Cultivating Marginal Areas*. Cambridge University Press, Cambridge

Glantz, M.H., Katz, R.W. and Nicholls, N. (eds) 1991 *Teleconnections: linking worldwide climate anomalies*. Cambridge University Press, Cambridge

Goldsmith, E. and Hildyard, N. (eds) 1984 *The Social and Environmental Effects of Large Dams*, Vol. 1, Case Studies, Wadebridge Ecological Centre, Cornwall

——1986 *The Social and Environmental Effects of Large Dams*, Vol. 2 Case Studies, Wadebridge Ecological Centre, Cornwall

Golladay, F. 1983 Meeting the needs of the poor for water supply and waste disposal, World Bank paper, *Appropriate Technology for Water Supply and Sanitation* 13, Washington, DC

Gordon, A., Grace, W., Schwerdtfeger, P and Bryon-Scott, R. 1998 *Dynamic Meteorology*. John Wiley, New York.

Gould, J. and Nissen-Petersen, E. 1999 *Rainwater Catchment Systems for Domestic Supply*. Intermediate Technology, London

Gould, M. 1981 A Water Quality Assessment of Development in the Senegal River Basin. *Water Resources Bulletin, American Water Resources Association* 17(3): 466–73

Graf, W. L. 1999 Dam nation: A geographic census of American dams and their large-scale hydrologic impacts: *Water Resources Research* 3: 1305–11

Gregory, K.J. 2000 *River Basin Planning*, in Thomas, D.S.G. and Goudie, A. (eds) *The Dictionary of Physical Geography*. Blackwell, Oxford

Gregory, J. M. and J. A. Lowe 2000 Predictions of global and regional sea level rise using AOGCMs with and without flux adjustment, *Geophysical Research Letters* 27(19): 3069–72

Gregory, S. (ed.) 1988 *Recent Climate Change*. Belhaven Press, London

Gribben, J. and Gribben, M. 1996 The greenhouse effect. *New Scientist* 151(2037): 1–4.

GRID-Arendal 2009 Trends in global water use by sector. http://www.grida.no/publications/vg/water2/page/3229.aspx

Griffith-Jones, S., Hedger, M. and Stokes, L. 2009 *The Role of Private Investment in Increasing Climate Friendly Technologies in Developing Countries*. UN-DESA. New York

Guardian 2008 Is there life in outer space? 1/8/2008 http://www.guardian.co.uk/science/poll/2008/aug/01/spaceexploration.spacetechnology

—— 2009 Obama Administration criticized over failure to disclose coal dump locations 18/6/09 http://www.guardian.co.uk/environment/2009/jun/18/coal-dump-location-obama-usa accessed 05/07/2009

Guillemin, F., Henry, P., Uwechue, N. and Monjour, L. 1991 Faecal contamination of rural water supply in the Sahelian area. *Water Resources* 25(8): 923–27

Guo S., Wang J., Xiong, L., Ying, A. and Li, D 2002 A macro-scale and semi-distributed monthly water balance model to predict climate change impacts in China. *Journal of Hydrology* 268: 1–15

GWI 2005 Global Water Intelligence: 100 largest desalination plants planned, in construction, or in operation, prepared by N. Wangnick GWI 41(5). http://www.globalwaterintel.com/

Gyau-Boakye, P. and Ampomah, B.Y. 2003 Water pricing and sector reforms informs information study in Ghana (International Water Resources Association). *Water International* 28(1): 11–18

Haas, J.E. 1974 Sociological aspects of weather modification, in Hess, W.N. (ed.) *Weather and Climate Modification*. J. Wiley, London. pp. 787–811

Hagen, R.M. 1988 Water issues, in Whitefield, E.E. et al. (eds) *Arid Lands Today and Tomorrow*. Westview Press, Boulder, pp. 9–10

Hardin, G. 1968 The tragedy of the commons. *Science* 162: 1243–48

Harrison, P. 1987 *The Greening of Africa*. Paladin, London

Hayami, J. and Ruttan, V. 1985 *Agricultural Development. An International Perspective*. Baltimore: Johns Hopkins University Press

Hayes, M.J., Svoboda, M.D., Wilhite, D.A. and Vanyarkho, O.V. 1999 Monitoring the 1996 drought using the standardized precipitation index. *Bulletin American Meteorological Society* 80(3): 429–38

Harvey, L.D.D. 2000 *Climate and Global Environmental Change*. Prentice Hall, Harlow

HDR 2006 *Beyond Scarcity*. Human Development Report 2006. New York: United Nations Development Programme

Heathcote, R.L. 1983 *The Arid Lands: Their Use and Abuse*. Longman, London.

Helmes, S.W. 1981 *Jawa: Lost City of the Black Desert*. Methuen, London

HESS 2007 Reprint of H.L. Penman's 1973 address, a view from the watershed. *Hydrology and Earth System Science (HESS)* 11(1): 12–16

Heimann, M. and Reichstien, M. 2008 Terrestrial ecosystem carbon dynamics and climate feedbacks. *Nature* 451(Jan.): 289–92

Henson, R. 2005 The heat was on in 2005 *Nature* 438(Dec.): 1062

Hewitson, B.C. and Crane, R.G. 2006 Consensus between GCM climate change projections with empirical downscaling: precipitation downscaling over South Africa. *International Journal of Climatology* 26: 1315–37

Hickey, S. and Mohan, G. 2004 *Participation. From Tyranny to Transformation?* London: Zed Books

Hills, E.S. (ed.) 1966 *Arid Lands*. Methuen and Co., London

Hirsch, T. 2003 Giant Talking Shop. http://news.bbc.co.uk/1/hi/sci/tech/2879125. stm (accessed 11 February 2010)

Hjort-as-Ornas, A. 2008 *Turning Hydropower Social*. Springer-Verlag, Berlin

Hlavinek, P., Kukharchyk, T., Marsalek, J. and Mahrikova, I. 2006 *Integrated Urban Water Resources Management*. Springer, Dordrecht

Hoekstra, A.Y. and Chapagain, A.K. 2007 Water footprints of nations: Water use by people as a function of their consumption pattern. *Water Resources Management* 21: 35–48

Holden, J. 2008 *An Introduction to Physical Geography and the Environment*, 2nd edn. Pearson Education, Harlow

Holling, C.S. 1973 Resilience and stability of ecological systems. *Annual Review of Ecology and Systematics* 4: 1–23

Hooijer A., Klijn, F., Pedroli, G. and van Os, Ad. 2004 Towards Sustainable Flood Risk Management in the Rhine and Meuse river basins: synopsis of the findings of IRMA-SPONGE. *River Research and Applications* 20: 343–57

Hoornaert, P. 1984 *Reverse Osmosis*. Pergamon Press, Oxford

Hope, R., Porras, I., Borgoyary, M., Miranda, M., Agarwal, C., Tiwari, S. and Amezaga, J. 2007 *Negotiating Watershed Services*. International Institute for Environment and Development, London

Hophmayer-Tokich, S. and Krozer, Y. 2008 Public participation in rural area water management: experiences from the North Sea countries in Europe. *Water International* 33(2): 243–57

Hornburg, C.D. 1987 Desalination for remote areas, in *Conference on Water and Engineering for Developing Countries, Developing World Water*, Grosvenor Press International, Hong Kong, pp. 230–32

Hough, M.N. and Jones, R.J.A. 1997 The UK Meterological Office rainfall and evaporation calculation system: MORECS version 2.0 an overview. *Hydrology and Earth System Sciences* 1(2): 227–39.

Houghton, J.T. 1997 *Global Warming, the Complete Briefing*, 2nd edn Cambridge University Press, Cambridge

—— 2004 *Global Warming, the Complete Briefing*, 3rd edn. Cambridge University Press, Cambridge

—— 2009 *Global Warming, the Complete Briefing*. 4th edn. Cambridge University Press, Cambridge

Houghton, J.T., Jenkins, G.J. and Ephraums, J.J. (eds) 1991 *Climate Change: The IPCC 1990 Scientific Assessment*. Cambridge University Press, Cambridge

Howard, G. and Bartram, J. 2003 Domestic Water Quantity, Service Level and Health WHO Report WHO/SDE/WSH/03.02 WHO, Geneva http://www.who.int/water_sanitation_health/diseases/WSH03.02.pdf

Howe, C.W. and Easter, K.W. 1971 *Interbasin transfers of water: economic issues and impacts*. John Hopkins University Press, Baltimore

Howe, J. and White. 2004 The mismanagement of surface water. *Applied Geography* 24: 261–80

Howell, P. and Allan, J.A. (eds) 1994 *The Nile: Sharing a Scarce Resource*. Cambridge: Cambridge University Press

Howell, P. and Lock, M. 1994 The control of the swamps of southern Sudan: Drainage schemes, local effects and environmental constraints on remedial development. In Howell, P. and Allan, J.A. (eds) *The Nile: Sharing a Scarce Resource*. Cambridge: Cambridge University Press. pp. 243–80

Howitt, R. and Sunding, D. 2003 Water infrastructure and water allocation in California, in Siebert, J. (ed.) *California Agriculture Dimensions and Issues*. Giannini Foundation of Agricultural Economics. University of California, Berkeley

Huggett, R., Lindley, S., Gavon, H. and Richardson, K. 2004 *Physical Geography, a Human Perspective*. Arnold, London

Hulme, M. 1996 *Climate change and Southern Africa: An Exploration of Some Potential Impacts and Implications in the SADC region*. Climatic Research Unit, UEA and WWF International.

Hulme, M. 1996 Recent climatic change in the world's drylands. *Geophysical Research Letters* 23(1): 61–64

Hulme, M. 2001 Climatic perspectives on Sahelian desiccation 1973–98. *Global Environmental Change* 11: 19–29

—— 2009a *Why We Disagree about Climate Change: Understanding Controversy, Inaction and Opportunity*. Cambridge University Press, Cambridge

—— 2009b The science and politics of climate change. *Wall Street Journal* 2/12/2009

Hult, J.L. 1982 Water supply from imported Antarctic icebergs, in *Conference on Alternative Strategies for Desert Development and Management Vol III:Water*. Pergamon Press, Oxford. pp.712–719

Hundecha, Y. and Bardossy, A. 2008 Statistical downscaling of extremes of daily precipitation and temperature and construction of their future scenarios. *International Journal of Climatology* 28(5): 589–610

Hunt, C.E. 2004 *Thirsty Planet: Strategies for sustainable water management.* Zed Books, London

Hunter, L. 2006 Household strategies in the face of resource scarcity in coastal Ghana: are they associated with development priorities? *Population Resources Policy Review* 25: 157–74

Hurriyet Daily News 2009 World Water Forum opens in Turkey with riot police quelling protest (cited text from 17 March 2009). http://www.hurriyet.com.tr/english/domestic/11218082.asp

Hussain, I. 2005 *Poverty in Irrigated Agriculture. Realities, Issues and Options with Guidelines. Pro-Poor Intervention Strategies in Irrigated Agriculture in Asia.* Colombo: International Water Management Institute

ICPR, 2009 International Commission for the Protection of the Rhine, Water Quality Webpage. http://www.iksr.org/index.php?id=134&L=3 (accessed 28 July 2009)

Intsiful, J. and Kunstman, N.H. 2008 Upscaling of land-surface parameters through inverse stochastic SVAT-modelling. *Boundary-Layer Meteorology* 129: 137–58

IPCC 1990 *First Assessment Report.* Intergovernmental Panel on Climate Change, Geneva

——— 1995 *Second Assessment Report: Climate Change.* Intergovernmental Panel on Climate Change, Geneva

——— 2001 *Third Assessment Report: Climate Change.* Intergovernmental Panel on Climate Change, Geneva. http://www.ipcc.ch/ipccreports/tar/vol4/index.php?idp=0

——— 2007 *Fourth Assessment Report on Climate Change.* IPCC, Geneva. Climate change synthesis report. Intergovernmental Panel on Climate Change, Geneva. http://www.ipcc.ch/publications_and_data/publications_ipcc_fourth_assessment_report_synthesis_report.htm

——— 2007 *Climate Change Fourth Assessment:* Working Group 2: Impacts, vulnerability and adaptation. http://www.ipcc.ch/ipccreports/ar4-wg2.htm

IPCC 2008 *Fourth Assessment on Climate Change.* IPCC, Geneva. http://www.ipcc.ch/publications_and_data/publications_ipcc_fourth_assessment_report_synthesis_report.htm

IRN 2003 Dammed rivers, dammed lies. International Rivers Network (online, http://www.internationalrivers.org/)

Jackson, C. 1993 Doing what comes naturally? Women and environment in development, *World Development* 21 (12): 1947–63

James, P. 1948 The São Francisco Basin: A Brazilian Sertão *Geographical Review* 38 (4): 658–61

Jansen, A. and Schulz, C. 2006 Water demand and the urban poor: a study of the factors influencing water consumption among households in Cape Town, South Africa. *South African Journal of Economics* 74(3): 593–609

Jarrah, M.A. 2005 Assessing runoff potential for water harvesting generation in Jordan. PhD thesis, University of Manchester

Jeffrey, P. and Geary, P. 2005 Consumer reactions to water conservation policy instruments. In Butler, D. and Memon, F.A. (eds) *Water Demand Management.* IWA, London. pp. 305–30

Jehangir, W. Masih, T., Ahmed, S., Gill, M., Ahmad, M., Mann, R., Chaudhary, M., Qureshi, A. and Turral, H. 2007 Sustaining crop water productivity in rice-wheat systems of South Asia: a case study from the Punjab, Pakistan. *Working Paper 115* Colombo: International Water Management Institute.

Jha, R. and Bhanu Murthy, K.V. 2006 *Environmental Sustainability: a Consumption Approach*. Routledge, London

Johnson, C., Penning-Rowsell, E. and Parker, D. 2007 Natural and imposed injustices: the challenges in implementing fair flood risk management policy in England. *The Geographical Journal* 173(4): 374–90

Jones, J.A. 1997 *Global Hydrology*. Longman, Harlow

Jones, P.D. and Hulme, M. 1996 Calculating regional time series for temperature and precipitation: methods and illustrations. *International Journal of Climatology* 16: 361–77

Jones, T. 2003 Pricing water. *Observer Spotlight* (OECD) No. 236 (March): 11–13

Johns, T.C., Gregory, J.M., Ingram, W.J., Johnson, C.E., Jones, A., Lowe, J.A., Mitchell, J.F.B., Roberts, D.L., Sexton, D.M.H., Stevenson, D.S. Tett, S.F.B. and Woodage, M.J. 2003 Anthropogenic climate change for 1860 to 2100 simulated with the HadCM3 model under updated emissions scenarios. *Climate Dynamics* 20(6): 583–612

Johnson, C., Penning-Rowsell, E. and Parker, D. 2007 Natural and imposed injustices: the challenges in implementing fair flood risk management policy in England. *The Geographical Journal* 173(4): 374–90

Jothityangkoon, C., Sivapalan, M. and Farmer, D.L. 2001 Process controls of water balance variability in a large semi-arid catchment: downward approach to hydrological model development. *Journal of Hydrology* 254: 174–98

Jowit 2008 Environment: Huge increase in spending on water urged to avert global catastrophe. *Guardian* 11 September 2008. http://www.guardian.co.uk/environment/2008/sep/11/water.climatechange

Kahinda, J.M., Lillie, E.S.B., Taigbenu, A.E., Taute, M. and Boroto, R.J. 2008 Developing suitability maps for rainwater harvesting in Southern Africa. *Physics and Chemistry of the Earth* 33: 788–99

Kamal, I. 2008 Myth and reality of the hydrid desalination process. *Desalination* 230: 269–80

Karshenas, M. 2004 Urban bias, intersectoral resource flows and the macroeconomic implications of agrarian relations: the historical experience of Japan and Taiwan. *Journal of Agrarian Change* 4 (1&2): 170–89

Kebbekus, B.B. and Mitra, S. 1998 *Environmental Chemical Analysis*. Blackie Academic and Professional, London

Keenlyside, N.S., Latif, M., Jungclaus, J., Kornblueh, L. and Roeckner, E. 2008 Advancing decadal-scale climate prediction in the North Atlantic sector. *Nature* 453: 84–88

Kenny, D.S., Goemans, C., Klein, R., Lowrey, J. and Reidy, K. 2008 Residential water demand management: Lessons from Aurora, Colorado. *Journal of American Water Resources Association* 44(1): 192–207

Kenny, J.F., Barber, N.L., Hutson, S.S., Linsey, K.S., Lovelace, J.K. and Maupin, M.A. 2009, Estimated use of water in the United States in 2005: *U.S. Geological Survey Circular* 1344. http://pubs.usgs.gov/circ/1344/pdf/c1344.pdf

Kerr, R.A. 2007a Global warming is changing the world. *Science* 316: 188–90
—— 2007b How urgent is climate change? *Science* 318: 1230–31

Keyantash, J. and Dracup, J.A. 2002 The quantification of drought: an evaluation of indices. *Bulletin of American Meteorological Society* (August): 1168–80

Khan, A.H. 1986 *Desalination Processes and Multistage Flash Distillation Practice*, Elsevier, Oxford

Kim, J.-W., Chang, J.-T., Baker, N.L., Wilks, D.S. and Gates, W.L. 1984 The statistical problem of climate inversion: determination of the relationship between local and large scale climate. *American Meteorological Society, Monthly Weather Review* 112: 2069–77

Kim, M.-K, Lau, W., Kim, K.-M and Lee, W.-S. 2007 A GCM study of effects of radiative forcing of sulfate aerosol on large scale circulation and rainfall in East Asia during boreal spring. *Geophysical Research Letters* 34: L24701, doi: 10.1029/2007GL031683.

Klawitter, S. 2006 Water resources at stake: The mountain aquifer beneath the occupied West Bank, Palestinian Territories. *Human Development Report.* New York: United Nations Development Programme.

Kolokytha, E.G. and Mylopoulos, N.A. 2004 Evaluating demand management aspects of urban water policy. The city of Volos case, Greece. *Water, Air and Soil Pollution* 4: 263–77.

Kondolf, G.M. 1997 Hungry Water: Effects of Dams and Gravel Mining on River Channels. *Environmental Management* 21(4): 533–551

Köppen, W. 1931 *Die Klimate der Erde: Grundriss der Klimakunde*, Berlin

Kruger, A.C. 2006 Observed trends in daily precipitation indices in South Africa: 1910–2004. *International Journal of Climatology* 26: 2275–85

Laffoley, D. and Grimsditch, G. (eds) 2009 The management of natural coastal carbon sinks. IUCN. Gland, Switzerland. http://www.naturalengland.org.uk/Images/carbon-sinks-report_tcm6–15112.pdf

Lancaster, W. and Lancaster, F. 1999 *People, Land and Water in the Arab Middle East,* Harwood, Amsterdam

Lahlou, Z.M. 2001 *Leak detection and water loss control.* Technical Brief. National Drinking Water Clearinghouse Fact Sheet (May)

Lankford, B. 2004 Resource-centred thinking in river basins; should we revoke the crop water requirement approach to irrigation planning? *Agricultural Water Management* 68: 33–46

Lattemann, S. and Hopner, T. 2008 Environmental impact and impact assessment of seawater desalination. *Desalination* 220: 1–15

der Leeden, V. 1975 Water resources of the world, Water Information Centre, New York.

Lemos, C. and Oliveira, J. 2004 *Can Water Reform Survive Politics? Institutional Change and River Basin Management in Ceará, Northeast Brazil. World Development,* 32(12) Elsevier

Lenton, R. and Muller, M. 2009 *Integrated Water Resources Management in Practice*. Earthscan, London

Ligon, F. K., W. E. Dietrich and W. J. Trush, 1995 Downstream ecological effects of dams: a geomorphic perspective, *Bioscience*, 45(3): 183–192

Lloyd-Hughes, B. and Saunders, M.A. 2002 A drought climatology for Europe. *International Journal of Climatology* 22: 1571–92

Lloyd, B. and Helmer, R. 1991 *Surveillance of Drinking Water Quality in Rural Areas*. Longman Scientific, Harlow

Lomborg, B. ed. 2009 *Global Crises, Global solutions*, 2nd edn. Cambridge University Press, Cambridge

Luckin, D. and Sharp, L. 2004 Remaking local governance through community participation ? The case of the UK community waste sector. *Urban Studies* 41(8): 1485–1505

Lundqvist, J. 2000 A global perspective on water and the environment. *Physics and Chemistry of the Earth* 25(3): 259–64

Lundqvist, J. Tortajada, C. Varis, O. and Biswas, A. 2005 Water management in megacities. *Ambio* 34(3): 267–68

McCaffrey, S.C. 1993 Water, politics and international law, in Gleick, P.H. (ed.) *Water in Crisis*. Oxford University Press, Oxford. pp. 92–104

McCulloch, J. 2007 All our yesterdays: a hydrological perspective. *Hydrology and Earth System Sciences (HESS)* 11(1): 3–11

McGuffie, K. and Henderson-Sellers, A. 1997 *A Climate Modelling Primer*. 2nd edn, J. Wiley & Sons, Chichester

McGuire, M.J. 2006 Eight revolutions in the history of UD drinking water disinfection. *Journal of American Water Works Association* 98(3): 123–49

McIlveen, R. 1998 *Fundamentals of Weather and Climate*. Routledge, London

McKee, T.B., Doesken, N.J. and Kleist, J. 1995 Drought monitoring with multiple time scales. Proceedings of the Ninth Conference on Applied Climatology. American Meteorological Society, Boston. pp. 233–36

McKibbin, W.J. and Wilcoxen, P.J. 2004 Climate policy and uncertainty: the roles of adaptation versus mitigation. *Brookings Discussion Papers in International Economics* 61: 1–15

Mackintosh, G. and Colvin, C. 2003 Failure of rural schemes in South Africa to provide potable water. *Environmental Geology* 44: 101–5

McNeill, D. 1985 The appraisal of rural water supplies. *World Development* 13(10/110): 1175–78

Maddaus, M.L., Maddaus, W., Torre, M. and Harris, R. 2008 Innovative water conservation supports sustainable housing development. *American Water Works Association Journal* 100(5): 105–110

Magiera, P., Taha, S. and Nolte, L. 2006 Water demand management in the Middle East and North Africa. *Management of Environmental Quality* 17(3): 289–298

Makoni, F.S., Manase, G. and Ndamba, J. 2004 Patterns of domestic water use in rural areas of Zimbabwe, gender roles and realities. *Physics and Chemistry of the Earth* 29: 1291–94

Mansfield, B. 2007 Privatisation: property and the re-making of nature–society relations. *Antipode* 39(3) 393–405

Mara, D. and Broome, J. 2008 Sewerage: a return to basics to benefit the poor. *Municipal Engineer* 161: 231–37

Mara, D. and Kramer, A. 2008 the 2006 WHO guidelines for wastewater and greywater use in agriculture: a practical interpretation, Chapter 1 in Al Baz, I. Otterphohl, R. and Wendland, C. (eds) *Efficient Management of Wastewater.* Springer-Verlag, Berlin. pp. 1–17

Marget, J. 2009 Water resources, in Woodward, J. (ed.) *The Physical Geography of the Mediterranean.* Oxford University Press, Oxford. pp. 583–97

Marino, E., White, D., Schweitzer, P., Chambers, M. and Wisniewskp, J. 2009 Drinking water in northwestern Alaska: using or not using centralized water systems in two rural communities. *Arctic* 62(1)75–83

Martinez-Espineira, R. 2002 Residential water demand in northwest Spain. *Environmental and Resource Economics* 21: 161–87

Marsalek, J., Jimenez-Cisneros, B.E., Malmquist, P.A., Karamouz, M., Goldenfum, J. and Chocat, B. 2006 *Urban water cycle processes and interactions. International Hydrological Programme Technical Documents in Hydrology* No. 78. UNESCO, Paris

Marsh T.J. 2001 The 2000/01 floods in the UK, a brief overview. *Weather* 56: 103–10

Marshall, R. and Plumb, A. 2007 *Atmosphere, Ocean and Climate Dynamics: An Introductory Text.* Academic Press, London

Mason, S.L. 1995 Sea surface temperatures and South African rainfall associations. *International Journal of Climatology* 15: 119–35

Matlock, W.G. 1988 The case for small scale water management systems in developing countries, in Whitehead, E. et al. (eds) *Arid Lands Today and Tomorrow,* Westview Press, Boulder. pp. 935–42

May, P.J. and Williams, W. 1986 *Disaster Policy Implementation,* Plenum Press, New York

Meadows, D.H., Meadows, D.L., Randers, J. and Behrens III, W.W. 1972 *Limits to Growth.* Universe Books, New York

Mehta, L. 2001 The manufacture of popular perceptions of scarcity: Dams and water-related narratives in Gujarat, India. *World Development* 29(12): 2025–41

Meinshausen, M. 2006 What does a 2°C target mean for greenhouse gas concentrations? A brief analysis based on multi-gas emission pathways and several climate sensitivity uncertainty estimates, in Schellnhuber, H.J., Cramer, W., Nakicenovic, N., Wigley., T, and Yohe, G. (eds) A*voiding Dangerous Climate Change.* Cambridge University Press, Cambridge. pp. 253–79

Meir, M.F. Dyurgerov, M.B. Rick, U.K. O'Neel, S. Pfeffer, W.T. Anderson, R.S. Anderson, S.P. and Glazovsky, A. 2007 Glaciers dominate eustatic sea level rise in 21st century. *Science* 317: 1064–67

Merrington, G., Winder, L., Parkinson, R. and Redman, M. 2002 *Agricultural Pollution.* Spon Press, London

Meteorological Office (UK) 2010 Climate change. http://www.metoffice.gov.uk/climatechange/

Metwally, A.M., Ibrahim, N.A., Saad, A. and Abu el-Ela, M.H. 2006 Improving the roles of rural women in health and environmental issues. *International Journal of Environmental Health Research* 16(2): 133–44

Micklin, P.P. 1971 Soviet plans to reverse the flow of rivers: the Kama-Vychegda-Pechora project, in Detwyler, T. (ed.), *Man's Impact on Environment*, McGraw-Hill, New York. pp. 302–18

Mitlin, D. 2003 Addressing urban poverty through strengthening assets. *Habitat International* 27: 393–406

—— 2005 *Beyond Second Best: The Whys, Hows and Wherefores of Water Subsidies*. Research Centre on Competition and Regulation Working Paper No. 93, Manchester: University of Manchester

Molden, D. ed. 2007 *Water for Food, Water for Life*. A comprehensive assessment of water management in agriculture. Earthscan, London and IWMI, Colombo.

Mollinga, P. 2003 On the Waterfront. Wageningen University Water Resources Series. Hyderabad: Orient Longman

Mollinga, P.P., Meinzen-Dick, R.S. and Merrey, D.J. 2007 Politics, plurality and problemsheds: A strategic approach for reform of agricultural water resources management. *Development Policy Review* 25(6): 699–719

MO 2010 The UK Meteorological office http://www.metoffice.gov.uk/index.html

Monneveux, P., Sanchez, C., Beck, D. and Edmeades, G.O. 2005 Drought tolerance improvement in tropical maize source populations: evidence of progress. *Crop Science* 46: 180–91

Monteith, J.L. and Unsworth, M.H. 1990 *Principles of Environmental Physics*. Edward Arnold, London

Montginoul, M. 2007 Analysing the diversity of water pricing structures: the case of France. *Water Resources Management* 21: 861–71

Moorehead, R. 1989 Changes taking place in common-property resource management in the inland Niger delta of Mali, in Berkes, F. (ed.) *Common Property Resources*. Belhaven, London

Moriarty, P. and Butterworth, J. 2003 *The Productive Use of Domestic Water Supplies*. IRC International Water and Sanitation Centre (www.irc.nl)

Morid, S. Smakhtin, V. and Bagherzadeh, K. 2007 Drought forecasting using artifical neural networks and time series of drought indices. *International Journal of Climatology* 27(15): 2103–11

Moris, J. and Thom, D. 1985 *African Irrigation Overview – summary. Water Management Synthesis II Project Report 37*. Logan: Utah State University

Morison, J.I.L., Baker, N.R., Mullineaux, P.M. and Davies, W.J. 2008 Improving water use in crop production. *Philosophical Transactions of Royal Society B* 363: 639–58

Morley, D. 2002 *Perspectives on Freshwater. Issues and Recommendations of NGOs*. London: United Nations Environment and Development Forum

Mortimore, M. 1989 *Adapting to Drought, Farmers, Famines and Desertification in West Africa*. Cambridge University Press, Cambridge

Morton, F.I. 1983 Operational estimates of evapotranspiration and their significance to the science and practice of hydrology. *Journal of Hydrology* 66: 1–76

—— 1994 Evaporation research, a critical review and its lessons for environmental science. *Critical Reviews in Environmental Science and Technology* 24(3): 237–80

Moseman, A. 2009 Does cloud seeding work? *Scientific American* February19, http://www.scientificamerican.com/article.cfm?id = cloud-seeding-china-snow

Moser, C. and Dani, A. (eds) 2008 *Livelihoods, Assets and Social Policy.* Washington, DC, World Bank

Moser, C. and Satterthwaite, D. 2008 T*owards Pro-Poor Adaptation to Climate Change in the Urban Centres of Low and Middle Income Countries.* International Institute for Environment and Development, London

Mpelasoka, F., Hennessy, K., Jones, R. and Bates, B. 2007 Comparison of suitable drought indices for climate change impacts assessment over Australia towards resource management. *International Journal of Climatology* 28(10): 1283–92

Nakayama, M. and Fujikura, R. 2006 Issues in World Commission on dams report and development: inconsistencies between the facts found and the guidelines. *Hydrological Processes* 20: 1263–72

Nature 2006 A scramble for Africa: Large dams benefit contractors and corrupt governments more than they aid the African people (Editorial). *Nature* 440: 383–84

NBI 2008 Nile Basin Initiative. http://www.nilebasin.org/index.php?option = com_content&task = category&sectionid = 6&id = 55&Itemid = 113

NEAA 2003 Netherlands Environment Assessment Agency (http://www.pbl.nl/en/index.html)

Neumann, D. 2002 Ecological rehabilitation of a degraded large river system – considerations based on case studies of macrozoobenthos and fish in the Lower Rhine and its catchment area. *International Review of Hydrobiology* 87: 139–50

New Scientist 2008 China's dams in the danger zone (Editorial). *New Scientist* 2659 (4 June)

Newsom, M. 1994 *Hydrology and the River Environment.* Clarendon Press, Oxford

Nickson, A. and Vargas, C. 2002 The Limitations of Water Regulation: the Failure of the Cochabamba Concession in Bolivia. *Bulletin of Latin American Research* 21(4): 99–120

Nicol, A. 2000 Adopting a sustainable livelihoods approach to water projects: implications for policy and practice, *Working Paper,* No. 133, London: Overseas Development Institute (ODI)

NID 2009 National Inventory of Dams. United States Army Corps of Engineers, was at http://crunch.tec.army.mil/nidpublic/webpages/nid.cfm; then replaced January 2009 by https://nid.usace.army.mil. see also ASCE 2010 http://www.infrastructurereportcard.org/sites/default/files/RC2009_dams.pdf

Nienhuis, P.H. 2008 *Environmental History of the Rhine-Meuse Delta: An ecological story on evolving human–environmental relations coping with climate change and sea-level rise.* Springer-Verlag, Berlin

Nir, D. 1974 *The Semi Arid World*, Longman, London

Nixon, W., Ghidaoui, M.S. and Kolyshkin, A.A. 2006 Range of validity of the transient dampling leakage detection method. *Journal of Hydraulic Engineering-ASCE* 132(9): 944–57

NRA 1993 *Ecologically Acceptable Flows*. National Rivers Authority, Bristol

Ntale, H.K. and Yew Gan, T. 2003 Drought indices and their application to East Africa. *International Journal of Climatology* 23: 1335–57

Nyong, A.O. and Kanaroglou, P.S. 2001 A survey of household domestic water-use patterns in rural semi-arid Nigeria. *Journal of Arid Environments* 49: 387–400

OECD 2003 *Social Issues in the Provision and Pricing of Water Services*. Organisation for Economic Cooperation and Development. OECD Publishing, Paris

OECD 2009 *Managing Water for All. An OECD Perspective on Pricing and Financing*. Organisation for Economic Cooperationand Development. OECD Publishing, Paris

OFWAT 2007 Security of supply 2006/07 report (http://www.ofwat.gov.uk/)

OFWAT 2010 UK Water Services Regulation Authority (http://www.ofwat.gov.uk/)

Oke, T.R. 1987 *Boundary Layer Climates*. Routledge, London

Oldfield, F. 2005 *Environmental Change: Key Issues and Alternative Approaches*. Cambridge University Press, New York

Oliver, M.A. 1987 Geostatistics and its application to soil science. *Soil Use and Management* 3: 8–20

OMVS 2003 L'organisation pour la mise en valeur du fleuve Senegal (OMVS) Un exemple reussi de gestion d'un grand bassin transfrontalier en Afrique de l'ouest http://www.inter-reseaux.org/IMG/pdf/OMVS.pdf

ONS 2010 Office of National Statistics (UK) (http://www.statistics.gov.uk/) STATBASE/xsdataset.asp?vlnk = 245)

Oppenheimer, M., O'Neill, B.C., Webster, M. and Agrawala, S. 2007 Climate change: The limits of consensus. *Science* 317: 1505–6

Oren, Y. 2008 Capacitive deionisation (CDI) for desalination and water treatment-past, present and future. *Desalination* 228: 10–29

Ortolano, L. and Cushing, K.K. 2002 Grand Coulee dam 70 years later, what can we learn ? *Water Resources Development* 18(3): 373–90

Osborn, T.J., Hulme, M., Jones, P.D. and Basnett, T.A. 2000 Observed trends in the daily intensity of UK precipitation. *International Journal of Climatology* 20: 347–64

Osborn, T.J. Hulme, M. 2002 Evidence for trends in heavy rainfall events over the UK. *Philosophical Transactions of the Royal Society London (A)* 360, 1313–25

Ostrom, E. 1990 *Managing the Commons*. Cambridge: Cambridge University Press

Ould Sidi Mohamed, Z. 1992 Mali: Une Reforme à Reformer, in Le Roy, E. (ed.) *La Mobilisation de la Terre dans les Strategies de Developpement en Afrique Noire Francophone*. Laboratoire d'Anthropologie Juridique de Paris, Université de Paris I.

Overman, M. 1976 *Water Solutions to a Problem of Supply and Demand*. Open University Press, Milton Keynes

Pachauri, R.K. and Reisinger, A. 2008 *Fourth Assessment of the Intergovernmental Panel on Climate Change*. IPCC, Geneva. (http://www.ipcc.ch/publications_ and_data/publications_ipcc_fourth_assessment_report_synthesis_report.htm)

Page, B. 2005 Paying for water and the geography of commodities. *Transactions of the Institute of British Geographers* NS 30: 293–306

Panigrahi, B., Panda, S.N. and Mull, R. 2001 Simulation of water harvesting potential in rain fed ricelands using water balance model. *Agricultural Systems* 69: 165–82

Parry, M., Arnell, N., Berry, P., Dodman, D., Fankhauser, S. Hope, C., Kovats, S., Nicholls, R., Satterthwaite, D., Tiffin, R. and Wheeler, T. 2009 *Assessing the Costs of Adaptation to Climate Change. A Review of the UNFCCC and Other Recent Estimates*. IIED, London

Patrick, E. 2000 Determining runoff potential for assessing suitability for water harvesting. PhD thesis, University College London

—— 2002 Researching crusting soils: themes, trends, recent development and implications for managing soil and water resources in dry areas. *Progress in Physical Geography* 26 (3): 442–61

PBS 2009 *Great Wall Across the Yangtze, Three Gorges Dam – Facts and Figures*. http://www.pbs.org/itvs/greatwall/dam1.html (accessed 7 April 2009)

Pearce, F. 1995 Fiddling while earth warms. *New Scientist* 145(1970): 14–15

—— 1997 Global warming chills out over the Pacific. *New Scientist* 153(2070): 16

—— 2004 Israel lays claim to Palestine's Water. *New Scientist* (2449): 6, 29 May

—— 2006 Mega-dams back on the agenda. *New Scientist* 191(2569): 10

Pemberton, B. 1987 *Nile Water Study, Conference on Developing World Water*. Grosvenor International, London. pp. 301–2

Penman, H. 1948 Natural evaporation from open water, bare soil and grass. *Proceedings of the Royal Society A* 193. pp. 120–45

—— 1963 *Vegetation and Hydrology*. Commonwealth Bureau of Soils, Technical Communication 53, Harpenden

Penman, H.L. Pereira, H.C. Nash, J.E. and Nixon, M. 2007 (retrospective) A view from the watershed. *Hydrology and Earth System Sciences (HESS)* 11(1): 12–25

Pereira, H.C. 1973 *Land Use and Water Resources*. Cambridge University Press, Cambridge

Pereira, L.S. 2007 Drought impacts in agriculture: water conservation and water saving practices and management. Chapter 17 in Rossi, G. et al. *Methods and Tools for Drought Analysis and Management*. Springer-Verlag, Berlin. pp. 349–73

Peter, G. 2006 Gender roles and relationships: Implications for water management *Physics and Chemistry of the Earth* 31: 723–30

Petts, G. 1984 *Impounded Rivers: Perspectives for Ecological Management*. John Wiley and Sons, Chichester, UK

Petts, G. Crawford, C. and Clarke, R. 1996 *Determination of Minimum Low Flows*. NRA, Bristol

Phillips, M., Page, S., Saratsi, E., Tansey, K. and Moore, K. 2008 Diversity, scale and green landscapes in the gentrification process: Traversing ecological and social science perspectives. *Applied Geography* 28: 54–76

Pinter, N., van der Ploeg, R., Schweigert, P. and Hoefer, G. 2006 Flood magnification on the River Rhine. *Hydrological Processes* 20: 147–64

Pittock, A.B. 2005. *Climate Change, Turning up the Heat*. Earthscan, London

Pittock, J., Lehner, B. and Lifeng, L. 2006 River basin management to conserve wetlands and water resources. Chapter 8 in Bobbink, R., Beltman, B., Verhoeven, J.T.A. and Whigham, D.F. (eds) *Wetlands: Functioning, Biodiversity Conservation and Restoration. Ecological Studies*, vol. 191, Springer-Verlag, Berlin. pp. 169–96

Ponte, L. 1976 *The Cooling*. Prentice-Hall, New Jersey.

Popkin, R. 1969 *Desalination, Water for the World's Future*. Pall Mall Press, London

Postel, S. 1992 *The Last Oasis, Facing Water Scarcity*. Earthscan, London

Prudhomme, C., Reynard, N. and Crooks, S. 2002 Downscaling of global climate models for flood frequency analysis: where are they now? *Hydrological Processes* 16: 1137–50

Pumphrey, R.G., Edwards, J.A. and Becker, K.G. 2008 Urban and rural attitudes towards municipal water controls: a study of a semi-arid region with limited water supplies. *Ecological Economics* 65: 1–12

Qdais, H.A. 2008 Environmental impacts of the mega desalination project: the Red–Dead Sea conveyor. *Desalination* 220: 16–23

Rahmstorf, S. 1997 Ice cold in Paris. *New Scientist* 153(2068): 26–30

Ramsar 2004 http://www.ramsar.org/wwd/4/wwd2004_rpt_mali_press_e.pdf

Rao, S.M. and Mamatha, P. 2004 Water quality in sustainable water management. *Current Science* 87(7): 942–47

Rathgeber, E. 1996 Women, men, and water-resource management in Africa, in Rached, E., Rathgeber, E. and Brooks, D.B. (eds) *Water Management in Africa and the Middle East: Challenge and Opportunities*, Ottawa, Ontario, Canada: International Development and Research Center

Rautanen, S.-L. and Baaniya, U. 2008 Technical work of women in Nepal's rural water supply and sanitation. *Water International* 33(2): 202–13

Ray, I. 2008 Equity re-examined: a study of community based rainwater harvesting in Rajasthan, India. *World Development* 37(2): 435–44

Raynaut, C. (ed.) (with Grégoire, E., Janin, P., Koechlin, J., Lavigne Delville, P.) 1997 *Societies and Nature in the Sahel*. Routledge, London

Reisner, C. 1993. *Cadillac Desert: The American West and its Disappearing Water*. London: Pimlico

Renzetti, S. 2000 An empirical perspective on water pricing reforms. In Dinar, Ariel (ed.) *The Political Economy of Water Pricing Reforms*. World Bank: Washington, pp. 123–40

Renzetti, S. 2002 *The Economics of Water Demands*. Kluwer, Dordrecht

Revadekar, J.V. and Kulkarni, A. 2008 The El Niño–Southern Oscillation and winter precipitation extremes over India. *International Journal of Climatology* 28(11): 1445–52

Riddell, J. 1982 *Land Tenure Issues in West African Livestock and Range Development Projects. LTC Research Paper 77*. Land Tenure Center, University of Wisconsin: Madison

Rietveld, L.C. Haarhoff, J. and Jagals, P. 2008 A tool for technical assessment of rural water supply systems in South Africa. *Physics and Chemistry of the Earth* 34: 43–49

Robertson, A.W. and Frankignoul, C. 1990 The tropical circulation: a simple model versus a general model. *Quarterly Journal of Royal Meteorological Society* 116: 69–87

Robertson, M. 2004 The Neoliberalisation of Ecosystem Services: Wetland Mitigation Banking and Problems in Environmental Governance. *Geoforum* 35: 361–73

Rockstrom, J. 2003 Resilience building and water demand management for drought mitigation. *Physics and Chemistry of the Earth* 28: 869–77

Rodda, J.C., Downing, R.A. and Law, F.M. 1976 *Systematic Hydrology*. Newnes-Butterworths, London

Ron, Z.D. 1985 Development and management of irrigation systems in mountain regions of the Holy Land. *Transactions of the Institute of British Geographers* 10: 149–69

—— 1986 Ancient and modern developments of water resources in the Holy Land and the Israeli–Arab conflict – a reply. *Transactions of the Institute of British Geographers* 11: 360–69

Roth, M. 2007 Review of urban climate research in (sub) tropical regions. *International Journal of Climatology* 27(14): 1859–73

Rothwell, J.J., Taylor, K.G., Ander, E.L., Evans, M.G. and Allott, T.E.H. 2009. Arsenic retention and release in ombrotrophic peatlands. *Science of the Total Environment* 407(4): 1405–1417

Rothwell, J.J., Evans, M.G., Daniels, S.M. and Allott, T.E.H. 2008. Peat soils as a source of lead contamination to upland fluvial systems. *Environmental Pollution* 153, 582–89

Rowley, G. 1986 Irrigation systems in the Holy Land, a comment. *Transactions of the Institute of British Geographers* 11: 356–59

Royal Society The 2008 Climate change controversies. http://royalsociety.org/uploadedFiles/Royal_Society_Content/News_and_Issues/Science_Issues/Climate_change/Climate_booklet_RS1420_reprint_Dec08.pdf

Russell, C.S., Clark, C.D. and Schuk, E.C. 2007 Economic instruments for water management in the Middle East and North Africa. *International Journal of Water Resources Development* 23(4): 659–77

Ryan, P. 2004 *Scaling up – A Literature Review*. IRC International Water and Sanitation Centre (www.irc.nl)

Safrai, I. and Zask, A. 2008 Reverse osmosis desalination plants–marine environmentalist regulator point of view. *Desalination* 220: 72–84

Salathe Jr, E.P., Mote, P.W. and Wiley, M.W. 2007 Review of scenario selection and downscaling methods for the assessment of climate change impacts on hydrology in the United States Pacific Northwest. *International Journal of Climatology* 27(12): 1611–21

Sanchez, L.D., Sanchez, A., Galvis, G. ande Latorre, J. 2006 *Multi-stage Filtration*. IRC International Water and Sanitation Centre (www.irc.nl)

SAP 2004 *Strategic Action Program for the Integrated Management of the São Francisco River Basin and its Coastal Zone*. Final Report. Brasilia, Agência Nacional de Águas

Scatena, F.N., Ortiz-Zayas, J.R. and Blanco-Libreros, J.F. 2008 Helping HELP with limited resources: the Luquillo experience. *Water South Africa* 34(4): 497–503

Schiermeier, Q. 2010 The real holes in climate science. *Nature* 463(7279): 284–87

Schiettecatte, W., Ouessar, M., Gabriels, D., Tanghe, S., Heirman, S. and Abdelli, F. 2005 Impact of water harvesting techniques onsoil and water conservation: a case study on a micro catchment in southeastern Tunisia. *Journal of Arid Environments* 61: 297–313

Schongart, J. and Junk, W. 2007 Forecasting the flood pulse in central Amazonia by ENSO indices. *Journal of Hydrology* 335: 124–32

Schmidt, J.J. 2007 Pricing water to death. *Alternatives Journal* 33(4): 29–31

Schultz, B. 2002 Role of dams in irrigation, drainage and flood control. *Water Resources Development* 18(1) 147–62

Scudder, T. 1973 The human ecology of big projects: river basin development and resettlement. *Annual Review of Anthropology* 2, (1973): 45–55

SDN 2009 Science and Development Network. Africa facing climate data shortage.11 November. http://www.scidev.net/en/news/africa-facing-climate-data-shortage.html

Selby, J. 2003 *Water, Power and Politics in the Middle East*. I.B. Tauris, London

Sewell, W.R.D. 1973 Weather modification: social concerns and public policies, in Sewell, W.R.D. (ed.) *Modifying the Weather*. Western Geographical Series 19. University of Victoria. pp. 1–49

Shafir, G. 1989 *Land, Labor and the Origins of the Israeli–Palestinian Conflict 1882–1914.* Cambridge University Press, Cambridge

Shakir, A. Ghosh, N.C. and Singh, R. 2008 Evaluating best evaporation estimate model for water surface evaporation in semi-arid India. *Hydrological Processes* 22(8): 1093–1106

Shah, T. 2007 The groundwater economy of South Asia: an assessment of size, significance and socio-ecological impacts. In: Giordano, M. and Villholth, K. (eds) *The Agricultural Groundwater Revolution. Opportunities and Threats to Development.* CABI, Wallingford

Sharp, L. 2006 Water demand management in England and Wales: constructions of the domestic water user. *Journal of Environment Planning and Management* 49(6): 869–89

Shaw, E. 1994 *Hydrology in, Practice*. Routledge, London

Sheppard, E. and McMaster, R.B. (eds) 1994 *Scale and Geographic Enquiry, Nature Society and Method*. J. Wiley and Sons, Chichester

Silver, R.S. 1978 *Steam Plant Aspects of Sea Water Distillation*. Burlington Press, Foxton

Silverman, B.A. 2001 A critical assessment of glaciogenic seeding of convective clouds for rainfall enhancement. *Bulletin of American Meteorological Society* 82(5): 903–23

—— 2003 A critical assessment of hygroscopic seeding of convective clouds for rainfall enhancement. *Bulletin of American Meteorological Society* (Sept.) 1219–30

Singh, V.P. and Xu, C.-Y 1997 Evaluation and generalization of 13 mass transfer equations for determining free water evaporation. *Hydrological Processes* 11: 311–23

Sivaplan, M., Bloschl G., Zhang L. and Vertessy R. 2003 Downward approach to hydrological prediction. *Hydrological Processes* 17: 2101–11

SIWI 2006 Stockholm International Water Institute (SIWI) *Water Scarcity Challenges in the Middle East and North Africa (MENA)* (prepared by Tropp, H. and Jagerkog, A.). *Human Development Report 2006* Occasional Paper. United Nations Development Programme, New York

Smith, M. 1992 *CROPWAT: A Computer Program for Irrigation Planning and Management. FAO Irrigation and Drainage Paper 46*, Food and Agriculture Organization of the United Nations. Rome

Smith, M.S. 2008 FAO (UN) methods for calculating evaporation and crop water requirements http://www.fao.org/ag/agl/aglw/webpub/REVPUB.htm (accessed 28 January 2008)

Sobolowski, S. and Frei, A. 2007 Lagged relationships between North American snow mass and atmospheric teleconnection indices. *International Journal of Climatology* 27: 221–31

Solomon, S., Alley, R., Gregory, R., Lemke, P. and Manning, M. 2008 A closer look at the IPCC report. *Science* 319 (Jan.): 409–10.

Sosland, J. 2007 *Cooperating Rivals: The Riparian Politics of the Jordan River Basin*. Albany (NY): State University of New York Press

Sprague, L.A., Mueller, D.K., Schwarz, G.E. and Lorenz, D.L. 2009 *Nutrient Trends in Streams and Rivers of the United States 2009. U.S. Geological Survey Scientific Investigations Report 2008–5202*.

Srivastava, R.C. 2001 Methodology for design of water harvesting system for high rainfall areas. *Agricultural Water Management* 47: 37–53

Stern, N. 2006 *The Economics of Climate Change*. Cambridge University Press, Cambridge

Stone, I. 1984 *Canal Irrigation in British India*. Cambridge University Press, Cambridge

Stott, P.A., Tett, S.F.B., Jones, G.S., Allen, M.R., Mitchell, J.F.B. and Jenkins, G.J. 2000 External control of 20th century temperature by natural and anthropogenic forcings. *Science* 290(5499): 2133–37

Strahler, A. and Strahler, A. 2006 *Introducing Physical Geography*, 4th edn. J. Wiley & Sons, Chichester

Sultana, F. 2009 Fluid lives: subjectivities, gender and water in rural Bangladesh, *Gender, Place & Culture* 16(4): 427–444

Sumner, D. and Brunke, H. 2003 Commodity policy and California agriculture. In Siebert, J. (ed.) *California Agriculture Dimensions and Issues*. Giannini Foundation of Agricultural Economics. University of California, Berkeley

Sutton, A. 2004 The Three Gorges Dam project on the Yangtze River, *Geography*, 89(2): 111–26

Swyngedouw, E. 1999 Modernity and Hybridity: nature, regeneracionismo and the production of the Spanish waterscape 1890–1930. *Annals of the American Association of Geographers* 89(3): 443–65

—— 2004 *Social Power and the Urbanization of Water: Flows of Power*. Oxford University Press, Oxford

—— 2006 *Power, Water and Money: Exploring the Nexus. Human Development Report 2006* Occasional Paper. New York: United Nations Development Programme

Sylla, O. 2006 *Decentralised Management of Irrigation Areas of the Sahel: Water Users Associations in the Senegal River Valley. Human Development Report 2006* Occasional Paper, New York, United Nations Development Programme

Symons, G.E. 2006 Water treatment through the ages. *Journal of American Water Works Association* 98(3): 87–98

Takakasi, Y. 2004 Dams, environment and regional development in Japan. *Water Resources Development* 20(1): 35–45

Tebaldi, C. Mearns, O. Nychka, D. and Smith, R.L. 2004 Regional probabilities of precipitation change: a Bayesian analysis of multimodel simulations. *Geophysical Research Letters* 31 online

Tebbutt, T.H.Y. 1998 *Principles of Water Quality Control*. Butterworth Heinemann, Oxford

Terblanche, D.E., Mittermaier, M.P. Burger, R.P. deWaal, K. and Ncipha, X.G. 2005 The South African rainfall enahcement programme 199702001. *Water South Africa* 31(3): 291–98

TERI 2003 *Coping with Global Change. Vulnerability and Adaptation in Indian Agriculture*. The Energy and Resources Institute (TERI), New Delhi. http://www.iisd.org/pdf/2004/climate_coping_global_change.pdf

Therkildsen, O.1988 *Watering White Elephants?* Scandinavian Institute of African Studies, Uppsala

Thompson, J., Porras, I., Tumwine, J., Mujwahuzi, M., Katui-Katua, M., Johnstone, N. and Wood L. 2001 *Drawers of Water II*. International Institute for Environment and Development, London.

Thorne, W. 1963 *Land and Water Use*. American Association for the. Advancement of. Science 73, Washington

Thornthwaite, C.W. 1948 An approach towards a rational classification of climate. *Geographical Review* 38: 55–94

Thornton, P.K., Jones, P.G., Owiyo, T., Kruska, R.L., Herrero, M., Kristjanson, P., Notenbaert, A., Bekele, N. and Omolo, A. (with contributions from Orindi, V., Otiende, B., Ochieng, A., Bhadwal, S., Anantram, K., Nair, S., Kumar, V. and Kulkar, U. 2006.) *Mapping Climate Vulnerability and Poverty in Africa. Report to the Department for International Development*, ILRI, PO Box 30709, Nairobi 00100, Kenya. pp. 171

Tleimat, B.W. 1980 Freezing methods, in Spiegler, K.S. and Laird, A.D.K. eds. *Principles of desalination part B*. Academic Press, London 359–400

Tollefson, J. 2010 Missed 2050 climate targets will reduce long term options. *Nature* (11 Jan.).

Tropp, H. 2005 Building new capacities for improved water governance. CSIR International Symposium on Ecosystem Governance, South Africa (cited UNWWD, 2006, p. 85)

Tropp, H. and Jagerkog, A. 2006 *Stockholm International Water Institute (SIWI): Water Scarcity Challenges in the Middle East and North Africa (MENA) Human Development Report 2006 Occasional Paper*. United Nations Development Programme, New York

Tsubo, M., Walker, S. and Hensley, M. 2005 Quantifying risk for water harvesting under semi-arid conditions. Part 1: Rainfall intensity generation. *Agricultural Water Management* 76: 77–93

TVA 2009 Tennessee Valley Authority website: http://www.tva.com (accessed 16 April 2009)

Twort, A.C., Ratnayaka, D.D. and Brandt, M.J. 2002 *Water Supply*, 5th edn. Butterworth-Heinemann, Oxford

UNCED 1992 United Nations Conference on Environment and Development, The Earth Summit. Rio de Janeiro. http://www.un.org/geninfo/bp/enviro.html

UKCIP 2002 UK Climate Impacts Programme. http://www.ukcip.org.uk/index. php?option=com_content&task=view&id=161&Itemid=287

UKCIP 2010 Climate Projections (version 2009) UK Climate Impacts Programme. DEFRA. http://ukclimateprojections.defr

UNDP 2004 Decentralised governance for development: A combined practice note on decentralization. United Nations Development Programme. New York

UNDP 2010 Women's empowerment. United Nations Development Programme. http://www.undp.org/women/

UNEP 2008 *Vital Water Graphics – An Overview of the State of the World's Fresh and Marine Waters*, 2nd edn. UNEP, Nairobi, Kenya

UNESCO 1977 *Map of the distribution of arid regions. MAB Technical Note No. 7*. UNESCO, Paris

UNESCO 2000 *Rain Water Harvesting and Artificial Recharge to Ground Water* http://www.unesco.org/water/ihp/publications/water_harvesting.pdf

UNESCO 2009 *World Water Assessment Programme* (http://www.unesco.org/water/wwap/)

UNESCO 2009 *World Water Development Report 2009*. Paris

UN 2010 *United Nations: The Dublin Statement on Water and Sustainable Development 1992* http://www.un-documents.net/h2o-dub.htm

United Nations Office for Human Affairs 2007 *Israeli-Palestinian Fatalities since 2000*. OCHA Special Focus, August 2007. http://domino.un.org/unispal.nsf/

UN-MDG 2009 *United Nations Millennium Development Goals 2009 report*. http://www.un.org/millenniumgoals/pdf/MDG_Report_2009_ENG.pdf

UNWWDR 2006 (United Nations World Water Development Report 2006) *Water a Shared Responsibility*. UNESCO, Paris (available at http://www.unesco.org/water/wwap/wwdr/wwdr2/table_contents.shtml)

USAID 2010 *Women in Development*. http://www.usaid.gov/our_work/cross-cutting_programs/wid/

US census 2009. http://www.census.gov/population/www/documentation/twps0027/twps0027.html (accessed 20 July 2009)

USDA 2005 www.fas.usda.gov/pecad/highlights/2005/09/uganda

USGS 2010 *Water Resources in the United States*. http://water.usgs.gov/

Van der Lee, G., Olde Venterink, H. and Asselman, N. 2004 Nutrient retention in floodplains of the Rhine distributaries in the Netherlands. *River Research and Applications* 20: 315–25

Van Dijk, G., Martein, E. and Schulte-Wulwer-Leidig, A. 1995 Ecological rehabilitation of the river Rhine: plans, progress and perspectives. *Regulated Rivers: Research and Management* 11: 377–88

Van Slobbe, E., Morris, E.D., Rohling, N., Torenbeek, R., Broker, K. and Heering, H. 2006 Social learning in wetland development, Chapter 12 in Bobbink, R. Beltman, B. Verhoeven, J.T.A. and Whigham, D.F. (eds) *Wetlands: Functioning, biodiversity conservation and restoration. Ecological Studies vol. 191*. Springer-Verlag, Berlin, pp. 269–83

Varis, O., Biswas, A., Tortajada, C. and Lundqvist, J. 2006 Megacities and water management. *Water Resources Development* 22(2) 377–94

Vecchi, G.A. and Soden, B.J. 2007 Global warming and the weakening of the tropical circulation. *Journal of Climate* 20(17): 4316–40

Venot, J-P, Turral, H., Samad, M., and Molle, F. 2007 *Shifting Waterscapes. Explaining Krishna Basin Closure in the Lower Krishna Basin, South India. Research Report 121*. Colombo: International Water Management Institute

Viessman, W. and Hammer, M.J. 1998 *Water Supply and Pollution Control*, 6th edn. Addison-Wesley Longman, Meno Park, California

Vink, R. and Behrendt, H. 2002 Heavy metal transport in large river systems: heavy metal emissions and loads in the Rhine and Elbe river basins. *Hydrological Processes* 16: 3227–44

Vohra, B.B. 1975 No more gigantism, *CERES* 8(4): 33–35

Voropaev, G.V. 1979 The scientific principles of large scale areal redistribution of water resources in the USSR. In Golubev, G. and Biswas, A.K. (eds) *Interregional Water Transfers*, Pergamon Press, Oxford. pp. 90–101

Wade, N. 1985 Choice of desalination and power plants. *Water and Sewage* 9(5): 3–6

Wakimizu, K., Nishiyama, K., Suzuki, Y., Tomine, K., Yamazaki, M., Isimaru, A., Ozaki, M., Itano, T., Naito, G. and Fukuta, N. 2002 A low level penetration seeding experiment of liquid carbon dioxide in a convective cloud. *Hydrological Processes* 16: 2239–53

Walker, G. and King, D. 2008 *The Hot Topic: How to Tackle Global Warming and Keep the Lights On*. Bloomsbury, London

Walker, S., Tsubo, M. and Hensley, M. 2005 Quantifying risk for water harvesting under semi-arid conditions Part II Crop yield simulation. *Agricultural Water Management* 76: 94–107

Walker, W.R. 1989 *Guidelines for Designing and Evaluating Surface Irrigation Systems. FAO Irrigation and Drainage Paper 45*, Food and Agriculture Organization of the United Nations. Rome, Online at http://www.fao.org/docrep/T0231E/t0231e00.htm#Contents

Walski, T.A. 2006 A history of water distribution. *American Water Works Association Journal* 98(3): 110–21

Wamsler, C. 2007 Bridging the gaps: stakeholder based strategies for risk reduction and financing for the urban poor. *Environment and Urbanization* 19(1): 115–42

Ward, B. 1966 *Spaceship Earth*. Hamilton, London

Ward, R. and Robinson, M. 1999 *Principles of Hydrology* 4th edn. McGraw-Hill Higher Education

Warren, A. and Agnew, C.T. 1988 *An Assessment of Desertification and Land Degradation in Arid and Semi-Arid Areas*. IIED, London

Warren, A. and Khogali, M. 1992 *Assessment of Desertification and Drought in the Sudano-Sahelian Region 1985–91*. UNSO, New York

Warrick, R.A., Barrow, E. and Wigley, T. (eds) 1993 *Climate and Sea Level Change: Observations, Projections and Implications*. Cambridge University Press, Cambridge

Washington, R., Harrison, M., Conway D., Black, E., Challinor, A., Grimes, D., Jones, R., Morse, A., Kay, G. and Todd, M. 2006 African climate change. *Bulletin of American Meteorological Society* (Oct.): 1355–66

Watson, E., Adams, W. and Mutiso, S. 1998 Indigenous irrigation, agriculture and development, Marakwet, Kenya. *Geographical Journal* 164: 67–84

WCD 2000 *World Commission on Dams Report: Dams and Development*, Earthscan. (http: //www.dams.org/report/earthscan.htm)

WCD 2009 World Commission on Dams (http://www.dams.org/)

WCED 1987 *Our Common Future* The World Commission on Environment and Development. Oxford University Press, Oxford

WDR 2008 World Development Report 2008. Washington, DC: World Bank

Webster, P.J., Holland, G.J., Curry, J.A. and Chang, H.R. 2005 Changes in tropical cyclone: number, duration, and intensity in a warming environment. *Science* 309(5742): 1844–46

WEDO 2010 Women's and Environment Development Organisation http://www.wedo.org/

Wei, H. Jian-Long, L. and Tian-Gang, L. 2005 Study on the estimation of precipitation resources for rainwater harvesting agriculture in semi-arid land of China. *Agricultural Water Management* 71: 33–45

Were, A., Villagarcia, L., Domingo, F., Alados-Arboledas, L., and Puigdefabregas, J. 2007 Analysis of effective resistance calculation methods and their effect on modelling evapotranspiration in two different patches of vegetation in semi-arid SE Spain. *Hydrology and Earth System Sciences* 11: 1529–42

Were, E., Roy, J. and Swallow, B. 2008 Local organisation and gender in water management: a case study from the Kenya highlands. *Journal of International Development* 20: 69–81

White, I. 2008 The Absorbent City: urban form and flood risk management *Proceedings of the Institution of Civil Engineers*: *Urban Design and Planning December* (DP4), 151–161

Whittington, D., Hanemann, W.M., Sadoff, C. and Jeuland, M. 2009 Water and Sanitation challenge. In Lomborg, B. (ed.) *Global Crises, Global Solutions*, 2nd edn Cambridge University Press, Cambridge

WHO 1985 *Guidelines for Drinking Water Quality*. World Health Organization, Geneva

——— 1989 *Health Guidelines for the Use of Waster Water in Agriculture and Aquaculture*. World Health Organization Technical report 778, Geneva

——— 2000 *Global Water Supply and Sanitation Assesment*. World Health Organization Geneva

——— 2006 *Guidelines for the Safe Use of Waster Water*. World Health Organization Geneva

——— 2006 *World Health Organization. Guidelines for Drinking Water Quality*, 3rd edn (http://www.who.int/water_sanitation_health/dwq/gdwq3rev/en/index.html)

——— 2009 *World Health Organization: WHO Guidelines for Drinking-water Quality* 3rd edition World Health Organization Geneva http://www.who.int/water_sanitation_health/dwq/gdwq3rev/en/index.html

WHO-UNICEF 2006 *Meeting the MDG drinking water and sanitation target*. WHO and UNICEF, Geneva

Wigley, T.M.L. and Raper, S.C.B. 1987 The global expansion of sea water associated with global warming. *Nature* 330(6144): 127–31

Wilby, R.L., Beven, K.J., Reynard, N.S. 2007 Climate change and fluvial flood risk in the UK: more of the same? *Hydrological Processes* 22(14): 2511–23

Wilhite, D.A. (ed.). 2000 *Drought: A Global Assessment. Natural Hazards and Disasters Series*. Routledge, London

Wilkinson, J.C. 1977 *Water and Tribal Settlement in S.E. Arabia*, Clarendon Press, Oxford

Williams, C.J.R., Kniveton, D.R. and Layberry, R. 2007 Climatic and oceanic associations with daily rainfall extremes over Southern Africa. *International Journal of Climatology* 27: 93–108

Wilson, Z. and Pfaff, B. 2008 Religious, philosophical and environmentalist perspectives on potable wastewater reuse in Durban, South Africa. *Desalination* 228: 1–9

Wirfa, L., Campbell, A and Rea, N. 2008 Implications of gendered environmental knowledge in water allocation processes in central Australia. *Gender, Place and Culture* 15(5): 505–18

Wittfogel, K. 1957 *Oriental Despotism: A Comparative Study of Total Power*. Random House, New York

Wolf, A. 1995 *Hydropolitics Along the Jordan River*. Tokyo: United Nations University Press

——— 2006 *Conflict and Cooperation Over Transboundary Waters*. Human Development Report Office Occasional Paper

Wolff, G., Cooley, H., Palaniappan, M., Samulon, A., Lee, E., Morrison, J. and Katz, D. Gleick, P. 2007 The world's water 2006–7. The biennal report on freshwater resources. Island Press

Wood, B.D. 2009 The role of scaling laws in upscaling. *Advances in Water Resources* 32: 723–36

Woodhouse, P. 1995 Water rights and rural restructuring in South Africa. A case study from the Eastern Transvaal. *Water Resources Development* 11(4): 527–45

—— 2003 African Enclosures: a default mode of development. *World Development* 31(10): 1705–20

—— 2008 *Water Rights in South Africa. Insights from Legislative Reform. BWPI Working Paper 36*. Manchester: Brookes World Poverty Institute. http://www.bwpi.manchester.ac.uk/resources/Working-Papers/bwpi-wp-3608.pdf

Woodhouse, P. and Ndiaye, I. 1990 *Structural Adjustment and Irrigated Food Farming in Africa: The "Disengagement" of the State in the Senegal River Valley. DPP Working Paper 20*, Open University, Milton Keynes

Woodhouse, P., Bernstein, H. and Hulme, D. 2000 *African Enclosures? The Social Dynamics of Wetlands in Drylands*. James Currey, Oxford

Woodley, W.L. and Rosenfeld, D. 2004 The development and testing of a new method to evaluate the operational cloud seeding programs in Texas. *Journal of Applied Meteorology* 43(2): 249–63

Woods, R., Sivapalan, M. and Duncan, M. 1995 Investigating the representative elementary area concept: an approach based on field data., *Hydrological Processes* 9(3/4): 291–312

Woolley, J., Cook, S.E., Molden, D. and Harrington, L. 2009 Water, food and development: the CIGAR challenge programme on water and food. *Water International* 34(1): 4–12

World Bank 1994 *Implementing the Water Resource Mandate of Agenda 21: The promise and the challenges for OECD countries*. Transportation, Water and Urban Development Department, World Bank: Washington, DC

—— 2006 *Water Supply and Sanitation Data*. World Bank, Washington DC. http://web.worldbank.org/WBSITE/EXTERNAL/TOPICS/EXTWAT/0, contentMDK: 21706928~menuPK: 4602430~pagePK: 148956~piPK: 216618~theSitePK: 4602123, 00.html

—— 2007 *World Development Report 2007: Development and the Next Generation*. World Bank, Washington, DC http://econ.worldbank.org/WBSITE/EXTERNAL/EXTDEC/EXTRESEARCH/EXTWDRS/EXTWDR2007/0, contentMDK: 20601683~pagePK: 64167702~piPK: 64167676~theSitePK: 1489834, 00.html

—— 2008 *World Development Report 2008: Agriculture for Development*. World Bank, Washington, DC. http://econ.worldbank.org/WBSITE/EXTERNAL/EXTDEC/EXTRESEARCH/EXTWDRS/EXTWDR2008/0, menuPK: 2795178~pagePK: 64167702~piPK: 64167676~theSitePK: 2795143, 00.html

—— 2009, 2010 *Water Data and Statistics*. World Bank. http://go.worldbank.org/VNZNZY9KZ0

—— 2010 *Rural water supply and sanitation*. http://web.worldbank.org/WBSITE/EXTERNAL/TOPICS/EXTWAT/0, contentMDK: 21630630~menuPK: 4618329~pagePK: 210058~piPK: 210062~theSitePK: 4602123, 00.html

—— 2010b *Gender and Development* http://web.worldbank.org/WBSITE/EXTERNAL/TOPICS/EXTGENDER/0, contentMDK: 21983335~pagePK: 210058~piPK: 210062~theSitePK: 336868, 00.html

World Resource Institute 1987, 1989, 1990, 2007, 2009, 2010 See WRI entries

Worrall, F., Burt, T.P., Adamson, J.K., Reed, M., Warburton, J., Armstrong, A. and Evans, M. 2007 Predicting the future carbon budget of an upland peat catchment. *Climatic Change* 85: 139–58

WRI 1987, 1989, 1990 *World Resource Institute: World Resources*. Basic Books Inc. New York

WRI 2007, 2009 World Resource Institute. *Earth trends*. http://www.wri.org/
—— 2010 World Resource Institute. *Earth Trends*. http://www.wri.org/

Wu, H., Hayes, M.J., Weiss, A. and Hu, Q. 2001 An evaluation of the standardized precipitation index, the Chinba-Z index and the statistical Z score. *International Journal of Climatology* 21: 745–58

Wu, H., Svobada, M.D., Hayes, M.J., Wilhite, D.A., and Wen, F. 2007 Appropriate application of the standardised precipitation index in arid locations and dry seasons. *International Journal of Climatology* 27: 65–79

WWAP 2006 *Water a Shared Responsibility*. World Water Assessment Programme 2nd UN World Water Development Report. UNESCO, Paris (see also WWDR)
—— 2009 *Water a Shared Responsibility*. World Water Assessment Programme 3rd UN World Water Development Report. UNESCO, Paris (see also WWDR 2009)

WWC 2000 World Water Council 2nd World Water Forum. The Hague, The Netherlands, March: http://www.worldwatercouncil.org/index.php?id=1594 &L=0%2Fst...usuarios.arnet.com.ar%2Fl%20target%3D%20target%3D%20 target%3D%20target%3D
—— 2006 World Water Council 4th World Water Forum, Mexico. Synthesis and final report http://portal.worldwaterforum5.org/wwf5/en-us/Forum KnowledgeBase/3rd%20World%20Water%20Forum/Communication/ Announcements/Final%20Report%202nd%20World%20Water%20 Forum.pdf
—— 2010 World Water Council WRC (http://www.worldwatercouncil.org/ index.php?id = 25)

WWDR 2006 *UN World Water Development Report, 2nd report: Water a Shared Responsibility*. UNESCO, Paris. http://www.unesco.org/water/wwap/ wwdr/wwdr2/
—— 2009 *The UN World Water Development Report, 3rd report: Water in a changing world*. UNESCO, Paris. http://www.unesco.org/water/wwap/wwdr/ wwdr3/

WWF 2008 World Wildlife Fund. Living planet report. http://assets.panda.org/ downloads/living_planet_report_2008.pdf

WWUK 2010 Water Works UK (Ltd.) http://www.wwuk.co.uk/

WWV 2000 *World Water Vision Commission Report: A Water Secure World*. World Water Council: http://www.worldwatercouncil.org/fileadmin/wwc/ Library/Publications_and_reports/Visions/CommissionReport.pdf

Wyseure, G.C.L. Gowing, J.W. and Young, M.D.B 2002 Parched-Thirst: an agrohydrological model for planning rainwater harvesting systems. In semi-arid areas. Chapter 10 in Singh, V.P. and Frevert, D.K. *Mathematical Models of Small Watershed Hydrology and Applications*. Water Resources Publications.

Xiong, L. and Guo, S. 1999 A two parameter monthly water balance model and its application. *Journal of Hydrology* 216: 111–23

Xu, C., Tunemar L., Chen, Y.D. and Singh, V.P. 2006 Evaluation of seasonal and spatial variations of lumped water balance model sensitivity to precipitation data errors. *Journal of Hydrology* 324: 80–93

Xu, C.-Y. and Singh, V.P. 2005 Evaluation of three complementary relationship evapotranspiration models by water balance approach to estimate actual regional evapotranspiration in different climatic regions. *Journal of Hydrology* 308: 105–21

Young, G., Dooge, J. and Rodda, J. 1994 *Global Water Resource Issues*. Cambridge University Press, Cambridge

Young, M.D.B., Gowing, J.W., Wyseure, G.C.L. and Hatibu, N. 2002 Parched-Thirst: development and validation of a process based model of rainwater harvesting. *Agricultural Water Management* 55: 121–40

Yu, F., Chen, Z., Ren, X. and Yang, G. 2009 Analysis of historical floods on the Yangtze River, China: characteristics and explanations. *Geomorphology* 113(3–4): 210–16

Zanchettin, D., Franks, S.W., Traverso, P. and Tomasino, M. 2008 On ENSO impacts on European wintertime rainfalls and their modulation by the NAO and the Pacific multi-decadal variability described through the PDO index. *International Journal of Climatology* 28(8): 995–1006

Zhao, H. and Moore, G.W.K. 2008 Trends in the boreal summer regional Hadley and Walker circulations as expressed in precipitation records from Asia and Africa during the latter half of the 20th century. *International Journal of Climatology* 28: 563–78

Zubair, L., Siriwardhana, M., Chandimala, J. and Yahiya, Z. 2008 Predictability of Sri Lankan rainfall based on ENSO. *International Journal of Climatology* 28: 91–101

Index